Microbial Population Genetics

Edited by

Jianping Xu

Department of Biology
McMaster University
Hamilton, ON
Canada

Caister Academic Press

Caister Academic Press
Norfolk, UK

www.caister.com

British Library Cataloguing-in-Publication Data
A catalogue record for this book is available from the British Library

ISBN: 978-1-904455-59-2

Cover image adapted from Fig. 3.2.

Printed and bound in Great Britain

Contents

Contributors

Fernando Gonzalez Candelas
Instituto Cavanilles de Biodiversidad y Biologia Evolutiva
University of Valencia
Valencia
Spain

fernando.gonzalez@uv.es

Yujun Cui
Beijing Institute of Microbiology and Epidemiology
Beijing
China

cuiyujun.lam@gmail.com

Bertrand D. Eardly
Department of Biology
Eberly College of Science
Pennsylvania State University
Reading, PA
USA

bde1@psu.edu

Beile Gao
Department of Biochemistry
McMaster University
Hamilton, ON
Canada

gaob@mcmaster.ca

G. Brian Golding
Department of Biology
McMaster University
Hamilton, ON
Canada

golding@mcmaster.ca

Radhey S. Gupta
Department of Biochemistry
McMaster University
Hamilton, ON
Canada

gupta@mcmaster.ca

Weilong Hao
Department of Biology
Indiana University
Bloomington, IN
USA

haow@indiana.edu

Deirdre A. Joy
Parasitology and International Programs Branch
Division of Microbiology and Infectious Diseases
NIAID/NIH/DHHS
Bethesda, MD
USA

djoy@mail.nih.gov

Yanjun Li
Beijing Institute of Microbiology and Epidemiology
Beijing
China

navylyj@163.com

Kui Lin
College of Life Sciences
Beijing Normal University
Beijing
China

linkui@bnu.edu.cn

Yingqin Luo
College of Life Sciences
Beijing Normal University
Beijing
China

yingqin.luo@asu.edu

Scott R. Miller
Division of Biological Sciences
The University of Montana
Missoula, MT
USA

scott.miller@mso.umt.edu

Thomas G. Mitchell
Department of Molecular Genetics and Microbiology
Duke University Medical Center
Durham, NC
USA

tom.mitchell@duke.edu

Teresa E. Pawlowska
Department of Plant Pathology and Plant-Microbe Biology
Cornell University
Ithaca, NY
USA

tep8@cornell.edu

Jim Provan
School of Biological Sciences
Queen's University Belfast
Belfast
UK

j.provan@qub.ac.uk

Rafael Sanjuán
Instituto Cavanilles de Biodiversidad y Biologia Evolutiva
University of Valencia
Valencia
Spain

rafael.sanjuan@uv.es

Jianping Xu
Department of Biology
McMaster University
Hamilton, ON
Canada

jpxu@mcmaster.ca

Yanfeng Yan
Beijing Institute of Microbiology and Epidemiology
Beijing
China

yanyf1983@163.com

Ruifu Yang
Beijing Institute of Microbiology and Epidemiology
Beijing
China

ruifuyang@gmail.com

Preface

Population genetics investigates the spatial and temporal patterns of genetic variation among individuals and populations of organisms, including the mechanisms for such patterns. This field has a rich history, started with Charles Darwin in 1859 with his seminal publication *On the Origin of Species by Means of Natural Selection, or the Preservation of Favoured Races in the Struggle for Life*. His revolutionary idea of Evolution by Natural Selection revealed the fundamental issues of population genetics: the existence of heritable variations among individuals, the contributions of these variations to the differential survival and reproduction among individuals, the accumulation of advantageous variants in a population over time, and the divergence among populations due to the accumulation of different variants. For most of the 19th and the early 20th century, there was very little agreement within the scientific community about the genetic basis of phenotypic variation among individuals in natural populations. However, this period laid some of the most important foundations of biology, including population genetics. The early pioneers in genetics and specifically population genetics include Francis Galton, Karl Pearson, Gregor Mendel, Thomas Morgan, Herman Nilsson-Ehle, Udny Yule, Ronald Fisher, J.B.S. Haldane, Sewall Wright, Sergei Chetverikov, and Theodore Dobzhansky.

Most early population geneticists used mathematical models to examine the factors that might help explain phenotypic differences among individuals and populations. When the principles of inheritance, the Mendelian laws, were rediscovered at the beginning of the 20th century, rapid advancements in population genetics soon followed, leading to the formulation of the New Synthesis of the theory of evolution. However, for the first half of the 20th century, most evolutionary biologists and population geneticists studied plants and animals. Few studies examined microbial populations.

The era of molecular population genetics began in 1966 when Lewontin and Hubby, and Harris, introduced the technique of protein electrophoresis to population genetics. Surprisingly abundant genetic variations were found in virtually all organisms examined, including some microorganisms. However, many of these variants had little phenotypic effect on the survival and reproduction of organisms. These findings have led to the emergence of the neutral theory of molecular evolution. The application of DNA sequencing technology into population genetics since the early 1980s further expanded our understanding of the extent of genetic variation among individuals and within and between populations. While most molecular population genetic studies from the 1960s to the early 1990s were still focused on macroorganisms such as plants and animals, the next phase of technical innovation, the development and application of high-throughput DNA sequencing and microarray technologies in the mid-1990s, were led by studies of microorganisms. These and other developments are now allowing unprecedented access to genetic materials from populations and communities of micro- and macro-organisms in their natural habitats. Indeed, comparative genomics, phylogenomics, phylodynamics, evolutionary genomics, population genomics, metagenomics, and systems biology are now burgeoning fields of scientific investigations.

The objective of this book is to bring up-to-date research advances in broad areas of microbial population genetics and genomics. It reviews the application of various molecular tools in our understanding of the patterns and the potential mechanisms for genetic variation of microorganisms at a broad range of scales, from those within micro-niches to among populations from different continents. The book introduces both fundamental concepts and recent molecular population genetic tools and data from SNP surveys, whole-genome DNA sequences, and microarray hybridizations etc. It covers broad groups of microorganisms including viruses, bacteria, archaea, fungi, protozoa, and algae.

The 12 chapters in this book contain a wide range of topics. Chapter 1 introduces recent advances in microbial systematics, with a special focus on the unique utility of insertions/deletions in helping to resolve the phylogenetic relationships among bacterial and archaeal phyla and orders. Chapter 2 reviews comparative microbial genomics and the impact of whole genome sequences in helping us understand the large-scale population and evolutionary issues. Chapter 3 summarizes our current understanding of horizontal gene transfer in bacteria, including the patterns, rates, and fate of horizontally transferred genes. Chapter 4 describes the molecular diversity of a deadly group of human bacterial pathogens and discussed how such information could be used for source tracking and rapid identification of human pathogenic microbes in general. Chapters 5 to 10 review the population genetics of large representative groups of microorganisms that include the nitrogen-fixing bacteria (Chapter 5), the photosynthetic cyanobacteria (Chapter 6), the eukaryotic microalgae (Chapter 7), the fungal mutualists (Chapter 8), the human fungal pathogens (Chapter 9), the human malaria parasites (Chapter 10), and human pathogenic viruses (Chapter 11). Chapter 12 presents an overview of the broad field of metagenomics and its wide-ranging impact on our understanding of microbial diversity, function, inter-relationships, and population genetics, from organisms ranging from viruses to bacteria, archaea, and eukaryotic microbes.

Microbial population genetics is a rapidly advancing field of investigation with relevance to many other theoretical and applied areas of scientific investigations. The theoretical issues that many of the chapters touched upon include the origins and evolution of species, of sex and recombination, and of life (Chapters 1, 2, 3, 11, and 12). On the applied side, population genetics lays the foundations for tracking the origin and evolution of antibiotic resistance and deadly infectious pathogens (Chapters 3, 4, 7, 9, 10, and 11). Population genetics is also an essential factor for devising strategies for the conservation and better utilization of beneficial microbes (Chapters 5, 6, 7, 8, and 12).

I want to thank Hugh Griffin at Horizon Scientific Press for his support and patience with this project. I am grateful to all the authors for their contributions. It has been a great privilege and honour to work with this group of scientists.

Jianping (JP) Xu, PhD
McMaster University
Hamilton, Ontario, Canada

Recent Advances in Understanding Microbial Systematics

Radhey S. Gupta and Beile Gao

Abstract

The higher taxonomic groups within Prokaryotes are presently distinguished mainly on the basis of their branching in phylogenetic trees. In most cases, no molecular, biochemical or physiological characteristics are known that are uniquely shared by species from these groups. Analyses of genome sequences are leading to discovery of novel molecular characteristics that are specific for different groups of *Bacteria* and *Archaea* and provide more precise means for identifying and circumscribing these groups of microbes in clear molecular terms and for understanding their evolution. These new approaches and their limited applications for clarifying microbial systematics are described here. Because of their taxa specificities, further studies on these newly discovered molecular characteristics should lead to discovery of novel biochemical and physiological characteristics that are unique to different groups of microbes.

Introduction

An understanding of the evolutionary history of life, which spans a period of more than 3.5 billion years (Ga), constitutes one of the most fascinating problems in life sciences (Schopf, 1978; Woese *et al.*, 1990; Margulis, 1993; Gould, 1994; Gupta, 1998a; Cavalier-Smith, 2002). The seminal work of Charles Darwin (Darwin, 1859) provided evidence that all living organisms shared a common ancestry and it also revealed key insights as to how the evolutionary process works to generate different life forms. Although Darwin's theory of evolution is applicable to all forms of life, his work primarily dealt with the macroscopic life forms, covering a period of < 1.0 Ga. There was almost no information available at that time about prokaryotic organisms, which based upon available evidence were the sole inhabitants of this planet for at least the first 1.5–2.0 Ga history of life (Schopf, 1978; Kasting, 1993). In the past 25–30 years, much has been learnt about the diversity of prokaryotic organisms (Woese, 1987; Olsen and Woese, 1993), but many critical issues remain unresolved (Gupta and Griffiths, 2002; Gupta, 2005a). However, with the dawn of genomic era, the prospects of gaining an understanding of these critical issues regarding the evolutionary relationships among prokaryotic organisms are beginning to emerge (Doolittle, 1999; Gupta, 2002; Ciccarelli *et al.*, 2006). This chapter describes some critical unresolved issues in prokaryotic phylogeny and new approaches that are proving helpful in understanding of microbial systematics and phylogeny.

Evolutionary relationships among prokaryotes: critical issues that need to be understood

The prokaryotic organisms are presently divided into two main domains, *Bacteria* and *Archaea* (Woese *et al.*, 1990; Ludwig and Klenk, 2005). Of these, *Bacteria* constitute the vast majority (>98%) of known prokaryotic organisms; hence, an understanding of the relationship among them constitutes a major part of prokaryotic phylogeny. Our current understanding of the evolutionary relationships among prokaryotes is mainly based on 16S rRNA sequences (Woese *et al.*, 1990;

Ludwig and Klenk, 2005). Based on branching in the 16S rRNA trees, the cultivable *Bacteria* are presently divided into about 24 main groups or phyla. These groups include *Thermotogae*, *Aquificae*, green non-sulphur bacteria or *Chloroflexi*, *Deinococcus-Thermus*, *Cyanobacteria*, low G+C Gram-positive (*Firmicutes*), high G+C Gram-positive (*Actinobacteria*), *spirochaetes*, green sulphur bacteria (*Chlorobi*), *Bacteroidetes*, *Chlamydiae*, *Planctomycetes*, *Proteobacteria*, *Thermodesulfobacteria*, *Thermomicrobia*, *Chrysiogenetes*, *Deferribacteres*, *Dictyoglomi*, *Fusobacteria*, *Acidobacteria*, *Fibrobacteres*, *Nitrospira*, *Flexistipes* and *Verrucomicrobia* (Ludwig and Klenk, 2005). Some of these phyla, namely *Thermodesulfobacteria*, *Thermomicrobia*, *Chrysiogenetes*, *Deferribacteres*, *Dictyoglomi*, *Fusobacteria*, *Acidobacteria*, *Fibrobacteres*, *Nitrospira* and *Flexistipes*, consist of only a few species, whereas other phyla such as *Proteobacteria*, *Cyanobacteria*, *Actinobacteria*, *Firmicutes* and *Bacteroidetes* contain thousands of species accounting for more than 90–95% of all known bacteria (Ludwig and Klenk, 2005).

The division of *Bacteria* into these 24 or so main groups is quite arbitrary and it currently has little evolutionary or taxonomic significance (Stackebrandt, 2006; Ludwig and Klenk, 2005). The main problem in this regard is that currently there are no objective criteria as to what constitute a phylum or other higher taxonomic levels such as Class, Order or Family (Stackebrandt, 2006; Ludwig and Klenk, 2005). The arbitrariness of the present bacterial classification is illustrated by the example of the *Proteobacteria* phylum. Proteobacteria comprise the largest group within prokaryotes accounting for nearly 50% of all cultured bacteria (Maidak *et al.*, 2001; Ludwig and Klenk, 2005; Kersters *et al.*, 2006). Based on their branching in the 16S rRNA trees, they have been further divided into five classes or divisions, named α, β, γ, δ, and ε (Maidak *et al.*, 2001; Ludwig and Klenk, 2005; Kersters *et al.*, 2006). Of these, α-, β- and γ-divisions harbour approximately 12%, 8% and 26% of all cultured bacteria (Maidak *et al.*, 2001). The species from these subgroups can be clearly distinguished from each other and from all other bacteria based on large numbers of molecular characteristics (Gupta, 2000b, 2005b, 2006; Kersters *et al.*, 2006; Ciccarelli *et al.*, 2006; Gupta and Sneath, 2007; Gupta and Mok, 2007; Gao *et al.*, 2009). However, despite their phylogenetic and molecular distinctness, these large groups of *Bacteria* are presently not recognized as distinct phyla, whereas numerous other poorly studied bacteria consisting of only a few species are recognized as separate phyla of bacteria.

It is important to point out that when these main groups of *Bacteria* were first described, only a limited number of sequences were available and these groups could be clearly distinguished based on long internal branches that separated them in the 16S rRNA trees (Woese *et al.*, 1985; Woese, 1987). However, with the enormous increase in the number of sequences, boundaries between these groups have become blurred making it difficult to clearly demarcate these groups in phylogenetic terms (Ludwig and Klenk, 2005). Further, except for their branching pattern in phylogenetic trees, for most bacterial groups, no molecular, biochemical or physiological characteristics are known that are unique to them. Hence, a central issue of fundamental importance to microbiology that remains to be understood and resolved is: 'In what aspects do different main groups of bacteria differ from each other and do species from these groups share any unique molecular, biochemical, structural or physiological characteristics that are distinctive of each group?' Another central issue in bacterial phylogeny is to understand how different main groups within *Bacteria* are related to each other and evolved from a common ancestor. Phylogenetic trees based on rRNA and other gene/protein sequences have not been able to resolve these relationships leading to notion that this important problem is insolvable (Doolittle, 1999; Ludwig and Klenk, 2005).

Based on this brief overview, it should be evident that in order to develop a reliable understanding of microbial systematics and phylogeny it is necessary at first to develop *new well-defined (molecular or biochemical) criteria* for identifying all of the main groups or divisions within *Bacteria* in a precise and definitive manner. These new criteria or properties should be such that they should enable identification and circumscription of all of the major taxa (at various taxonomic levels) in clear molecular and/or biochemical terms (Gupta and Griffiths, 2002). Further, it is also of central importance to understand how different groups of *Bacteria* are related to each other and have branched off from a common ancestor (Gupta, 2001).

New molecular markers for systematic and evolutionary studies

The availability of genome sequences from large numbers of microbes in recent years has opened up new windows of opportunities for discovering novel molecular characteristics that are unique for different groups of bacteria and can be used for their identification as well as for biochemical and functional studies (Nelson *et al.*, 2001; Korbel *et al.*, 2005). Comparative genomics provides the primary means for mining information from genomic sequences. These studies are leading to identification of different kinds of molecular markers that are proving of great value for understanding microbial systematics and phylogeny (Gupta, 1998a; Lerat *et al.*, 2005; Gupta and Griffiths, 2006). The ideal markers for such studies should have the following characteristics: '*These markers should be homologous apomorphic characters that evolved only once (synapomorphy) but not by convergence*' (Stackebrandt, 2006). Such markers also should not be affected by factors such as multiple changes at a given site, long-branch attraction effect, differences in evolutionary rates between and among species, lateral gene transfers, etc., which confound the inferences from phylogenetic trees (Delsuc *et al.*, 2005). Our recent work in this area describes two different types of molecular markers or rare genetic changes that generally satisfy these characteristics.

The first of these newly discovered molecular markers consist of conserved insertions and deletions (indels) in gene/protein sequences. The indels that provide useful phylogenetic markers are generally of defined size and they are flanked on both sides by conserved regions to ensure that they are reliable characteristics (Gupta, 1998a, 2000b, 2004, 2005b; Gupta and Griffiths, 2002; Gupta *et al.*, 2003; Gao and Gupta, 2005). Because of the highly specific nature of genetic changes that give rise to a given conserved indel, such changes are less likely to arise independently in different groups or taxa by either convergent or parallel evolution (i.e. homoplasy) (Gupta, 1998a; Rokas and Holland, 2000). Hence, when a conserved signature indel (CSI) of defined size is uniquely found in a phylogenetically defined group(s) of species, the simplest explanation for this observation is that the genetic change responsible for this CSI occurred once in a common ancestor of this group of species and then passed on to various descendents. Because the presence or absence of a given CSI in different species is not affected by factors such as differences in evolutionary rates, CSIs, which are restricted to particular clade(s), have generally provided good phylogenetic markers of common evolutionary descent. In addition, genetic changes leading to CSIs could be introduced at various stages during evolution, it is possible to identify CSIs in gene/protein sequences at different phylogenetic depths corresponding to various high taxonomic groupings (e.g. phylum, order, family or genus). Such CSIs, in turn, can provide well-defined markers for identifying different taxonomic groups of bacteria in molecular terms. Indeed, identified CSIs that are commonly shared by species from a number of different phyla have provided valuable information regarding branching order and interrelationships among different main groups of bacteria (Gupta, 2001, 2003; Gupta and Griffiths, 2002; Griffiths and Gupta, 2004b).

The second kind of molecular markers that have proven very useful for systematic and phylogenetic studies are whole proteins that are uniquely found in particular groups or subgroups of bacteria (Gupta and Lorenzini, 2007; Gupta and Mok, 2007). Comparative analyses of genomic sequences have indicated that such conserved signature proteins (CSPs), which are also referred to as ORFans (i.e. ORFs that have no known homologues), are also present at different phylogenetic depths (Siew and Fischer, 2003; Daubin and Ochman, 2004; Lerat *et al.*, 2005; Dutilh *et al.*, 2008). Recent studies show that many of these CSPs are uniquely present in all species from particular groups (Gao and Gupta, 2007; Gupta and Lorenzini, 2007; Gupta and Mok, 2007); hence it is likely that genes for these proteins evolved once in a common ancestor of these groups and then retained by all of its descendants. Because of their taxa specificity, these CSPs, again provide valuable molecular markers for identifying different groups of species in molecular terms. Further, similar to the CSIs, based upon species distribution patterns of these CSPs, it is again possible to draw robust phylogenetic inferences regarding interrelationships among various bacterial groups.

Our recent work in this area has led to identification of large numbers of CSIs and CSPs that are distinctive characteristics of various main groups within *Bacteria*. Based on these characteristics all of the main groups within *Bacteria* including α-, β-, γ-, δ and ε-*Proteobacteria* (Gupta, 2000b; Gupta, 2005b; Gupta, 2006; Gupta and Sneath, 2007; Gupta and Mok, 2007; Gao *et al.*, 2009), *Cyanobacteria* (Gupta *et al.*, 2003; Gupta, 2009), *Deinococcus-Thermus* (Griffiths and Gupta, 2004a; Griffiths and Gupta, 2007a), *Chlamydiae–Verrucomicrobia* (Griffiths *et al.*, 2005; Griffiths *et al.*, 2006; Gupta and Griffiths, 2006; Griffiths and Gupta, 2007b), *Fibrobacter--Chlorobi–Bacteroidetes* (Gupta, 2004; Gupta and Lorenzini, 2007), *Actinobacteria* (Gao and Gupta, 2005; Gao *et al.*, 2006), *Aquificales* (Griffiths and Gupta, 2004b; Griffiths and Gupta, 2006) and *Firmicutes* (Gupta and Gao, 2009), as well as *Archaea* (Gao and Gupta, 2007), can now be described and circumscribed in molecular terms. Information for some of the CSIs and CSPs for some of these microbial groups is provided in Table 1.1. These newly discovered molecular markers also provide powerful means for the discovery of novel biochemical and physiological characteristics that are unique and distinguishing characteristics of different groups of bacteria. To illustrate the usefulness of these new approaches for understanding microbial phylogeny and systematics, some of the work that has been done in this regard on *Archaea* and *Actinobacteria* is reviewed here.

Molecular markers for Archaea and its main groups

Archaea are widely regarded as one of the three domains of life (Woese, 1987, 1998; Woese *et al.*, 1990; Doolittle, 1999; Ludwig and Klenk, 2005), although a number of observations indicate this group of prokaryotes exhibits a close relationship to the Gram-positive bacteria (Mayr, 1998; Gupta, 1998a,b, 2000a; Koch, 2003; Skophammer *et al.*, 2007). Archaeal species were earlier believed to inhabit only extreme environments such as extremely hot, extremely saline, or very acidic or alkaline conditions (Woese, 1987; Woese *et al.*, 1990). However, recent studies provide evidence that they are widespread in different environments (Pace, 1997). Presently, very few molecular characteristics are known that are uniquely shared by either all archaea or different main groups within archaea (Woese, 1987; Woese *et al.*, 1990). The phylogenetic analyses of cultivable *Archaea* have led to their division into two major groups or phyla designated as *Crenarchaeota* and *Euryarchaeota* (Woese *et al.*, 1990; Ludwig and Klenk, 2005; Gribaldo and Brochier-Armanet, 2006). The species from both these groups, particularly *Euryarchaeota*, are highly diverse in terms of their metabolism and physiology. Based on their metabolic and physiological characteristics and other unique features, five functionally distinct groups within *Euryarchaeota* are currently recognized: methanogens, sulphate reducers, extreme halophiles, cell wall-less archaea, and extremely thermophilic sulphur-metabolizing archaea (Ludwig and Klenk, 2005; Gribaldo and Brochier-Armanet, 2006). The methanogens form the largest group within the *Euryarchaeota* and they are distinguished from all other prokaryotes by their ability to obtain all or most of their energy via the reduction of CO_2 to methane, the process of methanogenesis. However, methanogens are polyphyletic in different phylogenetic trees (Brochier *et al.*, 2005; Bapteste *et al.*, 2005; Gribaldo and Brochier-Armanet, 2006).

We have carried out comprehensive analyses on all available sequenced archaeal genomes to search for CSPs that are unique to either all archaea or its main subgroups (Gao and Gupta, 2007). These studies have identified >1400 proteins that are distinctive characteristics of *Archaea* and its various subgroups and whose homologues are not found in any other organisms. Six of these identified proteins are unique to all *Archaea*, 11 proteins are specific for *Crenarchaeota* and seven proteins are only found in various *Euryarchaeota* (Gao and Gupta, 2007). Additionally, many other proteins are specific for various subgroups within the *Archaea* (e.g. *Sulfolobales, Halobacteriales, Thermococci, Thermoplasmata*, all methanogenic archaea or particular groups of methanogens). Based upon the species distributions of these proteins, the evolutionary stages where the genes for these proteins have probably evolved are shown in Fig. 1.1. These proteins provide novel

Table 1.1 Summary of molecular markers (CSIs and CSPs) for some of the major bacterial phyla

Phylum	CSPs/CSIs	Reference
Actinobacteria	Phylum-specific: 5CSPs/CSIs (CoxI; CTPS; GluRS; 23S rRNA) Subgroups: *CMN subgroup*: 13CSPs/CSI (CarA) *Micrococcineae*: 8CSPs/CSI (S3)	Gao and Gupta (2005), Gao and Gupta (2006)
Deinococcus-Thermus	Phylum-specific: 65CSPs/CSIs (σ70, RpoC, ThrRS, L1, UvrA, Ffh, SerRS) Subgroups: *Deinocococci*: 206CSPs	Griffiths and Gupta (2004), Griffiths and Gupta (2007)
Cyanobacteria	Phylum-specific: CSIs (UvrD, SecA, EF-Tu, S1, Pol I, IMPDH, FtsH, GlgC, PSY, σ70) Subgroups: *Cyano Clade A*: CSI (EF-G) *Other Cyano expcept Clade A*: CSIs (Pol I, DnaX, TrpRS, TSB) *Synechococcophycidae*: CSIs (Pol I, RpoB, KgsA, TyrRS, RpoC)	Gupta (2003, 2009), Gupta *et al.* (2003)
Bacteroidetes, Chlorobi and Fibrobacter	Phylum-specific: 1CSPs/CSIs (RpoC; SHMT) Subgroups: *Bacteroidetes*: 27CSPs/CSIs (GyrB; SecA) *Chlorobi*: 51CSPs/CSIs (DnaE; AlaRS) *Bacteroidetes and Chlorobi*: 5CSPs/CSIs (ATPsyn; FtsK; UvrB)	Gupta and Lorenzini (2007), Gupta (2004), Griffiths and Gupta (2001)
Chlamydiae	Phylum-specific: 59CSPs/CSIs (RpoA; EF-Tu; GyrB; EF-P; LysRS; MgtE; MurA; TrmD) Subgroups: *Chlamydiaceae*: 79CSPs *Chlamydophila*: 20CSPs *Chlamydia*: 20CSPs	Griffiths *et al.* (2005, 2006), Gupta and Griffiths (2006)
Aquificae	Phylum-specific: 10CSPs/CSIs (GidA; RpoC; PolA; EF-Tu; SecA)	Griffiths and Gupta (2004, 2006)
α-Proteobacteria	Phylum-specific: 6CSPs/CSIs (Ctag; PurC; SAICAR Synthetase; DnaB; ATP1, etc. >13) Subgroups: *α-Proteobacteria except Rickettsiales*: 10CSPs/CSIs (Cox I; AlaRS; MutS) *Rickettsiales*: 3CSPs/CSIs (XerD integrase; Lap) *Rickettsiaceae*: 4CSPs/CSIs (Mfd; L19; FtsZ; σ70; ExoVII) *Anaplasmataaceae*: 5CSPs/CSIs (RP-314; Tgt) *Rhodobacterales, Caulobacter and Rhizobiales*: CSIs (DnaA; RP-057) *Rhodobacterales and Caulobacter*: CSI (AsnB) *Rhizobiales*: 6CSPs/CSI (TrpRS) *Rhizobiaceae, Brucellaceae and Phyllobacteriaceae*: CSIs (KGD; LytB; Lep A; SucC; GyrA) *Bradyrhizobiaceae*: 62CSPs/CSIs (SerS; LIG1)	Gupta and Mok (2007), Kainth and Gupta (2005), Guputa (2000, 2005)
γ-Proteobacteria	Phylum-specific: 4CSPs/CSI (PurH) Subgroups: *Enterobacteriales,Pasteurellales, Vibrionales, Aeromonadales and Alteromonadales*: 20CSPs/CSIs (RpoB; L16)	Gao and Gupta (2009), Gupta (2000)
ε-Proteobacteria	Phylum-specific: 49CSPs/CSIs (UvrB; PheRS; RecA; FtsH; RpoC) Subgroups: *Wolinella and Helicobacter:* 11CSPs/CSI (RpoB/RpoC) *Campylobacter:* 18CSPs/CSI (RpoC)	Gupta (2000, 2006)

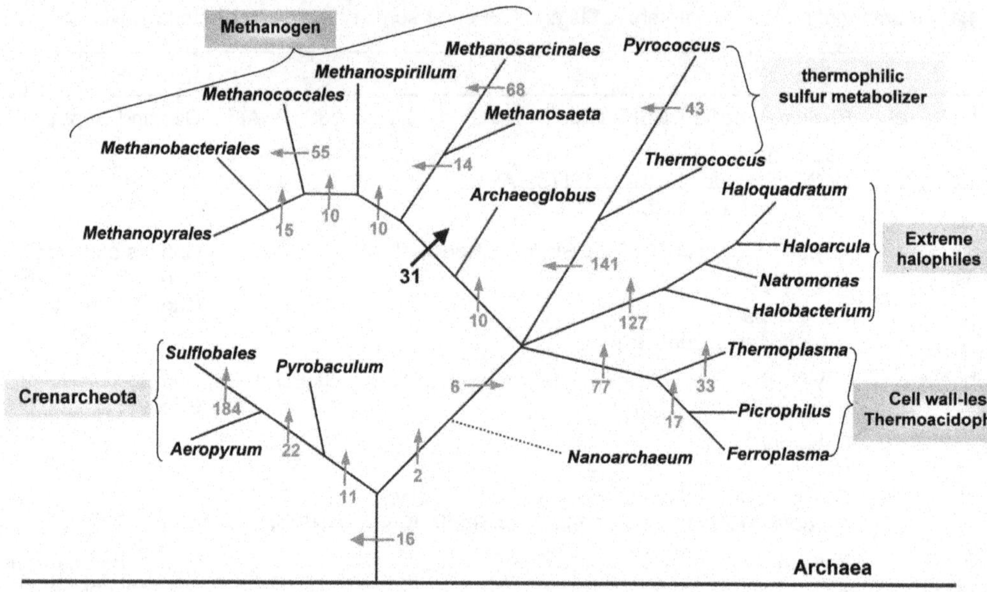

Figure 1.1 A summary diagram showing the species distribution patterns of various *Archaea*-specific proteins. The arrows mark the suggested evolutionary stages where proteins that are uniquely shared by the indicated groups (numbers indicated correspond to the CSPs that are specific for these groups) were introduced. The branching pattern shown here is based upon the species distribution patterns of these proteins and it is unrooted. The dotted line for *Nanoarchaeum* indicates that its placement within *Euryarchaeota* is uncertain. Modified from Gao and Gupta (2007).

molecular markers or signature proteins (CSPs) that are distinctive characteristics of *Archaea* and all of its major subgroups. Most of these proteins are of unknown function (Gao and Gupta, 2007) and further studies on them should lead to discovery of novel biochemical and physiological characteristics that are unique to these groups.

Among the archaea-specific proteins (Gao and Gupta, 2007), of particular significance is the observation that 31 proteins are uniquely present in virtually all methanogens including *Methanopyrus kandleri* (Fig. 1.1). As indicated above, in phylogenetic trees based on 16S rRNA and various proteins sequences, the methanogenic archaea form at least two distinct clusters, with *M. kandleri* branching distinctly from both these clusters (Brochier *et al.*, 2005; Bapteste *et al.*, 2005; Gribaldo and Brochier-Armanet, 2006). The methanogenic archaea in these trees are also interspersed by other groups of non-methanogenic archaea such as *Halobacteriales*, *Archaeoglobus*, *Thermoplasmatales* and *Thermococcales* (Brochier *et al.*, 2005; Bapteste *et al.*, 2005; Gribaldo and Brochier-Armanet, 2006). This has led to important questions concerning the origin of methanogenesis. To account for these results, it has been suggested that methanogenesis evolved once in a common ancestor of different methanogenic archaea, *Halobacteriales*, *Archaeoglobus*, *Thermoplasmatales* and also possibly *Thermococcales*; this was followed by loss of various genes involved in methanogenesis from various other archaeal groups except the methanogens (Brochier *et al.*, 2005; Bapteste *et al.*, 2005; Gribaldo and Brochier-Armanet, 2006). According to this scenario, a methanogenic archaea was the common ancestor of different physiologically and metabolically distinct groups within *Euryarchaeota* and this capability was subsequently independently lost in all other lineages.

In contrast to this proposal, our results showing the presence of 31 proteins that are uniquely found in all methanogens strongly suggest that this group of archaea form a monophyletic lineage exclusive of other archaea (Gao and Gupta, 2007). Importantly, our analyses have also identified 10 additional proteins that are uniquely shared by various methanogens and *Archaeoglobus fulgidus* (Fig. 1.1). In contrast to *A. fulgidus*, no protein was identified that was uniquely shared by various methanogenic archaea and any of the *Halobacterales* or *Thermoplasmatales.* These observations are highly significant because they strongly suggest that *Archaeoglobus* and all of the methanogens shared a common ancestor exclusive of various other archaea (Gao and Gupta, 2007). In other words, the ancestral lineage that led to the origin of methanogenesis very probably evolved from the *Archaeoglobus* lineage (Fig. 1.1). It is also significant that among the proteins that are uniquely shared by *Archaeoglobus* and methanogens, several are parts of the complexes that are important for nitrogen assimilation and methanogenesis. These results support the view that these characteristics have their origin within the *Archaeoglobus* lineage (Gao and Gupta, 2007).

Molecular markers for the Actinobacteria phylum and its subgroups

Gram-positive bacteria with high G+C DNA content are currently recognized as a distinct phylum, *Actinobacteria,* on the basis of their branching in 16S rRNA trees (Embley and Stackebrandt, 1993; Stackebrandt and Schumann, 2000; Ludwig and Klenk, 2005). This phylum constitutes one of the largest groups among *Bacteria,* comprising 130 genera (Garrity *et al.*, 2005). Actinobacterial species exhibit high levels of diversity in terms of their morphology and physiology. They also play important roles in medicine, industry and environment; some genera such as *Streptomyces* are major antibiotic producers while many others (e.g. *Mycobacterium, Corynebacterium, Nocardia, Leifsonia, Tropheryma,* etc.) cause serious human, animal and plant diseases (Embley and Stackebrandt, 1993; Stackebrandt and Schumann, 2000; Ludwig and Klenk, 2005; Ventura *et al.*, 2007). However, except for their distinct branching in phylogenetic trees, until recently no other biochemical or molecular characteristics were known that could distinguish species of this group from all other bacteria (Stackebrandt and Schumann, 2000; Ludwig and Klenk, 2005).

Sequence alignments of various proteins from different bacterial species have identified a number of CSIs that are specific for either all actinobacteria or certain subgroups within them (Gao and Gupta, 2005). An example of a CSI that is specific for the entire *Actinobacteria* phylum is shown in Fig. 1.2. In the partial sequence alignment of cytochrome c oxidase subunit 1 (CoxI) that is presented in this figure, a 2-aa indel is present in a conserved region that is unique to various actinobacterial species, but not seen in any other bacteria (Gao and Gupta, 2005). The shared presence of this 2-aa indel in all actinobacteria strongly indicates that the genetic change leading to this occurred only once in a common ancestor of actinobacteria and passed on to all descendant species. Besides this signature indel, several other CSIs including a 4-aa indel in CTP synthetase, a 5-aa indel in glutamyl-tRNA synthetase (GluRS), and a large insert in the 23S rRNA that are also specific for actinobacteria have been identified (Roller C *et al.*, 1992; Gao and Gupta, 2005). The actinobacteria-specificity of several of these CSIs (viz. CoxI, GluRS and CTP synthase) has been examined by sequencing fragments of these genes from 23 actinobacterial species, covering many different families. All of these gene fragments, except two in GluRS, were found to contain these CSIs providing strong evidence that they are distinctive characteristics of the entire phylum (Gao and Gupta, 2005). In view of their actinobacteria-specificity, these CSIs provide good molecular markers for circumscribing the *Actinobacteria* phylum and distinguishing species of this group from all other bacteria.

In addition to these CSIs, which are specific for all *Actinobacteria,* we have also identified many other CSIs that are specific for certain subgroups within this large phylum. Some of these CSIs also provide information regarding the interrelationships among different subgroups. As an example, in the ribosomal protein S3, we have identified a 5-aa indel that is uniquely shared by various actinobacterial species belonging to the suborders *Micrococcineae* and *Bifidobacterineae* (Fig. 1.3). The species from these two suborders are quite diverse in terms of their phenotypic and physiological characteristics. However, the uniquely shared presence of this indel by various *Micrococcineae* and *Bifidobacterineae* species provide evidence that these two suborders are phylogenetically close and they shared a common ancestor exclusive of other actinobacteria. The CSIs in a number of other proteins, as well as the clustering of these two subgroups in phylogenetic trees based on combined sequences for multiple

Group	Species	Accession	326		368
	Escherichia coli	NP_752476	LSFIVWLHHFFTMG	AG	ANVNAFFGITTMIIAIPTGVKIFNWLF
Proteobacteria	*Pseudomonas aeruginosa*	ZP_00138941	-G-T----------	S-	GD--G---VA--L-S------L-----
	Vibrio vulnificus	NP_762523	---V--A--M--T-	MP	VFAEL--MYC--M--V---------VA
	Yersinia pseudotuberculosis	YP_069492	--------------	S-	---------A----S------------
	Agrobacterium tumefaciens	NP_353177	--YL----------	S-	--------------S----A-------
	Caulobacter crescentus	NP_420580	--Y-----------	S-	-S------------S----A-------
	Bordetella parapertussis	NP_886335	---L--A--M--T-	IP	VVGQL--MYA--L-S---------VA
	Burkholderia fungorum	ZP_00032266	---M--A--M-AT-	MP	VTGQL--MYA--L--V-----V---VA
Other bacteria	*Cytophaga hutchinsonii*	ZP_00309629	------A--M-VT-	MN	PFLGSI-MFL-L---V-SA--A--YIA
	Leptospira interrogans	NP_710424	V--L--G--M-VS-	QS	EFAGVL-SF---LVGV--AI-L---IS
	Prochlorococcus marinus	NP_895169	-GLV--A--MF-S-	TP	PWMRL--T-A-SF--V---I-F----A
	Gloeobacter violaceus	NP_925109	-GFL--A--M--S-	TP	DWLRM--MV-SFL--V---I-V-S--G
	Nostoc punctiforme	ZP_00110624	-GLI--A--M--S-	IP	GWLRM--M-------V---I---S--A
	Chloroflexus aurantiacus	ZP_00355855	-G-L--G--M-VSS	QS	VYAGLI-SFI--LV---SAI-V---TA
	Thermus thermophilus	YP_005640	-GTM--A--M--V-	ES	TLFQIA-AFF-AL--V-----L--IIG
	Deinococcus radiodurans	NP_296339	V-C------M-AV-	IP	EAWQIA-M-S-L-V-V--------LIG
Firmicutes	*Bacillus cereus*	NP_830510	---V----------	--	PA--S--S-S--A-S------------
	Exiguobacterium sp.	ZP_00182647	-GFM--V--M--V-	L-	PVA--I-AVA--A--V-----------
	Geobacillus kaustophilus	YP_149311	------V-------	--	PA--SA-S----A--------------
	Oceanobacillus iheyensis	NP_693175	---V--V-------	Q-	-LT-SI-S----A--V---I------L
	Listeria innocua	NP_469361	---L--V-------	S-	-L--S--S----M------I-------
	Staphylococcus aureus	YP_040448	---L--V-------	N-	-LI-S--S-S--L-G------L----L
Actinobacteria	*Mycobacterium tuberculosis*	NP_217559	--VA--A--M-AT-		-VLLP--SFM-YL--V---I-F---IG
	Corynebacterium efficiens	NP_739028	--MAV-A--MFVT-		-VLLP--SFM-FL-SV-----F---VG
	Nocardia corynebacterioides	AY876134	--IA--A--MYAT-		-VLLP--SFM-FL--V-----F---IG
	Rhodococcus erythropolis	NP_898717	--IA--A--MYAT-		-VLLPY-SFM-FL--V---------IG
	Gordonia westfalica	NP_954783	--VA--A--MYVT-		-VLLP--SFMTFL--V-----F---IG
	Tsukamurella paurometabola	AY876131	--VA--A--MYAT-		-VLLP--SFM-FL--V-----F---I-
	Williamsia murale	AY876139	--VA--A--MYAT-		-VLLP--SFM-F---V-----F---IG
	Streptomyces avermitilis	NP_827224	--VT--A--MYVT-		-VLLP--SFM-FL--V-----F---IG
	Streptosporangium roseum	AY876128	--IT--A--M-PT-		QVLLP--SFM-FL--V-----F---IG
	Microtetraspora niveoalba	AY876123	--MT--A--M-AT-		-ALLP--SMLSFL------I-F---TG
	Planobispora rosea	AY876132	--VA--A--M-PT-		QVLLP--SFM-FL--V-----F---IG
	Thermobifida fusca	ZP_00057854	--MT--A--M-PT-		-VLLP--SFM-FL--V-----F---IG
	Pseudonocardia halophobica	AY876129	--AA--A--MYAT-		-VLL---SF--LL------I-FV--IG
	Saccharopolyspora erythraea	AY876135	--VV--A--MYAT-		-VLLP--AF--FL--V---M-F---IG
	Nocardioides simplex	AY876138	--VA--A--M-VT-		-VNLP--SGM-FL--V-----F---IG

Figure 1.2 Partial alignment of cytochrome c oxidase subunit 1 sequences showing a 2-aa indel (boxed) that is specific for various actinobacterial species. Dashes in all sequence alignments indicate identity with the amino acid on the top line. Modified from Gao and Gupta (2005).

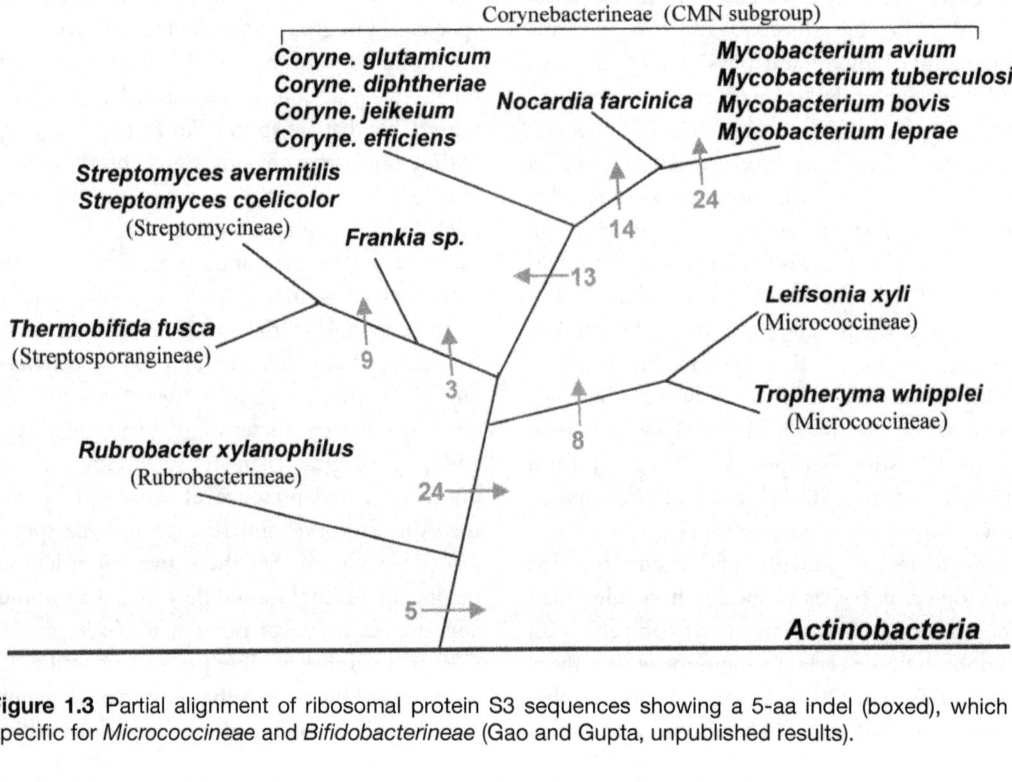

Figure 1.3 Partial alignment of ribosomal protein S3 sequences showing a 5-aa indel (boxed), which is specific for *Micrococcineae* and *Bifidobacterineae* (Gao and Gupta, unpublished results).

proteins, also support this inference. Thus, these CSIs provide useful molecular markers to define and circumscribe this large clade (*Micrococcinea e* + *Bifidobacterineae* suborders) of *Actinobacteria.*

Besides identifying CSIs that are specific for actinobacteria, comprehensive BLAST searches have been carried out on each ORF from four actinobacterial genomes (viz. *Mycobacterium leprae* TN, *Leifsonia xyli* subsp. xyli str. CTCB07, *Bifidobacterium longum* NCC2705 and *Thermobifida fusca* YX) to identify CSPs that are specific for actinobacteria or its various subgroups. In this study, a total of 233 actinobacteria-specific proteins were identified, homologues of which are generally not found in any other bacteria (Gao *et al.,* 2006). Based on their species distribution pattern, these proteins could be grouped as actinobacterial phylum-specific, suborder-specific, family-specific, genus-specific and so on (Gao *et al.,* 2006). For example, five CSPs are shared by all actinobacteria, including *Rubrobacter,* which is the deepest branch within this phylum; 16 CSPs are specific for the suborder *Corynebacterineae;* and 24 CSPs are unique to the genus *Mycobacterium* (Fig. 1.4). These proteins provide novel molecular means for defining and circumscribing the *Actinobacteria* phylum and its different subgroups (Gao *et al.,* 2006).

As is the case for the *Archaea,* the species distribution patterns of various actinobacteria-specific CSPs also provide valuable information regarding interrelationships and branching order among various actinobacterial subgroups (Gao *et al.,* 2006). For example, nine of the identified CSPs are uniquely shared by *Thermobifida fusca* and *Streptomyces* species, which for the first time suggests that the species from these two subgroups are closely related and they probably shared a common ancestor exclusive of other actinobacteria (Fig. 1.4). Within the *Corynebacterineae* (CMN subgroup), 17 proteins are shared only by *Mycobacterium* and *Nocardia,*

Group	Species	Accession			
Micrococcineae	*Arthrobacter sp.*	YP_832448.1	MGQKVNPHGFRLGITTDHVSHWFADST	KAGQR	YKDFVREDIRIR
	Renibacterium salmoninarum	YP_001625305	----------------N---------N	-----	---------K--
	Kocuria rhizophila	YP_001854475	----I--N--------------Y---N	QP---	---YI---VK--
	Clavibacter michiganensis	YP_001223356	-------Y-----------R--S---	-K---	-S-Y-A--V---
	Micrococcus luteus	ZP_02944562	----I--N------------------H	-E---	-A--LK--VK--
	marine actinobacterium	ZP_01129342	-------Y-----------R--S---	-P---	-S-Y-A--VK--
	Janibacter sp.	ZP_00993899	---------------SE-R-R------	-E---	-R-Y-K--VA--
	Leifsonia xyli	YP_0628481	-------Y-----------R--S---	-K---	-S-YLA--VK--
	Tropheryma whipplei	NP_7876761	----I--Y-L------------YS---	RP---	-A-Y-S---K--
	Brevibacterium linens	ZP_00379557	----I--N---------K-K------	-P---	-S-Y-L--VK--
	Kineococcus radiotolerans	YP_001360447	-------F----------R-R------	-T---	-A-Y-K--VA--
	Actinomyces odontolyticus	ZP_02043500	-------R----------R-R--S---	TK---	-A-Y-A--VA--
Bifidobacterineae	*Bifidobacterium longum*	ZP_00121721	----I--F-Y-----EN-R-K--S-N	--E--	-R---L--DQ--
	Bifidobacterium dentium	ZP_02917074	----I--F-Y-----EN-R-K--S-N	--E--	-R---L--DA--
	Bifidobacterium adolescentis	ZP_02028160	----I--F-Y-----EN-C-K--S-N	V-E--	-R---L--DA--
	Bifidobacterium animalis	ZP_02963319	----I--F-Y-----ES-R-K--S-N	V-E--	-R---L--DA--
Other Actinobacteria	*Thermobifida fusca*	YP_2906961	-------------V---FK-R----KL		---Y-K--VA--
	Streptomyces coelicolor	NP_6288671	-----------------FK-R-Y--KL		---Y-K--VA--
	Nocardia farcinica	YP_1169481	----I------------WK-R-Y--KQ		-A-Y-K--VA--
	Frankia sp.	YP_001510293	---------------SEFT-R-Y--KQ		--AY-G--VK--
	Mycobacterium tuberculosis	NP_2152211	----I------------WK-R-Y--KQ		-AEY-K--VA--
	Corynebacterium efficiens	NP_7371381	----IH---L-----S-WK---Y--QS		-A-Y-A---K--
	Salinispora arenicola	YP_001539081	-----H--------S--WK-R----KL		---YIG--VK--
	Propionibacterium acnes	YP_0565391	----I--------V----KTR-Y-EKQ		-AEL-G--VK--
Firmicutes	*Bacillus pumilus*	YP_001485368	-------V-L-I-VIR-WE-K---GKD		-A--LH--LK--
	Oceanospirillum sp.	ZP_01168436	-----H-T-I----VK--T-V-Y--KS		E-ASKLQN-LQV-
	Moorella thermoacetica	YP_4312791	-----H-K-L----IR-WD-R-Y-GKE		--TLLH--LK--
	Leuconostoc mesenteroides	YP_8177061	----I--T---V-VIR-WDAK----KA		D-ANQLH--L---
	Mycoplasma hyopneumoniae	YP_2875852	-------N---F---RN-NAI-Y--KN		KFSINLL--VK--
Other bacteria	*Deinococcus geothermalis*	YP_6053251	--N-I--N-------KGWN-R-Y-GKK		Q-SQLLK--EK--
	Prochlorococcus marinus	NP_8936691	--N-I--T-------QE-R-K---T-K		T-PTLLQ--DK--
	Nostoc sp.	NP_4882491	----IH-V-------QE-Q-R---EPS		--PELLQ--HKL-
	Leptosira terrestris	YP_001382203	-----H-L---V---KK-R-L--SN-K		Q-PQS-L--YV--
	Pseudomonas syringae	NP_7904791	-----H-I-I----VKE-T-V-Y--GR		T-A-YLFA-LKV-
	Sinorhizobium medicae	YP_001326680	----I--I------NRTWD-R----NA		E-GQLLH--LK--
	Neisseria meningitidis	NP_2732061	----I--T----AV-K-WA-K---K--		DFSTVLKQ--DV-
	Agrobacterium radiobacter	YP_002544211	----I--I------NRTWD-R----NA		E-GQLLH--LKM-
	Coxiella burnetii	ZP_01947564	-------V-M-I---R-WT-N-Y--KK		DFA-RLN--LNV-
	Bordetella petrii	YP_001633565	----IH-T----AV-RNWS-R----DK		AFGSMLA----V-
	Caulobacter sp.	YP_001683247	-------V-L----NRTWD-R----GE		Q-GKLLHQ-LAV-
Eukaryotes	*Arabidopsis thaliana*	NP_0510961	----I--L-----T-QS-H-L---QPK		K-SEGLE--KK--
	Lotus japonicus	NP_0848351	----I--L-----T-QS-D-I---QP-		N-SENLK--KK--
	Zea mays	NP_0430621	----I--L-----T-QN-H-F---QPK		N-SEGLQ--KK--

Figure 1.4 A summary diagram showing the species distribution patterns of various actinobacteria-specific proteins. The arrows mark the suggested evolutionary stages where the CSPs that are specific for different groups were probably introduced. Modified from Gao *et al.* (2006).

suggesting that these two genera are more closely related to each other than to *Corynebacterium*. Overall, the relationships suggested by the species distribution patterns of these CSIs and CSPs are consistent with the branching of these species in phylogenetic trees (Gao *et al.*, 2006). However, in many cases, these molecular characteristics have provided new insights into the branching order and interrelationships among different subgroups that were not resolved by phylogenetic trees (Gupta, 1998a; Gao *et al.*, 2006).

The above are just some examples of the conserved CSIs and CSPs that have been discovered by recent comparative genomic analyses. Table 1.1 gives an overview of the large numbers of such characteristics that have been discovered in recent years. Based on these molecular markers, all of the major groups within *Bacteria* as well as *Archaea* can now be identified in definitive terms. These studies have also been instrumental in the discovery of several CSIs that have provided valuable information regarding the branching order among the major bacterial phyla. An example of one such CSI in the Hsp70 protein is provided in Fig. 1.5. In this case, a 21–23 aa insert (boxed) is present in all of the Gram-negative bacteria that are bounded with two membranes (i.e. diderm prokaryotes) (Gupta and Golding, 1993; Gupta, 1998a). Evidence has been presented that this large indel, which is also lacking in various *Archaea*, is an insert in the Gram-negative bacteria (see Fig. 1.5) thereby indicating that the bacterial phyla lacking this indel (viz. *Firmicutes*, *Actinobacteria* and *Thermotogae*) are ancestral

Group	Species	Accession	61					123
Proteo-Bacteria	Escherichia coli	SP/ P04475	NPQNTLFAIKRLIG	RRF	QD	EEVQRDVSIMPFKIIAAD	NGDAWVEVKGQK	MAPPQISAEVLKKM
	Xylella fastidiosa	AAF85139	--K--FY-V-----	-K-	G-	A---K-LDLV-Y--TQH-	------ATADAK	L--QE---K--E--
	Pse. aeruginosa	AAG08147	------Y-V-----	---	EE	NV--K-IQMV-YS-VK--	-----------	-------------
	Vibrio cholerae	2623605	--------------	---	E-	------IK---Y--VK--	-------A----	--A--V-----L--
	Ral. solanacearum	16429657	--R---Y-V-----	-K-	EE	K---K-IGL--YT-SK--	--------RDK-	-----------R--
	Nei. meningitidis	H81185	-AK--IY-A-----	HK-	E-	K-----IES---E--K-N	------KAQ-KE	LS---------R--
	Mesorhizobium loti	13473981	--E--I--V-----	--Y	D-	PVTEK-KKLV-Y--VKG-	-------AG-K-	QS-S----MI-Q--
	Rickettsia conorii	15892156	--R--IY-V-----	-N-	I-	PM-RK-QG-V-YN-VK--	-------ADNN-	YS-S----FI-Q--
	Caulo. crescentus	GB/ M95799	--T-----------	-TA	S-	PV-EK-KGMV-YRSSR-R	A-----KAH-KD	YS-QEV--FI-Q--
	Myxococcus xanthus	1805284	--E--V--A-----	-K-	DS	P-GKKAIGVS---VASSP	-------IR-KG	YS--EV--I--M--
	Geo. metallireducens	ZP_00080430	--E-----------	-KY	DT	---RK-I--S----VK--	-------AR-KM	YSA-E---M--Q--
	Des. vulgaris	ZP_00130430	--ER-V--V---M-	--G	DA	P--G-WKEHS-YR-V-GA	----A---Q-RP	YSA-E---MI-G-L
	Campylobacter jejuni	CAB73024	--EK-IYS---IM-	LMI	NE	DAAKEAKNRL-YH -TER	--ACAI-IA-KI	YT-QE---K--M-L
	Helicobacter pylori	2072520	--EK-IYS---IM-	LM-	NE	DKAKEAEKRL-Y--VDRN	GA-CAI-IS-KI	YT-QE---KI-M-L
Other Gram-Negative Bacteria	Chl. trachomatis	SP/ P17821	--EK--AST--F--	-K-		S--ESEIKTV-Y-VAPNS	K---VFD-EQKL	YT-EE-G-QI-M--
	Chlam. pneumoniae	AAF38114	--EK--GS---F--	-KY		S--ASEIQTV-YTVTSGS	K---VF--D-KQ	YT-EE-G-QI-N--
	Prot. amoebophila	YP_008498	--E--ITSS--F--	-KY		Q--LSEIKTV-Y-VTNNN	----VF-IQ-KI	VT-EE-A-QI-I--
	Lentisphaera araneosa	ZP_01875441	--T--I-SA--FM-	-K-		A--KHEIDLV-YEVVE-K	----HIK-GDKV	YS--E---KIIA-L
	Cyt. aquatilis	AF013110	--TK-IAS---FM-	HT-		A-TTNESKRVSY-VVKGV	QQYST-DID-RL	YTAQEL--MT-Q--
	Fib. succinogenes	AY017382	--EK-IYS---FM-	-TA		G-CSAAEKN--Y-LVGTG	SDPVR-QIDDKQ	F---E---A--QT-
	Por. gingivalis	Q9ZAD3	--TK-IYS---FM-	ETY		DQ-S-E-ERV---VVRG-	-NTPR-DID-RL	YT-QE---MI-Q-L
	Chlrobium tepidum	AF130447	--K--I-S---FM-	-KY		D--PNEKKLASYDVVN-E	G-Y-K-KIGDKT	YS-QE---MI-Q--
	Hydro. marinus	AY188440	D----IYES--F--	-K-		D--KEEIKYV-Y-VV-D-	K---AFD-PNAGKIVR-EEVG-Q----L	
	Aquifex aeolicus	2983493	D-E--VYES--F--	-K-		N---KEAKRVSY-VVPDE	K---AFDIP-KL	VR-EEVG-H--R-L
	Treponema pallidum	3322484	--EH-IYS---F--	S--		N-LTGEAKKV-Y--VPQG	-D-VR---E-KL	YSTQE---FI-Q--
	Bor. burgdorferi	SP/ P28608	--E--IYS---FM-	---		---ASEIKMV-Y--EKGL	----R-NISNI-KQ-S--E---AT-T--	
	Lep. interrogans	AAN50903	-AV--IRSA--F--	--L		N-CESEMKHVSY-V-RSG	-EGVKF-TSAGE	FT-QE---R--M--
	Synechococcus PCC7942	GB/ D29668	--R--FAN---F--	--Y		D-LTDESKRV-YTVRRDP	E-NVRIVCPQLSREF--EEVA-MI-R-L	
	Synechocystis PCC6803	GB/ M57518	--G--FYSV--F--	-K-		D-ITNEATEVAYSVVKDG	--NVKLDCPA-GKQF--EE---Q--R-L	
	Anabaena variabilis	2073390	-----F--V--Y--	--Y		N-LSPESKRV-YT-RKD-	V-NIK-ARLNKE	F-AEE---M----L
	Cfx. aurantiacus	AF130446	--E---YSV--F--	-S-		D--TEERDV----VVKGP	RN-VRIY-PQTNKEY--QE---M--Q-L	
	Herpeto. aurantiacus	YP_001543093	--E--I-S---FM-	LE-		DKATQER-TV-YE-VNGP	R---R-R-PRTGKDYT-QE---MT-Q-L	
	D. radiodurans	A75557	--AA---EV--F--	--W		D--KEEAARS--TVKEGP	S-SVRI--N-KD	L--E-V-----R-L
	Thermus aquaticus	GB/ Y07826	--EG-I-E---F--	---		----EEAKRV-Y-VVPGP	D-GVR-----KL	YT-EE---MI-R-L
Thermotogae	Thermotoga maritima	3882431	--ER-IKS---KM-				T-YK-RIDDKE	YT-QE---FI---L
	Fervido. nodosum	YP_001410713	-SDR-IKS---KM-				S-YK-RIDDKE	YT-QE---YI---L
Actinobacteria	Mycobacterium leprae	SP/ P19993	-VDR-IRSV--HM-				S-WSI-ID-K-	YTAQE---R--M-L
	Str. coelicolor	GB/ 014499	-VDR-IRSV--HM-				T-WK-NLD-KD	FN-Q----F--Q-L
	Bifido. longum	AAN24348	-VDR-ISSV--HM-				S-WT-DID-K-	WT-QE--AQILMKL
	Cor. glutamicum	Q8NLY6	-VDR-IRSV--H--				T-WS-AIDDKN	YTSQE---RT-M-L
Firmicutes	Clo. acetobutylicum	118726	--DK-IIS---KM-				TAEK-AIDDKN	YT-QE---MI-Q-L
	Bacillus subtilis	GB/ U39711	-- --IMS---HM-				T-YK--IE-KD	YT-QEV--II-QHL
	Mycoplasma genitalium	1361531	-- --IVS----M-				TSNK-K-TTKE	LS-E-V--QI-SYL
	Strep. mutans	2145133	-- E-ILS--SKM-				TSEK-SANAKE	YT-QE---MI-QYL
	Lactococcus lactis	GB/ M98865	-- E-IIS--SKM-				TSEK-SAN-KE	YT-QE---MI-QNL
	Sta. aureus	GB/ X75428	-- --VQS---HM-				T-YK-DIE-KS	YT-QE---MI-QNL
Archaea	The. acidophilum	7522204	--EG-I--A--KM-				T-YKFK-FDKE	FT-Q----FI-Q-I
	Methanosarcina mazei	1169378	--D--VYS---HM-				EANYK-TLN-KD	YT-QE---MI-Q-L
	H. marismortui	232284	--DE-IQS---HM-				QD-YS--LD-EE	YT-E-V--MI-Q-I

Figure 1.5 Excerpts from DnaK sequence alignment showing a 21–23 insert (boxed) that is uniquely found in all Gram-negative (diderm) bacteria but absent in various Gram-positive bacteria and *Archaea* (Gupta and Golding, 1993; Gupta *et al.*, 1994; Gupta, 1998a). A 2-aa insert within this large insert is specific for various Proteobacteria (Gupta, 2000b). Dashes (–) indicate identity with the amino acid on the top line.

(Gupta and Golding, 1993; Gupta, 1998a; Lake *et al.*, 2007). The presence of this large insert in all of the eukaryotic Hsp70 homologues also provides strong evidence for their origin from an insert-containing Gram-negative bacterium (Gupta *et al.*, 1994; Gupta, 1998a). Within this large insert, a 2-amino-acid insert that is specific for various proteobacteria has also been identified (smaller box in Fig. 1.5) (Gupta, 1998a, 2000b). Based upon the species distribution patterns of many other CSIs of this kind in widely distributed proteins, it is now possible to logically deduce that the various main groups or phyla within the *Bacteria* have evolved in the following order; (from the earliest to the most recent) *Firmicutes* → *Actinobacteria* → *Fusobacteria–Thermotogae* → *Deinococcus–Thermus–Chloroflexi* → *Cyanobacteria* → *Spirochetes* → *Fibrobacteres–Chlorobi–Bacteriodetes* → *Chlamydiae- Verrucomicrobi–Planctomyctes* → *Aquificae* → *Proteobacteria* (δ and ε) → *Proteobacteria* (α) → *Proteobacteria* (β) and → *Proteobacteria* (γ) (Gupta, 1998a, 2001, 2003, 2005a; Gupta and Griffiths, 2002; Griffiths and Gupta, 2004b, 2007b).

Importance of CSIs and CSPs for discovering biochemical and physiological characteristics that are unique to different groups of bacteria

As indicated earlier, virtually all of the higher taxonomic groups within the *Bacteria* are currently identified only in phylogenetic terms. For most of them, no biochemical or physiological characteristics are known that are unique to them. In this context, our discovery of numerous CSIs and CSPs that are specific for different groups of bacteria is highly significant as they provide powerful new means for discovering novel biochemical and physiological characteristics that are unique to these groups. Hence, functional studies on these lineage-specific CSIs and CSPs should prove of great value in filling this important gap in our knowledge regarding different groups of microbes. In this regard it should be noted that most of the discovered CSIs are present in widely distributed proteins (viz. CoxI, Hsp60, Hsp70, RpoB, EF-Tu, CTP synthase, GluRS, RpoB, RpoC, Gyrase A and B subunits, etc.) that are involved in essential functions. The primary biochemical functions of these proteins are vital for cell survival and they are expected to remain the same in all organisms. Hence, the question arises: what is the functional significance of these evolutionarily preserved CSIs that are specific for different bacterial lineages? In our recent work, we have provided evidence that a number of conserved indels that are found in the Hsp60 (GroEL) and Hsp70 (DnaK) proteins are essential for the groups of species where they are found and deletions or most other changes in these CSIs resulted in failure to support cellular growth (Singh and Gupta, 2009). Because of the presence of these CSIs on the surfaces of various essential proteins, we have suggested that these CSIs confer some new functional capabilities (i.e. ancillary functions) on these essential proteins (Singh and Gupta, 2009). These ancillary functions are expected to be important for the lineages where these CSIs are found and they could include ability of the protein(s) to interact with some other cellular proteins or ligands (with the CSI serving as a docking site) that either modulate the activity, or confer some new function on these proteins. Hence, further studies on understanding how these CSIs modify the cellular functions of various proteins are of much interest as they will probably reveal novel biochemical and physiological properties that are unique to different groups of bacteria.

Unlike the lineage-specific CSIs, which are found in widely distributed proteins, virtually all of the lineage-specific CSPs are of unknown functions. The discovery of these proteins points to our lack of knowledge regarding many fundamental aspects of cells, particularly functions that are specific or unique for different bacterial groups. Hence, an important challenge for the future is to understand the cellular functions of these genes/proteins (Danchin, 1999; Galperin and Koonin, 2004; Roberts, 2004; Gupta and Griffiths, 2006), which should provide valuable insights into the biochemical and physiological characteristics that are unique to different groups of bacteria at various taxonomic levels.

Concluding remarks

The availability of genome sequence data is enabling identification of numerous rare genetic changes consisting of CSIs and CSPs that are specific for different groups of prokaryotes at various taxonomic levels. The discovery of these new molecular markers is proving very useful in understanding some of the critical issues in prokaryotic phylogeny and systematics. In particular, based on these molecular markers, it is now possible to identify and circumscribe most of the major phyla, as well as many of their subgroups, within *Bacteria* and *Archaea* in definitive molecular terms. Additionally, these molecular markers are also providing means to logically delineate the branching order of the main phyla within *Bacteria.* Because of their taxa specificity, functional studies on these newly discovered molecular markers hold much promise for discovering novel biological characteristics that are distinctive characteristics of different groups of prokaryotes, and potentially for eukaryotes as well.

References

Bapteste, E., Brochier, C., and Boucher, Y. (2005). Higher-level classification of the Archaea: evolution of methanogenesis and methanogens. Archaea *1*, 353–363.

Brochier, C., Forterre, P., and Gribaldo, S. (2005). An emerging phylogenetic core of Archaea: phylogenies of transcription and translation machineries converge following addition of new genome sequences. BMC Evol. Biol. *5*, 36.

Cavalier-Smith, T. (2002). The neomuran origin of archaebacteria, the negibacterial root of the universal tree and bacterial megaclassification. Int. J. Syst. Evol. Microbiol. *52*, 7–76.

Ciccarelli, F.D., Doerks, T., von Mering, C., Creevey, C.J., Snel, B., and Bork, P. (2006). Toward automatic reconstruction of a highly resolved tree of life. Science *311*, 1283–1287.

Danchin, A. (1999). From protein sequence to function. Curr. Opin. Struct. Biol. *9*, 363–367.

Darwin, C. (1859) The Origin of Species by Means of Natural Selection, or the Preservation of Favoured Races in the Struggle for Life (London: John Murray).

Daubin, V. and Ochman, H. (2004). Bacterial genomes as new gene homes: the genealogy of ORFans in *E. coli.* Genome Res. *14*, 1036–1042.

Delsuc, F., Brinkmann, H., and Philippe, H. (2005). Phylogenomics and the reconstruction of the tree of life. Nat. Rev. Genet. *6*, 361–375.

Doolittle, W.F. (1999). Phylogenetic classification and the universal tree. Science *284*, 2124–2128.

Dutilh, B.E., Snel, B., Ettema, T.J., and Huynen, M.A. (2008). Signature genes as a phylogenomic tool. Mol. Biol. Evol. *25*, 1659–1667.

Embley, T.M. and Stackebrandt, E. (1993). The molecular phylogeny and systematics of the actinomycetes. Annu. Rev. Microbiol. *48*, 257–289.

Galperin, M.Y. and Koonin, E.V. (2004). 'Conserved hypothetical' proteins: prioritization of targets for experimental study. Nucleic Acids Res. *32*, 5452–5463.

Gao, B. and Gupta, R.S. (2005). Conserved indels in protein sequences that are characteristic of the Phylum *Actinobacteria.* Int. J. Syst. Evol. Microbiol. *55*, 2401–2412.

Gao, B. and Gupta, R.S. (2007). Phylogenomic analysis of proteins that are distinctive of *Archaea* and its main subgroups and the origin of methanogenesis. BMC Genomics *8*, 86.

Gao, B., Mohan, R., and Gupta, R.S. (2009). Phylogenomics and protein signatures elucidating the evolutionary relationships among the Gammaproteobacteria. Int. J. Syst. Evol. Microbiol. *59*, 234–247.

Gao, B., Parmanathan, R., and Gupta, R.S. (2006). Signature proteins that are distinctive characteristics of *Actinobacteria* and their subgroups. Antonie van Leeuwenhoek *90*, 69–91.

Garrity, G.M., Bell, J.A., and Lilburn, T.G. (2005). The Revised Road Map to the Manual. In: Bergey's Manual of Systematic Bacteriology, Volume 2, Part A, Introductory Essays, pp. 159–220. Eds D. J. Brenner, N. R. Krieg, J. T. Staley (New York: Springer).

Gould, S.J. (1994). The Evolution of Life on the Earth. Sci. Am. Oct., 85–91.

Gribaldo, S. and Brochier-Armanet, C. (2006). The origin and evolution of Archaea: a state of the art. Philos. Trans. R. Soc. Lond. B Biol. Sci. *361*, 1007–1022.

Griffiths, E. and Gupta, R.S. (2004a) Distinctive protein signatures provide molecular markers and evidence for the monophyletic nature of the *Deinococcus-Thermus* phylum. J. Bacteriol. *186*, 3097–3107.

Griffiths, E. and Gupta, R.S. (2004b) Signature sequences in diverse proteins provide evidence for the late divergence of the order *Aquificales.* Int. Microbiol. *7*, 41–52.

Griffiths, E. and Gupta, R.S. (2006). Molecular signatures in protein sequences that are characteristics of the Phylum Aquificales. Int. J. Syst. Evol. Microbiol. *56*, 99–107.

Griffiths, E. and Gupta, R.S. (2007a) Identification of signature proteins that are distinctive of the Deinococcus–Thermus phylum. Int. Microbiol. *10*, 201–208.

Griffiths, E. and Gupta, R.S. (2007b) Phylogeny and shared conserved inserts in proteins provide evidence that Verrucomicrobia are the closest known free-living relatives of chlamydiae. Microbiology *153*, 2648–2654.

Griffiths, E., Petrich, A., and Gupta, R.S. (2005). Conserved indels in essential proteins that are distinctive characteristics of *Chlamydiales* and provide novel means for their identification. Microbiology *151*, 2647–2657.

Griffiths, E., Ventresca, M.S., and Gupta, R.S. (2006). BLAST screening of chlamydial genomes to identify signature proteins that are unique for the *Chlamydiales, Chlamydiaceae, Chlamydophila* and *Chlamydia* groups of species. BMC Genomics *7*, 14.

Gupta, R.S. (1998a) Protein phylogenies and signature sequences: A reappraisal of evolutionary relationships among Archaebacteria, Eubacteria, and Eukaryotes. Microbiol. Mol. Biol. Rev. *62*, 1435–1491.

Gupta, R.S. (1998b) What are archaebacteria: Life's third domain or monoderm prokaryotes related to

Gram-positive bacteria? A new proposal for the classification of prokaryotic organisms. Mol. Microbiol. *29*, 695–708.

Gupta, R.S. (2000a) The *Natural* evolutionary relationships among prokaryotes. Crit. Rev. Microbiol. *26*, 111–131.

Gupta, R.S. (2000b) The phylogeny of Proteobacteria: relationships to other eubacterial phyla and eukaryotes. FEMS Microbiol. Rev. *24*, 367–402.

Gupta, R.S. (2001). The branching order and phylogenetic placement of species from completed bacterial genomes, based on conserved indels found in various proteins. Int. Microbiol. *4*, 187–202.

Gupta, R.S. (2002). Phylogeny of Bacteria: Are we now close to understanding it? ASM News *68*, 284–291.

Gupta, R.S. (2003). Evolutionary relationships among photosynthetic Bacteria. Photosynth. Res. *76*, 173–183.

Gupta, R.S. (2004). The phylogeny and signature sequences characteristics of *Fibrobacters, Chlorobi* and *Bacteroidetes*. Crit. Rev. Microbiol. *30*, 123–143.

Gupta, R.S. (2005a) Molecular sequences and the early history of life. In: *Microbial Phylogeny and Evolution: Concepts and Controversies*, pp. 160–183. Ed. J. Sapp (New York: Oxford University Press).

Gupta, R.S. (2005b) Protein signatures distinctive of Alpha proteobacteria and its subgroups and a model for Alpha proteobacterial evolution. Crit. Rev. Microbiol. *31*, 135.

Gupta, R.S. (2006). Molecular signatures (unique proteins and conserved Indels) that are specific for the epsilon proteobacteria (Campylobacterales). BMC Genomics 7, 167.

Gupta, R.S. (2009). Protein signatures (molecular synapomorphies) that are distinctive characteristics of the major Cyanobacterial clades. Int. J. Syst. Evol. Microbiol. 59.

Gupta, R.S. and Gao, B. (2009). Phylogenomic analyses of clostridia and identification of novel protein signatures that are specific to the genus Clostridium sensu stricto (cluster I). Int. J. Syst. Evol. Microbiol. *59*, 285–294.

Gupta, R.S. and Golding, G.B. (1993). Evolution of HSP70 gene and its implications regarding relationships between archaebacteria, eubacteria, and eukaryotes. J. Mol. Evol. *37*, 573–582.

Gupta, R.S. and Griffiths, E. (2002). Critical issues in Bacterial phylogenies. Theor. Popul. Biol. *61*, 423–434.

Gupta, R.S. and Griffiths, E. (2006). Chlamydiae-specific proteins and indels: novel tools for studies. Trends Microbiol. *14*, 527–535.

Gupta, R.S. and Lorenzini, E. (2007). Phylogeny and molecular signatures (conserved proteins and indels) that are specific for the Bacteroidetes and Chlorobi species. BMC Evol. Biol. 7, 71.

Gupta, R.S. and Mok, A. (2007). Phylogenomics and signature proteins for the alpha Proteobacteria and its main groups. BMC Microbiol. 7, 106.

Gupta, R.S. and Sneath, P.H.A. (2007). Application of the character compatibility approach to generalized molecular sequence data: Branching order of the proteobacterial subdivisions. J. Mol. Evol. *64*, 90–100.

Gupta, R.S., Aitken, K., Falah, M., and Singh, B. (1994). Cloning of *Giardia lamblia* heat shock protein HSP70 homologs: implications regarding origin of eukaryotic cells and of endoplasmic reticulum. Proc. Natl. Acad. Sci. U.S.A. *91*, 2895–2899.

Gupta, R.S., Pereira, M., Chandrasekera, C., and Johari, V. (2003). Molecular signatures in protein sequences that are characteristic of Cyanobacteria and plastid homologues. Int. J. Syst. Evol. Microbiol. *53*, 1833–1842.

Kasting, J.F. (1993). Earth's early atmosphere. Science *259*, 920–926.

Kersters, K., Devos, P., Gillis, M., Swings, J., Vandamme, P., and Stackebrandt, E. (2006). Introduction to the Proteobacteria. In: The Prokaryotes: A Handbook on the Biology of Bacteria, M. Dworkin, S. Falkow, E. Rosenberg, K. H. Schleifer, E. Stackebrandt, eds (New York: Springer), pp. 3–37.

Koch, A.L. (2003). Were Gram-positive rods the first bacteria? Trends Microbiol. *11*, 166–170.

Korbel, J.O., Doerks, T., Jensen, L.J., Perez-Iratxeta, C., Kaczanowski, S., Hooper, S.D., Andrade, M.A., and Bork, P. (2005). Systematic association of genes to phenotypes by genome and literature mining. PLoS Biol. 3, e134.

Lake, J.A., Herbold, C.W., Rivera, M.C., Servin, J.A., and Skophammer, R.G. (2007). Rooting the tree of life using nonubiquitous genes. Mol. Biol. Evol. *24*, 130–136.

Lerat, E., Daubin, V., Ochman, H., and Moran, N.A. (2005). Evolutionary Origins of Genomic Repertoires in Bacteria. PLoS Biol. *3*, e130.

Ludwig, W. and Klenk, H.-P. (2005). Overview: A phylogenetic backbone and taxonomic framework for prokaryotic systamatics. In: Bergey's Manual of Systematic Bacteriology, D. J. Brenner, N. R. Krieg, J. T. Staley, G. M. Garrity, eds (Berlin: Springer-Verlag), pp. 49–65.

Maidak, B.L., Cole, J.R., Lilburn, T.G., Parker, C.T., Jr., Saxman, P.R., Farris, R.J., Garrity, G.M., Olsen, G.J., Schmidt, T.M., and Tiedje, J.M. (2001). The RDP-II (Ribosomal Database Project). Nucleic Acids Res. *29*, 173–174.

Margulis, L. (1993). Symbiosis in Cell Evolution. (New York: W.H. Freeman and Company).

Mayr, E. (1998). Two empires or three? Proc. Natl. Acad. Sci. U.S.A. *95*, 9720–9723.

Nelson, K.E., Paulsen, I.T., and Fraser, C.M. (2001). Microbial Genome Sequencing: A window into Evolution and Physiology. ASM News *67*, 310–317.

Olsen, G.J. and Woese, C.R. (1993). Ribosomal RNA: a key to phylogeny. FASEB J. *7*, 113–123.

Pace, N.R. (1997). A molecular view of microbial diversity and the biosphere. Science *276*, 734–740.

Roberts, R.J. (2004). Identifying protein function--a call for community action. PLoS Biol. 2, E42.

Rokas, A. and Holland, P.W. (2000). Rare genomic changes as a tool for phylogenetics. Trends Ecol. Evol. *15*, 454–459.

Roller C., Ludwig W., and Schleifer K.H. (1992). Gram-positive bacteria with a high DNA G+C content are characterized by a common insertion within their 23S rRNA genes. J. Gen. Microbiol. *138*, 167–175.

Schopf, J.W. (1978). The evolution of the earliest cells. Sci. Am. *239*, 110–120.

Siew, N. and Fischer, D. (2003). Analysis of singleton ORFans in fully sequenced microbial genomes. *Proteins* 53, 241–251.

Singh, B. and Gupta, R.S. (2009). Conserved inserts in the Hsp60 (GroEL) and Hsp70 (DnaK) proteins are essential for cellular growth. Mol. Genet. Genomics *281*, 361–373.

Skophammer, R.G., Servin, J.A., Herbold, C.W., and Lake, J.A. (2007). Evidence for a Gram-positive, eubacterial root of the tree of life. Mol. Biol. Evol. *24*, 1761–1768.

Stackebrandt, E. (2006). Defining Taxonomic Ranks. In: The prokaryotes, M. Dworkin, S. Falkow, E. Rosenberg, K.-H. Schleifer, and E. Stackebrandt, eds (New York: Springer), pp. 29–57.

Stackebrandt, E. and Schumann, P. (2000). Introduction to the Taxonomy of Actinobacteria. In: The Prokaryotes: An Evolving Electronic Resource for the Microbiological Community, M. Dworkin, *et al.*, eds (New York: Springer-Verlag).

Ventura, M., Canchaya, C., Tauch, A., Chandra, G., Fitzgerald, G.F., Chater, K.F., and van Sinderen, D. (2007). Genomics of Actinobacteria: tracing the evolutionary history of an ancient phylum. Microbiol. Mol. Biol. Rev. *71*, 495–548.

Woese, C.R. (1987). Bacterial evolution. Microbiol. Rev. *51*, 221–271.

Woese, C.R. (1998). The universal ancestor. Proc. Natl. Acad. Sci. U.S.A. *95*, 6854–6859.

Woese, C.R., Kandler, O., and Wheelis, M.L. (1990). Towards a natural system of organisms: proposal for the domains Archaea, Bacteria, and Eucarya. Proc. Natl. Acad. Sci. U.S.A. *87*, 4576–4579.

Woese, C.R., Stackebrandt, E., Macke, R.J., and Fox, G.E. (1985). A phylogenetic definition of the major eubacterial taxa. Syst. Appl. Microbiol. *6*, 143–151.

Comparative Microbial Genomics: Analytical Tools, Population Genetic Patterns and Evolutionary Implications

2

Yingqin Luo, Kui Lin and Jianping Xu

Abstract

Microbes are ubiquitous in the world in which we live. With the development of high throughput DNA sequencing technology, there has been an explosion of DNA sequence data on microbes. The major aim of future microbial genomics will be to identify the functional significances of individual gene and genomic fragments and to use the information to help improve human health and promote our society development. One current major undertaking to understand genomic information is the comparative analyses between genomes that are not only distantly related, but also closely related ones. Such comparative analyses between genomes that have diverged at different evolutionary time scales allow us to extract different types of information about biological functions and evolutionary processes. We review the tools and databases that have been established for comparative analyses of microbial genomes and discuss the implications of such analyses on our understandings of the common properties of life, the extent of genome plasticity and diversity within and between species, the processes and mechanisms underlie the observed genome diversity, and the origin and evolution of life.

Introduction of comparative microbial genomics

Microbes are ubiquitous in the world in which we live. They impact almost every aspect of the human welfare and are a vital component of the global biological diversity, including nutrient cycling. At the most fundamental level, microbes can offer clues on how life began and how it is sustained today. Recently, there has been an explosion of DNA sequence data on microbes. However, knowing the DNA sequence of microbes does not tell us directly how this genetic information leads to the observable phenotypes that we want to understand. The major aim of future microbial genomics will be to identify the functional significances of individual genome sequences and to use the information to help improve human health and promote our society development. Comparative analyses of genome sequences will be a major component of this effort.

While it is clear that genomics offers unprecedented opportunities, there are major areas of research as yet untouched that would increase our understanding of the broader microbial world, including its diversity and potential applications. At the time of this writing, more than 700 prokaryotic genomes have been sequenced, and the number of completely sequenced genomes has increased rapidly over the past few years and will continue to increase. For example, more than 3600 genome-sequencing projects are currently under way. The sequenced genomes offer opportunities for addressing a variety of questions. We believe it is important to conduct comparative analyses not only of distantly related genomes, but also of closely related genomes, since we can extract different types of information about biological functions and evolutionary processes from comparisons of genomes that have diverged at different evolutionary time scales. Among the 700 available prokaryotic genomic sequences, many of them are closely related to each other (Liolios *et al.*, 2008).

Extracting useful information from such a growing number of genomes is a major challenge in comparative genomics (Uchiyama, 2007). The role of comparative genomics becomes much more important to utilize these amounts of sequences not only for elucidating commonality in all of life, but also for understanding the evolutionary diversity within various groups, including understanding the evolutionary processes or mechanisms that generate such genome diversities.

Comparative genomics: the next wave of the genomics research

The basic principles of comparative genomics are straightforward: common features of two organisms will often be encoded within the DNA as conserved elements between the genomes. More precisely, the DNA sequences encoding functional proteins and RNAs that were conserved from the last common ancestor should be preserved in contemporary genome sequences. Likewise, the transcriptional factor binding motifs controlling the expression of genes that are regulated similarly in two related species may also be conserved, in particular between closely related organisms. Conversely, sequences that encode (or control the expression of) proteins and RNAs responsible for differences between species should be divergent (Hardison, 2003).

With so many completed genome sequences in hand, many interesting questions can be addressed by comparing these genomes at different phylogenetic distances. For instance, broad insights about the types of essential genes can be gleaned by genomic comparisons at very long phylogenetic distances, e.g. greater than 1 billion years since two genomes have separated. Obviously, over such very large distances, the order of genes and the sequences regulating their expression will unlikely be conserved. At moderate phylogenetic distances (roughly 70–100 million years of divergence), for example, both functional and non-functional DNA fragments can be found within the conserved DNA (Hardison, 2003). In these cases, the functional sequences often show signatures of purifying or negative selection. That is, the functional sequences would show lower levels of divergence than non-functional or neutral DNA sequences (Jukes and Kimura, 1984). Not only can comparative genomics discriminate conserved from divergent and functional from non-functional DNA, it can also contribute to identifying the general functional class of certain DNA segments, such as coding exons, non-coding RNAs, and some gene regulatory regions. Examples of analyses at this time scale have been carried out among enteric bacteria (McClelland *et al.*, 2000), and among several species of yeast (Cliften *et al.*, 2001; Kellis *et al.*, 2003). In contrast, very similar genomes are particularly apt for finding the key sequence features that may account for the differences between the organisms. Such features/diverged sequences may be indicators of the effects of differential adaptive evolution between the organisms. Comparative genomics helps identify such features (Kellis *et al.*, 2003). As genomic sequence data accumulate, their comparative analyses will become increasingly more powerful at addressing biological questions.

Basic questions of comparative microbial genomics

In 2003, in a pioneer study by Koonin and Galperin on comparative genomics studies, the following two basic questions were raised.

Question 1: How good is our current collection of genome sequences?

To address this issue, they superimposed the completely sequenced genomes over the taxonomy tree and examined how densely populated the main branches were. When this was done with the prokaryotic part of the taxonomy, the result seemed to be rather encouraging: the main bacterial and archaeal lineages were already represented by either a complete genome sequence or a genome project that was nearing completion. However, that result needs to be taken with a grain of salt because our knowledge of prokaryotic diversity was (and is still) quite limited, in particular for microbes from natural environments. Environmental molecular evolutionary studies have indicated that the great majority of bacterial and archaeal species is uncultivable with the current methods (Pace, 1997). Recent techniques aimed at growing these organisms (Kaeberlein *et al.*, 2002) might eventually result in a real revolution in the genomics of some of these uncultivable microbes, but it will probably take decades to unfold. A case in point is the report of a new archaeal phylum, the Nanoarchaea (Huber *et al.*, 2002). With these caveats, it is fair to say that, to the best of our knowledge, the current diversity of prokaryotes is reasonably well covered by completed genome sequences, and hence, significant insights could be gained for prokaryotic evolutionary genomics (Koonin and Galperin, 2003).

Question 2: Why does comparative genomics work to give us information on gene functions and evolution?

The fundamental principle for addressing this question was provided by the neutral theory of molecular evolution (Gillespie, 1984; Kimura, 1979). Neutral evolution can be very fast, as convincingly demonstrated, for example, by the rapid deterioration of pseudogene sequences. Therefore, whenever we detect sequence conservation among proteins or nucleic acids from species separated by a long time span (and this, in practical terms, involves any comparison between two species because they are typically separated by millions of years, time more than sufficient for a pseudogene to change, often beyond recognition), we can be sure that this conservation is due to the pressure of purifying selection driven by functional constraints (Gilligan and Venkatesh, 2004). Furthermore, we can reasonably assumed that conserved amino acids and nucleotides probably perform the same or similar functions, at least in structural and biochemical terms, in homologous protein, RNA, or DNA molecules (Huynen *et al.*, 2004). These fundamental principles have allowed (and will continue to lead) scientists revealing important signals on the functions and evolutionary relationships of the great majority of genes and their products between two or more genomes.

In summary, the theory of molecular evolution and, in particular, the simple assumed connection between evolutionary conservation and function form the core theoretical framework and the main methodology of comparative genomics and functional genomics. The comparative approach allows researchers to predict the functions of genes and/or their products by transferring annotation from known characterized genes of model organisms to their uncharacterized homologous in other organisms. More importantly, it enables us to delineate the functionally critical portions of protein (and RNA) molecules, such as catalytic or binding sites of enzymes. Naturally, the quality of these inferences depends on the sensitivity and robustness of computational methods employed by comparative genomics. We believe that comparative genomics can play increasingly important roles in help prioritizing targets for systematic experimental studies.

Analytic tools in comparative genomics

The rapid accumulation of bacterial genome sequences has opened up a new field of research, that of comparative genomics. Interpretation of raw DNA sequence data involves the identification and annotation of genes, proteins, and regulatory and/or metabolic pathways (Hardison, 2003). Therefore, there is a natural shift towards the creation of tools for viewing and manipulating data in a comparative genomics context. In addition, genome annotations need to be reprocessed on a regular basis to take into account the newly characterized functions of genes. Furthermore, large-scale functional analyses generate additional data that contribute to the interpretation of genomic data. These considerations are driving the research community to think about how to manage public collections of genomes in novel ways. One role of bioinformatics is to assist biologists in the extraction of biological knowledge from this flood of data. Consequently, software designed for the analyses and functional annotation of a single genome have evolved to tools for comparative genomics, detecting the relatively conserved information across many genomes simultaneously. Here we introduce several popular tools for bacterial genome annotation and comparative genomics.

Microbial genome data resources

A variety of specialized data resources manage the results of microbial genome data processing and interpretation at different stages. These stages correspond to different levels of microbial genome characterization. Draft and finished microbial genome data are continuously incorporated in various microbial genome data resources. Below are brief descriptions to the main microbial genome data resources.

GOLD

GOLD (Genomes Online Database) is a World Wide Web resource for comprehensive access to information regarding complete and ongoing genome projects, as well as metagenomes and metadata, around the world (Bernal *et al.*, 2001; Kyrpides, 1999; Liolios *et al.*, 2008; Liolios *et al.*, 2006). GOLD was created in 1997 with the aim to (i) monitor all genome sequencing projects from instigation to completion and (ii) provide the community with a centralized database integrating diverse information related to those projects in the form of hyper-text links to disparate web-based resources. Although several different types of statistics, related to each of the data fields, can be derived from the user at any point using the search engine, the database also provides readily available graphical overviews for specific data types.

Complete and ongoing projects and their associated metadata can be accessed in GOLD through pre-computed lists and a search page. As of March 2008, GOLD contains information on more than 3613 sequencing projects, out of which 731 have been completed and their sequence data deposited in the public databases (GOLD V2.0). GOLD continues to expand with the goal of providing metadata information related to the projects and the organisms/environments towards the Minimum Information about a Genome Sequence' (MIGS) guideline. GOLD is available at http://www.genomesonline.org and has a mirror site at the Institute of Molecular Biology and Biotechnology, Crete, Greece, at http://gold.imbb.forth.gr/.

ASAP

ASAP (a systematic annotation package for community analysis of genomes) is a relational database and has a web interface developed to store, update and distribute genome sequence data and their functional characterizations (Glasner *et al.*, 2003). ASAP facilitates ongoing community annotation of genomes and tracking of information as genome projects move from preliminary data collection through post sequencing functional analysis. The database includes multiple genome sequences at various stages of analysis, corresponding experimental data and access to collections of related genome resources. Its development was motivated by the need to more directly involve a greater community of researchers, with their collective expertise, in keeping the genome annotation current and to provide a synergistic link between up-to-date annotation and functional genomic data.

ASAP supports three levels of users: public viewers, annotators and curators. Public viewers can readily browse updated annotation information such as for *Escherichia coli* K-12 strain MG1655, genome-wide transcript profiles from more than 50 microarray experiments and an extensive collection of mutant strains and associated phenotypic data. Annotators worldwide are currently using ASAP to participate in a community annotation project for the *Erwinia chrysanthemi* strain 3937 genome. Curation of the *E. chrysanthemi* genome annotation as well as those of additional published enterobacterial genomes are under way and will be publicly accessible in the near future. ASAP is available at https://asap.ahabs.wisc.edu/annotation/php/ASAP1.htm.

CMR

CMR (Comprehensive Microbial Resource) is a tool that allows researchers to access all the bacterial genome sequences completed to date (Peterson *et al.*, 2001). It contains robust annotation of all completed microbial genomes and allows for a wide variety of data retrievals. For each genome not sequenced at The Institute of Genome Research (TIGR), two kinds of annotation are displayed: the Primary annotation taken from the genome sequencing centre and the TIGR annotation generated by an automated annotation process at TIGR. CMR thus allows access of all the information on all of the bacterial genomes or any subset of them. The bacterial information has been placed on the web at http://www.tigr.org/CMB for retrieval using standard web browsing technology. Retrievals can be based on protein properties such as molecular role assignments and taxonomy. The CMR also has special web-based tools to allow data mining using pre-run homology searches, whole genome dot-plots, batch downloading and traversal across genomes using a variety of data types. It is available at http://www.tigr.org/CMR.

IMG

The IMG (Integrated Microbial Genomes) system serves as a community resource for comparative analysis and annotation of all publicly available genomes from the three domains of life, in a uniquely integrated context (Markowitz *et al.*, *2006*, 2008a,b; Markowitz and Kyrpides, 2007). IMG provides tools and viewers for analysing and annotating genomes, genes and functions, individually or in a comparative context. An increasing number of eukaryotic genomes, viruses (including phages) and plasmids have also been added to IMG in order to increase its genomic context for comparative analysis. IMG's analytical tools have been gradually generalized and enhanced in terms of their usability, analysis flow and performance. These tools allow users to focus on a subset of genes, genomes and functions of interest, and conduct analysis using summary tables, graphical viewers and various methods for comparing genes, pathways and functions across genomes. IMG is provided by the DOE-Joint Genome Institute (JGI) and is available from http://img.jgi.doe.gov.

SEED

SEED is a software environment to support early phases in building design that has been adopted for comparative genomics research. Database support in SEED allows designers to store and retrieve different design versions, alternatives and past designs that can be reused and adapted in different contexts (case-based design in the terminology of Artificial Intelligence). In addition, the database stores recurring problem specifications and typical requirements for building types or functional areas common to many buildings. The database serves also as a main means of information exchange between modules, which do not communicate design decisions directly to each other. Current literature refers to this as information modelling or product and process modelling. The SEED is available at the web site: http://www.seedling.org/IJDC/DB/.

Genome-wide comparative alignment tools

Genome sequence comparison has been an important method for understanding gene function and genome evolution since the early days of gene sequencing. Alignment of DNA sequences is the core process in comparative genomics. In recent years, an important new sequence-analysis task has emerged: comparing an entire genome with another. Several powerful alignment algorithms have been developed to align two or more sequences.

MUMmer

MUMmer is a system for rapidly aligning entire genomes, whether in complete or draft form (Delcher *et al.*, 1999, 2002; Kurtz *et al.*, 2004). MUMmer can also align incomplete genomes; it can handle thousands of contigs from a shotgun sequencing project, and will align them to another set of contigs or a genome using the NUCmer program included within the system. If the species are too divergent for a DNA sequence alignment to detect similarity, then the PROmer program within the environment can generate alignments based upon the six-frame translations of both input sequences. The original MUMmer system, version 1.0, was described in a 1999 Nucleic Acids Research paper. Version 2.1 appeared a few years later and was described in a 2002 Nucleic Acids Research paper (Delcher *et al.*, 2002), and the most recent version MUMmer 3.0 was described in a 2004 Genome Biology paper (Kurtz *et al.*, 2004). Users can get the Mummer source code from the web site: http://mummer.sourceforge.net/.

BLAT

BLAT (The BLAST-Like Alignment Tool) is a new tool for sequence alignment, which is similar in many ways to BLAST (Kent, 2002). The program rapidly scans for relatively short matches (hits), and extends these into high-scoring pairs (HSPs). However, BLAT differs from BLAST in several significant ways. Specifically, where BLAST builds an index of the query sequence and then scans linearly through the database, BLAT builds an index of the database and then scans linearly through the query sequence. Where BLAST triggers an extension when one or two hits occur in proximity to each other, BLAT can trigger extensions on any number of perfect or near-perfect hits. Where BLAST returns each area of homology between two sequences as separate alignments, BLAT stitches them together into a larger alignment. The web interface is available at http://genome.ucsc.edu and a stand-alone BLAT is also available, which is more suitable for batch runs on one or more CPUs. Both the client/server and the stand-alone can do comparisons at the nucleotide, protein, or translated nucleotide level.

MegaBLAST

MegaBLAST uses the greedy algorithm of Zhang *et al.* (2000) for nucleotide sequence alignment search and concatenates many queries to save time spent scanning the database. This program is optimized for aligning sequences that differ slightly as a result of sequencing or other similar 'errors'. It is up to 10 times faster than more common sequence similarity search and alignment programs and therefore can be used to swiftly compare two large sets of sequences against each other. The source code of the MEGABlast can be obtained from the web site: http://www.ncbi.nlm.nih.gov/blast/megablast.shtml.

Genome comparison visualization tool

Comparative analysis is an increasingly important step in the annotation and analysis process of genome sequence data, allowing phenotypic differences between strains and species to be correlated with changes in the chromosomes. For example, comparative sequence analysis has enabled the identification of cis-regulatory regions and location of coding exons using purely computational means. Visual front-ends are necessary and important to make the process of viewing alignments intuitive and easy to facilitate discovery of conserved sequences for functionally significant regions. Below we describe a few visualization tools for genome comparisons.

PipMaker/MultiPipMaker

PipMaker is a web site for comparing two long DNA sequences to identify conserved segments and for producing informative, high-resolution displays of the resulting alignments (Schwartz *et al.*, 2000). One display is a per cent identity plot (pip), which shows both the position in one sequence and the degree of similarity for each aligning segment between the two sequences in a compact and easily understandable form. The web site also provides a plot of the locations of those segments in both species. PipMaker is appropriate for comparing genomic sequences from any two related species, although the types of information that can be inferred (e.g. protein-coding regions and cis-regulatory elements) depend on the level of conservation and the time and divergence rate since the separation of the species. PipMaker supports analysis of unfinished or 'working draft' sequences by permitting one of the two sequences to be in un-oriented and unordered contigs.

Similarly, MultiPipMaker allows the user to visualize relationships among more than two sequences (Schwartz *et al.*, 2003). All pairwise alignments with the first sequence are computed and then returned as interleaved pips. Moreover, MultiPipMaker can be requested to compute a true multiple alignment of the input sequences and return a nucleotide-level view of the results. Both the PipMaker and MultiPipMaker are available at the web site: http://bio.cse.psu.edu/pipmaker/.

ACT

ACT (Artemis Comparison Tool) is a DNA sequence comparison viewer, such as parsed BLAST alignments based on Artemis – an annotation tool (Carver *et al.*, 2005). Similar to other Artemis tools, ACT is written in Java and runs on Unix, GNU/Linux, Macintosh and MS Windows systems. It can read complete EMBL and GENBANK entries or sequence in FASTA or raw sequence format. Other types of readable sequence input files include EMBL, GENBANK and GFF formats. The sequence comparison displayed by ACT is usually the result of running a blastn or tblastx search. The ACT software is available at the web site: http://www.sanger.ac.uk/Software/ACT/.

VISTA

Vista (Visualization and Alignment Software for Comparative Genomics) is a visualization tool for alignments, which displays GLASS alignments (Carver *et al.*, 2005). It is a program to depict long alignments of DNA sequences from two or more organisms with various types of annotation in a clear and easily interpretable format. Originally it was developed to locate conserved sequences in syntenic regions of different genomes. The key features of the VISTA program are the following:

- clean graphical output, allowing for easy identification of sequence similarities and differences;
- easily configurable, enabling the visualization of alignments of up to several million bases at different levels of resolution;
- displays alignments of draft sequences;
- displays sequence annotations such as repeats, coding exons, UTRs and more.

The VISTA plot is based on moving a user-specified window over the entire alignment and calculating the per cent identity over the window at each base pair. VISTA can be accessed from the web site: http://www.lbl.gov/tt/techs/lbnl1690.html.

SynPlot

Synplot (displays DIALIGN and GLASS alignments) is an application program, written in Perl, for viewing global alignments of syntenic regions of genomic DNA sequence (Gottgens *et al.*, 2001). The alignment is used to calculate the percentage identity along the alignment within a sliding window, the width of which can be specified by the user. This information is used to draw a picture of the alignment in postscript format. The sequences are rendered as lines interrupted by spaces corresponding to the gaps introduced by the alignment, with a plot of the percentage identity underneath. Features can also be drawn on the sequence lines. This program uses a GFF format file output by ACeDB from the annotated genomic sequence, and a configuration file which specifies the colour, height and order in which the rectangles representing the features are drawn. Synplot can be downloaded at the web site: http://www.sanger.ac.uk/Users/jgrg/SynPlot/.

Microarray analyses technology

The advent of DNA microarray technology has greatly expanded our ability to monitor changes in the abundance of transcripts. Such a development has been a milestone in several areas of microbiology. In clinical microbiology, microarrays are used for microorganism detection and identification and gene-expression analysis. DNA microarrays have allowed us to monitor the effects of pathogens on host-cell gene expression in a much greater depth and on a significantly broader scale than previous single gene studies. The results generated by these studies are complex, and few systematic studies have been carried out to compare results among studies (Raoult *et al.*, 2004).

Comparative transcriptomics – whole genome mRNA transcript profiling using microarrays

Whole-genome microarrays from fully sequenced genomes are a powerful platform for identifying differences in gene content between organisms and for studying gene expression dynamics. The generation of messenger RNA expression profiles is referred to as transcriptomics, as these are based on the process of transcription. Given the inhospitable *in vivo* and the varied *ex vivo* environments encountered by most microbial pathogens, transcriptome analysis holds particular promise for identifying and determining the functions of differentially regulated, virulence associated genes (Wren, 2000). The basic principle of this technique involves extracting the mRNA expressed under a range of environmental conditions and hybridizing these sequences to a high-density gridded microarray of the DNA content of an organism. Such high-throughput analysis allows massive parallel gene expression and gene discovery studies to be undertaken (Brown and Botstein, 1999). DNA microarray analysis will complement other technologies such as *in vivo* expression technology and differential fluorescence analysis to identify and investigate which bacterial genes are differentially expressed in the host (Mahan *et al.*, 1993; Strauss and Falkow, 1997; Valdivia and Falkow, 1997).

The application of DNA microarrays to microbial pathogens

The study of the complete set of genes expressed and modified in a cell is an important and rapidly evolving discipline that is readily applicable to microbial pathogens (Wren, 2000). For example, strains of *Staphylococcus aureus* resistant to the antibiotic vancomycin present a potentially serious public health problem (Coenye, 2003). In the case, the Gaasterland group described the development of a multi strain *S. aureus* microarray. Pairwise comparisons of the available genomes of strains of *S. aureus* have revealed considerable variation in gene content across the epidemiological landscape. They identified changes in protein-coding potentials that correlate with antibiotic resistance by measuring differences in gene expression in vancomycin-sensitive and vancomycin-resistant pairs of *S. aureus* isolates (Sobral *et al.*, 2007; Utaida *et al.*, 2003). Philip Butcher's group used microarrays to help understand the complex pathophysiology of *Mycobacterium tuberculosis* infection (Butcher, 2004; Schroeder *et al.*, 2002; Stewart *et al.*, 2002). Besides discussing some methodological aspects of microarray work, they focused on the use of *M. tuberculosis* microarrays to investigate the intracellular lifestyle of this organism and its interaction with host macrophages (Bacon *et al.*, 2004; Brodin *et al.*, 2006). In the future, it would be exciting to integrate results from *in vitro* work like this with results from *in vivo* microarray work on mammalian hosts to provide a whole-genomic view of host–pathogen interactions (Coenye, 2003; Wren, 2000).

Comparison of metabolic pathways in microbial genomes

Understanding the regulatory mechanisms should allow the examination of engineering pathways with pre-determined expression patterns (i.e. expression is activated by a given compound or in a specific environmental or physiological condition). Metabolic pathways have evolved to execute their function efficiently, while tolerating perturbations, such as changes in environmental parameters or in the physiological status of the cell. Below we describe some of the databases and programs for integrated analyses of metabolic pathways.

KEGG

KEGG (Kyoto Encyclopedia of Genes and Genomes) is a database of biological systems that integrates genomic, chemical and systemic functional information (Kanehisa *et al.*, *2006*, 2008; Kanehisa and Goto, 2000). KEGG provides a reference knowledge base for linking genomes to life through the process of PATHWAY mapping. The PATHWAY database contains information about conserved subpathways (or pathway motifs), which are often encoded by positionally coupled genes on the chromosome and which are especially useful in predicting gene functions. The genomic information is stored in the GENES database, which is a collection of gene catalogues for all the completely sequenced genomes and some partial genomes with up-to-date annotation of gene functions. A third database in KEGG is LIGAND, which includes information about chemical compounds, enzyme molecules and enzymatic reactions. In addition, KEGG provides a reference knowledge base for linking genomes to the environment, such as for the analysis of drug–target relationships, through the process of BRITE mapping. KEGG BRITE is an ontology database representing functional hierarchies of various biological objects, including molecules, cells, organisms, diseases and drugs, as well as relationships among them. Additionally, the KEGG resource is being expanded to suit the needs for practical applications. KEGG DRUG contains all approved drugs in the US and Japan, and KEGG DISEASE is a new database linking disease genes, pathways, drugs and diagnostic markers.

KEGG provides Java graphics tools for browsing genome maps, comparing two genome maps, manipulating expression maps, as well as including computational tools for sequence comparison, graph comparison and path computation. The KEGG databases are updated daily and made freely available through http://www.genome.ad.jp/kegg/.

BioCyc

The BioCyc database collection is a set of 160 pathway/genome databases (PGDBs) for most eukaryotic and prokaryotic species whose genomes have been completely sequenced to date (Karp *et al.*, 2005). Each PGDB offers a wealth of genomic and metabolic information on certain microorganisms, including *P. aeruginosa* and *S. cerevisiae*. Each database provides information on a microorganism's annotated genome, on the biochemical reaction(s) that each gene product catalyses and on the organism's metabolic pathways, predicted from its annotated genome by a program called PathoLogic. The information from each database is comprehensive and complex. In addition, each bacterial PGDB includes predicted operons for the corresponding species. The BioCyc collection provides a unique resource for computational systems biology, namely global and comparative analyses of genomes and metabolic networks, and a supplement to the BioCyc resource of curated PGDBs. The Omics viewer available through the BioCyc website allows scientists to visualize combinations of gene expression, proteomics and metabolomics data on the metabolic maps of these organisms (Krummenacker *et al.*, 2005). The BioCyc is available at the web site: http://www.biocyc.org/.

MetaCyc

MetaCyc is a universal database of metabolic pathways and enzymes from all domains of life (Caspi *et al.*, 2008). The pathways in MetaCyc are curated from the primary scientific literature, and the small-molecule metabolic pathways are experimentally determined. Each reaction in a MetaCyc pathway is annotated with one or more well-characterized enzymes. Because MetaCyc contains only experimentally elucidated knowledge, it provides a uniquely high-quality resource for metabolic pathways and enzymes. MetaCyc stores pathways involved in both primary metabolism and secondary metabolism. MetaCyc also stores compounds, proteins, protein complexes and genes associated with these pathways. It is extensively linked to other biological databases containing protein and nucleic-acid sequence data, bibliographic data and protein structures. MetaCyc also contains objects for the genes that encode many enzymes within the DB. While it does not contains primary sequence data, MetaCyc does contain links to external sequence databases. The MetaCyc is available at the web site: http://metacyc.org/.

EcoCyc

EcoCyc is a bioinformatics database that describes the genome and the biochemical machinery of *E. coli* K-12 MG1655 (Karp *et al.*, 2007). The long-term goal of this project is to describe the molecular catalogue of the *E. coli* cell, as well as the functions of each of its molecular parts, to facilitate a system-level understanding of *E. coli*. EcoCyc is an electronic reference source for *E. coli* biologists, and for biologists who work with related microorganisms. EcoCyc contains the complete genome sequence of *E. coli*, and describes the nucleotide position and function of every *E. coli* gene. The annotation of the *Escherichia coli* K-12 genome in the EcoCyc database is one of the most accurate, complete and multidimensional genome annotations. EcoCyc information was derived from 15 000 publications. The database contains extensive descriptions of *E. coli* cellular networks, describing its metabolic, transport and transcriptional regulatory processes. Database queries to EcoCyc survey the global properties of *E. coli* cellular networks and illuminate the extent of information gaps for *E. coli*, such as dead-end metabolites. EcoCyc provides a genome browser with novel properties, and a novel interactive display of transcriptional regulatory networks. The EcoCyc is available at the web site: http://ecocyc.org/.

Phylogenetics and phylogenomics for microbial genomes

Generally, microorganisms, in particular prokaryotes often lack morphological and behavioural characters amenable to phylogenetic analysis. Such a lack of information in these areas makes gene sequence information the most prevalent source of data for phylogenetic analysis in pre-genomic era. Molecular phylogenetics based on single genes, in particular the small-subunit rRNA (SSU rRNA) (Woese, 1987), has laid the foundation for a modern classification system, conceptually represented by the 'universal tree of life' (Woese *et al.*, 1990). However, phylogenetic trees based on single genes or gene families may show conflict results (Teichmann and Mitchison, 1999) due to a variety of problem, such as mutational saturation of the single genes and horizontal gene transfer. Consequently, although SSU rRNA gene sequence continue to be considered as molecular criteria for species delineation, it is anticipated that much additional taxonomic information can be extracted from complete genome sequences.

Now that large-scale genome-sequencing projects are sampling many organismal lineages, it is becoming possible to compare large data sets of DNA or protein sequences to study speciation and evolution. The steady increase in the number of completely sequenced prokaryotic genomes has created a boom for bioinformatics. With more than 700 prokaryotic genomes completely sequenced, there has been an increasing interest in the use of various characters in whole genomes for prokaryotic genomes studies. This is giving birth to a brand new field of research – phylogenomics (Nelson *et al.*, 1999). Phylogenomics uses entire genomes to infer a species tree and has become the *de facto* standard for reconstructing reliable phylogenies (Ciccarelli *et al.*, 2006).

One major branch of phylogenomics involves the use of these data to reconstruct the evolutionary history of organisms. Access to large amount of genomic data could potentially alleviate problems associated with single-gene-based phylogenetics. This is because a large number of characters can now be used for phylogenetic analysis to avoid biases. With this increase, the emphasis of phylogenetic inference is shifting from the search for informative characters to the development of better reconstruction methods using genomic data. Existing models used in tree-building algorithms only partially take into account molecular evolutionary processes, and phylogenomic inference will benefit from an increased understanding of these mechanisms. Interestingly, phylogenomics is also providing the opportunity to use new 'morphological-like' characters that are based on genome structure, such as rare genomic changes (RGCs) (Deeds *et al.*, 2005; Gupta, 1998; Luo *et al.*, 2006; Rokas, 2006; Rokas and Holland, 2000).

The integration of genomics data into the phylogenetics is still at an early stage. Given the breadth of organismal diversity, the gene-scale era of phylogenetics is still an invaluable asset to the pursuit of the Tree of Life. Comparative genomics, with its ability and potential to vastly increase both the amount and type of molecular data available for a small but critical fraction of biodiversity, is bound to play an increasingly important role in efforts to assemble a robust picture of the Tree of Life (Rokas, 2006).

BPhyOG – a database for overlapping genes in prokaryotic genomes

BPhyOG (Bacterial Phylogeny based on Overlapping Genes) is an online interactive server for reconstructing completely sequenced bacterial genomes on the basis of their shared overlapping genes (Luo *et al.*, 2007). It provides two tree-reconstruction methods: neighbour joining (NJ) and unweighted pair-group method using arithmetic averages (UPGMA). Users can apply the desired method to generate phylogenetic trees, which are based on an evolutionary distance matrix for the selected genomes. The distance between two genomes is defined by the normalized number of their shared OG pairs. BPhyOG also allows users to browse the OGs that were used to infer the phylogenetic relationships. It provides detailed annotation for each OG pair and the features of the component genes through hyperlinks. Users can also retrieve each of the homologous OG pairs that have been determined among 177 genomes. BPhyOG is useful tool for analysing the tree of life and overlapping genes from a genomic standpoint. It currently includes 177 completely sequenced bacterial genomes containing 79,855 OG pairs, with their annotation and homologous OG pairs comprehensively integrated. The reliability of phylogenies and completeness of annotations make BPhyOG a comprehensive and powerful web server for genomic and genetic studies. Fig. 2.1 shows the flow chart of the phylogenetic tree of 30 gamma-proteobacterial genomes reconstructed on the basis of conserved OG pairs. BPhyOG is freely available at http://cmb.bnu.edu.cn/BPhyOG.

Examples for microbial genomics

CFGP

CFGP (Comparative Fungal Genomics Platform) is a web-based multifunctional informatics workbench (Park *et al.*, 2008). The CFGP comprises three layers, including the basal layer, middleware and the user interface. The data warehouse in the basal layer contains standardized genome sequences of 65 fungal species. The middleware processes queries via six analysis tools, including BLAST, ClustalW, InterProScan, SignalP 3.0,

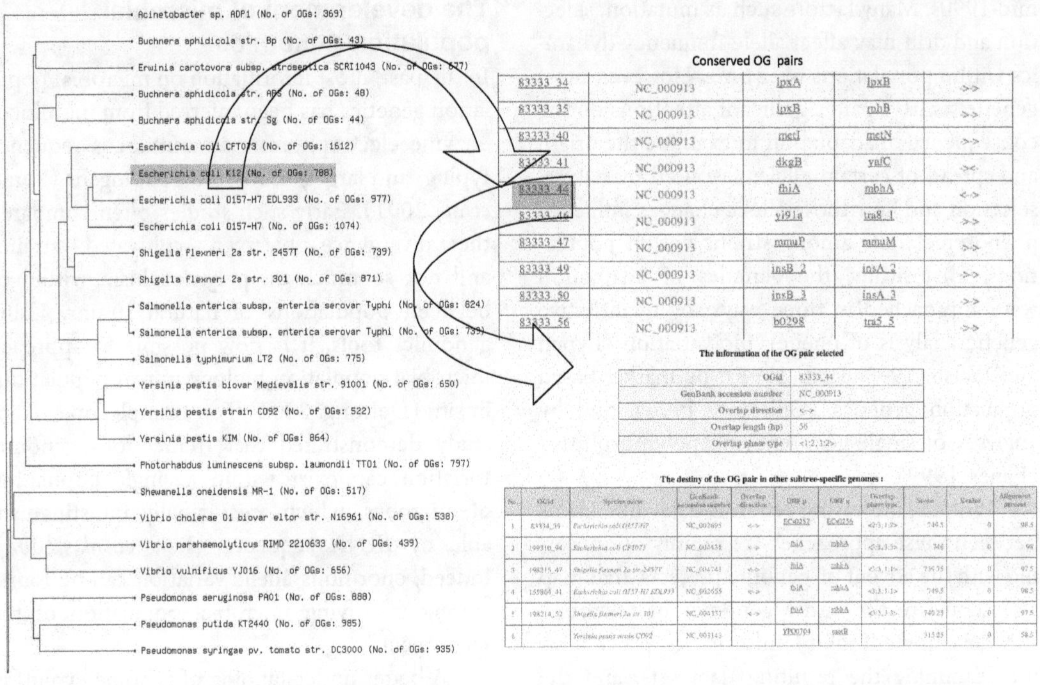

Figure 2.1 The flow chart for constructing the phylogenetic tree of 30 gamma-proteobacterial genomes on the basis of conserved OG pairs. The left panel shows a phylogenetic UPGMA tree for the 30 gamma-proteobacterial genomes based on the number of their shared OG pairs. Each branch is labelled by the species name and the total number of OG pairs in the genome in square brackets. The right upper panel shows all the conserved OG pairs in the genome of interest when the species name is clicked on the tree, e.g. *Escherichia coli K12*. The right lower panel describes the information about one OG pair (e.g. *83333_44*) and the list of its orthologous OG pairs within the subclade-specific genomes (Luo *et al.*, 2007).

PSORT II and a newly developed tool named BLASTMatrix. The BLASTMatrix permits the identification and visualization of genes homologous to a query across multiple species. The Data-driven User Interface (DUI) of the CFGP was built on a new concept of pre-collecting data and post-executing analysis instead of the 'fill-in-the-form-and-press-SUBMIT' user interfaces utilized by most bioinformatics sites. A tool termed Favorite, which supports the management of encapsulated sequence data and provides a personalized data repository to users, is another novel feature in the DUI. CFGP is available at the web site: http://cfgp.snu.ac.kr.

MicroScope

MicroScope is a microbial genomes annotation and comparative analysis platform, which was developed by the French National Sequencing Center located at Genoscope. It is made of three major components (Strous *et al.*, 2006): (i) a set of syntactic and functional annotation tools, (ii) a relational database, the Prokaryotic Genome DataBase, (PkGDB) which is linked to metabolic pathway databases (MicroCyc) created using the Pathway Tools software, and (iii) a graphical interface, the Magnifying Genome (MaGe), which allows performing relevant expert annotation that combine synteny results with metabolic network predictions.

The website for accessing MicroScope is http://www.genoscope.cns.fr/agc/mage/wwwpkgdb/MageHome/index.php?webpage=microscope.

Population genetic patterns

Recently, because sequencing thousands of genes of known function and position in a genome has become a routine, modern population genetics underwent a large paradigm shift since the mid-1990s. Many factors such as mutation, selection and drift may affect allele frequency dynamics within populations over time. However, most geneticists have paid significant attentions on the consequence of adaptation to examine the origin and spread of certain alleles favoured by natural selection and how those allelic changes influence phenotypic traits among strains within populations. Historically, the genetics of adaptation were approached in three ways: by quantitative genetic analysis of phenotypic variation (Lynch and Walsh, 1998; Roff, 1997), by marker-based population genetics (Gillespie, 1994), and by analyses of single-gene enzyme polymorphisms (Eanes, 1999).

With the increasing genomic data, differences between these approaches are rapidly disappearing with the advent of 'genome projects' that map and sequence the entire genomes of microbes, the development of bioinformatics procedures for examining the resulting data sets, and the creation of genetic maps that identify loci affecting phenotypic variation, i.e. quantitative trait loci (QTL). Population genetic theory provides statistical tools to infer the evolutionary forces acting on genes. These tools are now allowing the construction of an integrated strategy capable of simultaneously describing the genetic basis of phenotypic variation and the evolution of those genes that underlie that variation.

The emerging field of 'population genomics' attempts to describe the process of simultaneous sampling of numerous variable loci within a genome and the inference of locus-specific effects from the sample distributions (Black *et al.*, 2001; DeLong, 2002, 2004; Jorde *et al.*, 2001). The conceptual basis for population genomics depends on four assumptions (Black *et al.*, 2001). First and most important, genes and nucleotide substitutions evolve independently through extensive and continuous recombination. Second, migration, drift, and inbreeding influence all parts of the genome equally. Third, an investigator has the means to assay SNPs at many loci rapidly and inexpensively. Fourth, the investigator has access to information on the relative linkage and/or physical positions of genetic markers. It should be noted that some of these assumptions would be unrealistic in certain situations.

The development of microbial population genomics

In the past, most information on microbial population genetics has been inferred from multilocus enzyme electrophoresis or multilocus sequence typing, and largely focused on pathogens (Spratt *et al.*, 2001). Early such studies often compared the divergence between cultivated strains, and not so much on polymorphism within or between populations of natural strains. Using genomics tools, it is now possible to approach microbial population biology in real populations in situ (DeLong, 2002). For example, one recent study demonstrated that tremendous genomic variation can exist within a single population of microbes, which are virtually indistinguishable by rRNA sequence (Beja *et al.*, 2002). Indeed, enormous allelic variation can be found among free-living microbial population of the same species.

A better understanding of genome evolution will undoubtedly result from microbial population genomic studies. Environmental genomic (metagenomic) approaches are cultivation independent, and so they can provide relatively equal access to the collective genomes of abundant, yet uncultivated, microbial species. Genomic analyses of uncultured microbes can provide significant insight into the biological properties of individuals within microbial populations. Applied microbial population genomics is now facilitating predictions about phenotype from genotype of such uncultured microbes, broadening our knowledge of natural environmental processes, and extending our ability to exploit natural microbial products and processes (DeLong, 2002). Genomic windows into the natural microbial world now also provide unprecedented insight into microbial population structure, biodiversity and dynamics. Additionally, the genomic patterns that give rise to emergent properties of populations and their relationship to ecosystem function are bound to become increasingly more apparent (DeLong, 2002).

Microbial population genomics now promises to better define the relative importance of lateral gene transfer (see chapter by Hao and Golding in this volume), recombination and genetic drift on shaping the diversity of microbial species and genomes within and between populations. Thus, such studies should spur the development of new theory, and greatly improve our understanding of microbial evolution as it occurs in different organismal, ecological and population contexts (DeLong, 2002).

The nature and dynamics of bacterial genomes

The organization of bacterial genomes is simple yet elegant. These genomes are small, ranging from 500 to 10,000 kb, and are tightly packed with genes and other functional elements. The coding regions themselves are almost intronless and are short, averaging a scant 1 kb, and are positioned almost contiguously along the chromosome (Ochman and Davalos, 2006). The streamlining of bacterial genomes is commonly assumed to be the result of selection acting on replication efficiency and growth rates.

One of the interesting issues is to examine which forces have shaped bacterial genomes during the course of evolution. Early research showed bacterial genomes to be tightly packed with functional elements, but unprecedented discoveries from genome analyses have shown that the genetic information encoded within bacterial genomes decays over evolutionary time scales (Andersson *et al.*, 1998; Cole *et al.*, 2001; Parkhill *et al.*, 2001). The primary force countering the erosion of genomes is natural selection, which serves to maintain the functional regions. Genes less essential for survival are more susceptible to neutralizing and deletional mutations, whereas those critical for survival will resist such mutational meltdown. Moreover, the degree of selection acting on any particular gene can change over time and depend on a specific ecological context or respective environment (Ochman and Davalos, 2006). For example, the inactivation or loss of one or more genes can increase the value of others, and changes in bacterial ecology or lifestyle might keep some genes redundant (Moran and Plague, 2004).

The impact of these factors is evident from the contents and organizations of sequenced bacterial genomes. As expected, in very large populations, deleterious mutations in beneficial genes are not likely to become fixed by chance; but in small populations, even useful genes can be inactivated and deleted. Thus, organisms with small population sizes are more prone to gene decay and loss, even for genes that are usually considered beneficial (Ochman and Davalos, 2006). This is the case for pathogens, which typically require only small inocula to infect each new host, resulting in a reduction in the effective size of their populations. When compared with their free-living relatives, facultative bacterial pathogens often harbour substantial numbers of pseudogenes (Andersson and Andersson, 1999; Jin *et al.*, 2002) and have population structures that promote the maintenance of certain deleterious mutations (Andersson and Hughes, 1996; Cooper and Lenski, 2000; Mira *et al.*, 2001). Such pathogens represent an intermediate stage of genome erosion relative to the obligate host-associated lineages, whose genomes are highly reduced, having undergone massive inactivation and loss of non-essential genes (Ochman and Davalos, 2006).

Intraspecies genomic variation – the core genome and dispensable genome

Although genomics has achieved great successes in various biological research fields, several fundamental questions such as what is a bacterial species and how many genes there are in a species have not been satisfactorily addressed. Mathematical modelling predicts that new genes will be discovered even after sequencing hundreds of genomes per species (Medini *et al.*, 2005). Indeed, experimental data have shown that in some species new genes are discovered even after sequencing the genomes of several strains (Medini *et al.*, 2005). Although the processes of gene gain and gene loss have been appreciated for some time, the scope of genotypic variation – as measured by gene presence or absence – has been unveiled in several taxa for which multiple complete genomes are available, including multiple strains of almost 40 genera that represent eight bacterial divisions (Lawrence and Hendrickson, 2005). Many studies on comparative genomics have revealed large differences in gene content between even closely related strains. For example, comparison of the laboratory strain *Escherichia coli* K12 to both uropathogenic and enterohaemorrhagic strains revealed that a startlingly small proportion of genes (<40% of the total number of genes present) were shared among these three strains (Welch *et al.*, 2002). One appealing interpretation is that different strains exploit somewhat different ecological niches by virtue of metabolic differences imparted by different gene inventories. Therefore, a bacterial species can be described by its pan-genome, which is composed of a 'core genome' containing genes present in all strains, and a 'dispensable genome' containing genes present in two or more strains and genes unique to single strains (Lawrence and Hendrickson, 2005). The collection of genes shared among members of the same species–that is, the clade-specific metagenome – has been termed the 'pan-genome' (Medini *et al.*, 2005).

Multiple forces shape microbial genomes

Comparisons of available genome sequences have revealed wide variation in genome organization among bacterial pathogens. These studies have shown that multiple forces shaped the prokaryotic genome during its evolution. These forces include gene loss/genome reduction, genome rearrangement, expansion of functional capabilities through gene duplication, and acquisition of functional capabilities through lateral gene transfer (Fraser-Liggett, 2005).

Gene loss

Evidence has suggested that the transition to obligate intracellular parasites often involves deletions of large segments of DNA early on in the process, probably catalysed by genome instability mediated by insertion sequence (IS) elements or other mobile DNA that are themselves eventually lost from the genomes (Belda *et al.*, 2005; Moran and Mira, 2001; Moran and Plague, 2004). Is there any beneficial effect due to the loss of gene(s) in those bacteria during their exploitation of new ecological environments? This is true in the case of certain pathogens, including *Mycobacterium tuberculosis* (Tsolaki *et al.*, 2004), which has lost metabolic pathways and become more virulent. Similarly, *Shigella* spp. (Nakata *et al.*, 1993) and *Bordetella pertussis* (Parkhill *et al.*, 2003) have probably lost genes encoding cell surface antigens, a situation that may allow them enhanced ability to evade the host immune system attacks.

Genome rearrangements

Complete genome projects have revealed that, on one hand, free-living bacteria often contain large numbers of IS elements and repetitive DNA sequences that may mediate homologous recombination (Fraser-Liggett, 2005), and on the other hand, obligate intracellular organisms that exist in relative isolation generally contain few mobile elements, and their genomes tend to be stable over long periods of time. As we know, genome rearrangements mediated by IS elements can play a major role in genome plasticity and the extent of such rearrangements often reflects the lifestyle of the organism. One example of the role of IS elements in mediating genome-wide rearrangements comes from the comparative analysis of *Burkholderia mallei* (Nierman *et al.*, 2004), an obligate mammalian pathogen that causes *glanders*, and *Burkholderia pseudomallei* (Holden *et al.*, 2004), an environmental soil-dwelling organism. The *B. mallei* genome contains 171 complete or partial IS elements that collectively represent >3% of the genome sequence (Nierman *et al.*, 2004). The *B. mallei* genome is 1.4 Mb smaller than that of its closest relative, *B. pseudomallei*, and most of the synteny break points between the two genomes are bounded by IS elements. In addition, two syntenic portions of chromosome 1 in *B. pseudomallei* that are found on chromosome 2 in *B. mallei* are flanked by IS elements, lending further support to the idea that genome rearrangement can play an important role in genome structural alteration in certain species (Nierman *et al.*, 2004).

Gene duplication

Gene duplication and subsequent functional diversification is another mechanism for generating diversity within microbial genomes. Gene paralogous (genes related by duplication) can represent as much as 50% of the larger microbial genomes, and an interesting subset of such genes are lineage-specific duplications that presumably are responsible, in part, for species-specific biology (Fraser-Liggett, 2005). Based on the Clusters of Orthologous Groups (COGs) database (Tatusov *et al.*, 2003), Konstantinidis and Tiedje found two interesting results when analysing about 115 completed prokaryotic genome sequences: (i) that larger genomes are disproportionately enriched in genes encoding proteins involved in regulation, secondary metabolism, and transport; and (ii) that an inverse correlation was observed with proteins involved in translation and DNA processing (Konstantinidis and Tiedje, 2004). This analysis provides a possible explanation for why species with larger genomes are more apt to dominate environments where nutrients are scare, because they are more versatile in terms of their ability to sense and respond to changing environmental conditions (Fraser-Liggett, 2005).

Lateral gene transfer

Another source of genome variability that plays an important role in prokaryotic genome evolution is lateral gene transfer (LGT), which brings new genes into a genome and provides a means for rapid adaptations to changing demands on an organism (Boucher *et al.*, 2003). For example, acquisition of virulence determinants on pathogenicity islands appears to play a major role in pathogen evolution. Phylogenetic analysis of multiple strains of *Staphylococcus aureus* indicated that the diversification of the highly variable RD13 region, encoding putative pathogenesis-related proteins, has probably occurred by LGT and recombination (Fitzgerald *et al.*, 2003). Comparative analysis between *S. aureus* and its close relative, *Staphylococcus epidermidis*, demonstrated that genome islands in non-syntenic regions of the genome, probably acquired by LGT, were the primary source of variations in pathogenicity and virulence between these species (Gill *et al.*, 2005). LGT probably also plays an important role in generating diversity among environmental bacteria. Several lines of evidence have suggested extensive LGT from archaeal species to the hyperthermophilic bacteria, *Thermotoga maritima* (Mongodin *et al.*, 2005; Nelson *et al.*, 1999; Nesbo *et al.*, 2002).

Re-evaluating prokaryotic species in microbial population genetics, ecology and genomics

The concept of species for prokaryotes is still controversial, and currently, assignment of isolates to species is mainly based on measures of phenotypic or genome similarity. The recent methods for defining prokaryotic species are inadequate and incapable of keeping pace with the levels of diversity that are being uncovered in nature (Gevers *et al.*, 2005). We are witnessing the field of prokaryotic taxonomy continuously influenced by advances in microbial population genetics, ecology and genomics, and by the ease with which sequence data can be obtained (Gevers *et al.*, 2005).

Defining prokaryotic species by DNA–DNA hybridization

Since the 1970s, the cornerstone of genotypic characterization has been the measurement of overall genetic similarity among isolates, assessed by the degree to which their genomes hybridize under standard conditions (DNA–DNA hybridization or DDH). Prior to the availability of DNA sequence information, DDH provided a standardized means for identifying and classifying prokaryotes that lack well-defined morphological or other phenotypic characteristics (Gevers *et al.*, 2005). Nevertheless, DDH has serious limitations: it is a time-consuming procedure, carried out properly by few laboratories only, and it is ill-suited for rapid identification of prokaryotes. It is also unavailable for the classification of prokaryotes that are currently non-culturable, now recognized to be the majority of prokaryotes (>99%) in the biosphere (Amann *et al.*, 1995). More importantly, classification by DDH requires pairwise comparisons of two prokaryotic genomes, and individual strains cannot be analysed and compared with a database using a codified set of criteria to assign it to a known taxon or to propose a new taxon.

Defining prokaryotic species by an rRNA gene sequence approach

As pioneered by Woese (Woese, 1987; Woese and Fox, 1977), ribosomal RNA (rRNA) gene sequence similarity allows the classification of prokaryotes using a universally distributed trait. Importantly, classification of organisms in this manner employs sequence databases. By comparing the rRNA gene sequence of an isolate to those of all known prokaryotic species, microbiologists can rapidly identify prokaryotic strains. In addition, microbiologists who explore the vast diversity of prokaryotic life using molecular approaches can place their findings in a universal context even when the isolates cannot be cultured

(Gevers *et al.*, 2005). Although the advantages of the direct genotypic approaches are obvious, classification by rRNA gene sequence alone – an increasingly common practice – is unsatisfying for several reasons. For example, the rRNA gene sequences often lack resolution when compared with DDH. Consequently, near identity of rRNA gene sequences does not eliminate the need to apply other methods to further explore whether isolates are sufficiently similar to be assigned to the same species. Most importantly, despite the perceived reliability of 16S rRNA gene sequence as a phylogenetic marker, any single measure of sequence similarity is subject both to simple stochastic variation and to the influence of recombination or horizontal gene transfer (Boucher *et al.*, 2004; Gogarten *et al.*, 2002).

Defining prokaryotic species by a multilocus sequencing approach

Sequence-based approaches to defining species require loci that evolve more rapidly than rRNA genes, and multiple genes provide a buffer against the distorting effects of recombination inherent in the analysis using a single locus. Indeed, this is the premise that underlies the development of multilocus sequence typing (MLST), a method for the genotypic characterization of prokaryotes at the infraspecific level using the allelic mismatches of a small number (usually 7) of housekeeping genes (Maiden *et al.*, 1998). Groups of isolates with identical allelic profiles define strains or clones and form the basis for a genotypic classification system (Gevers *et al.*, 2005). Therefore, whereas the rRNA gene sequence might assign an isolate to a genus, MLST might assist in grouping isolates into the major genetic lineages within a species (Cooper and Feil, 2004). MLST refers to a specific tool designed for molecular epidemiology and for defining strains within named species. Such an intraspecific genotyping system should help define species and, indeed, has been used to identify cryptic species in many groups of organisms.

What are prokaryotic species?

The demarcation of familiar species could in principle be used to calibrate the approximate range of diversity that is expected to encompass newly characterized species (Gevers *et al.*, 2005). However, defining species limits by using levels of sequence similarity typically found within existing named species is clearly inappropriate as there are wide variations among species in the levels of genetic similarity among strains within a species. How can we define a widely accepted concept of prokaryotic species? A more attractive approach is to seek ecological, genomic or phenotypic differences among the major clusters resolved by MLSA or complete genomic variation analysis that would justify their separation into species. Ecotype may be useful in understanding the differences among similar prokaryotes. Ecotypes are defined as populations that are genetically cohesive and ecologically distinct. Moreover, ecotypes are expected to be irreversibly separate from one another. Ecotypes therefore hold all the quintessential properties of species as understood in systematics outside of microbiology (de Queiroz, 1998). Defining species as ecotypes might remain reserved for those groups whose ecological distinctness is obvious and important (for example, pathogens), as well as species yet to be discovered (Gevers *et al.*, 2005).

Community genomics in natural microbial communities – another new frontier

At present, relatively little is known about the evolutionary processes that structure and maintain microbial diversity. This is because, until recently, it was difficult to explore individual-level patterns of variation at the microbial scale. With the great advent of high-throughput DNA sequencing techniques, community genomics as a new molecular approach are making great advances towards answering fundamental questions about the evolutionary ecology of microorganisms. In the past few years, massive community-genomic sequence data have been generated, and these data are allowing us to assess variation at different levels across large segments of microbial genomes. Thus, population-genomic analysis of these data are being used to determine how selection and genetic exchange shape the evolution of new microbial lineages (Whitaker and Banfield, 2006). Once independent lineages have been identified, such analyses enable the identification of genome changes that drive niche differentiation and promote the coexistence of closely related lineages within the same environment. Whitaker and Banfield suggested that understanding the evolutionary ecology of natural microbial populations through population-genomic analyses will enhance our understanding of genome evolution across all domains of life (Whitaker and Banfield, 2006).

Community genomics can also provide information about uncultured microbial communities. Two studies published in 2004 demonstrated the power of this approach, particularly with simple microbial communities (Tyson *et al.*, 2004; Venter *et al.*, 2004). Using a whole-genome shotgun approach, Tyson *et al.* (2004) were able to reconstruct two almost complete genome sequences of *Leptospirillum group II* and *Ferroplasma type II* and the partial sequence of three other species from a low-complexity acid mine drainage biofilm growing underground within a pyrite ore body. A larger project published shortly thereafter by Venter and colleagues (Venter *et al.*, 2004) used a whole genome shotgun strategy to explore the diversity in the Sargasso Sea. A total of almost 1.5 billion base pairs of DNA were generated, and based on sequence relatedness, it was estimated that this single sample of seawater contained 1800 species. The project identified >1.2 million new genes, including >700 new rhodopsin- like photoreceptors that may be involved in a new form of phototrophy in the oceans (Allen and Banfield, 2005).

Community-genomic studies help to address several goals in microbial community ecology. Because of the important economic values of many groups of microorganisms, many projects focused on identifying novel enzymes, antibiotics and signalling molecules through functional screens (Riesenfeld *et al.*, 2004), including the diversity of metabolisms responsible for such compounds that can be recognized from the analyses of functional genes from the environment (Tringe *et al.*, 2005). Another goal was to uncover the community structures of microbial populations (including viral populations, e.g. Breitbart *et al.*, 2002); estimating community-level diversity based on sampling statistics of genomic libraries (Venter *et al.*, 2004); and assessing the types of organisms present within a sample using phylogenetic 'anchors' that are well represented in the public sequence databases (DeLong *et al.*, 2006; Tringe *et al.*, 2005; Venter *et al.*, 2004). Because microorganisms have small genomes, population-genomic analysis is possible on a larger scale (i.e. reconstructing more genomes from more individuals and sampling more sites) than is possible for organisms with larger genomes. Indeed, microbial community genomics is bringing us to an exciting era of microbial ecology and population biology.

Evolutionary implications of comparative genomics

Evolutionary biology has two sets of encompassing goals: One is to discover the history of life on earth and the other is to understand the causal processes of evolution. The first set of goals includes: (i) to determine the ancestor–descendant relationships among all species that have ever lived–their phylogeny; (ii) to determine the times at which they originated and for some, when they became extinct; and (iii) to determine the origin of and the rate and course of change in their characteristics. The second set of goals includes understanding the following issues: (i) the origins of hereditary variations; (ii) how various processes act to affect the fate of those variations; (iii) the relative importance of the many co-acting processes of change; (iv) how rapidly changes occur; (v) how processes such as mutation, natural selection, and genetic drift have given rise to the diverse molecular, anatomical, behavioural, and other characteristics of different organisms; and (vi) how populations become different species. Virtually all fields of biology deal with and impact on our understandings of the processes and causes of evolution.

Fundamental issues in evolutionary implications

Large amount of genomic sequencing data are ideal for understanding some of the most fundamental issues in evolutionary biology: suboptimality, homology, hierarchy, ancestry, novelties, the role of natural selection, and the relative importance of directional versus stabilizing selection (Kondrashov, 1999). Indeed, genomic data have provided the best available evidence for many of the above evolutionary phenomena, including the Darwinian evolutionary paradigm. Comparative genomics has played a major role in these efforts and understandings.

Causes of suboptimality: random drift and mutational pressure

Darwin discovered a spectacular paradox: information about past adaptive evolution is provided not by perfect adaptation of a modern species but by its suboptimality (Darwin, 1859). Indeed, natural selection performs local, greedy optimization: at any moment, population slowly moves within the space of genotypes (or phenotypes) in the direction of increasing fitness. Comparative genomics revealed two extra causes for suboptimality: random drift and mutational pressure. Both affect only very weakly selected mutations and, thus, can be significant only at the level of the whole genome. Random drift, important when it can overrule selection, can lead to temporary fixation of any allele. Mutational pressure, important when it can overrule selection, leads to fixation of the allele favoured by mutation, instead of selection (Kondrashov, 1999).

Evidence for homology: orthologous and paralogous

Comparative genomics have been used to discover many cases when proteins performing the same function (analogous) in different species are very dissimilar. Thus, profound similarity among proteins performing such a function in some other species can be taken as homologous (Galperin *et al.*, 1998). It is of particular interest whether similarity of the fundamental genetic processes are also homologous. Many analogous enzymes involved in another fundamental process –DNA replication –are profoundly dissimilar within some pairs of species, implying that their interspecific similarity is homologous and that replication perhaps originated twice (Leipe *et al.*, 1999).

However, independent parallel or convergent evolution from similar initial conditions under similar selective pressures can preserve and even reinforce initial homology. Preservation of homology has been demonstrated for some cases of protein evolution (Hunt *et al.*, 1998). Convergent evolution of function of even initially very different proteins can lead to some sequence similarity (Maruyama *et al.*, 1999; Shen and Jacobs-Lorena, 1999). In many other cases, moreover, such evolution never leads to similarity of sequences and even of structures, presumably because the same function can be performed by very similar proteins (Galperin *et al.*, 1998). Any interspecific similarity of functionless DNA fragments is homologous by definition and must have been inherited from a common ancestor because parallel or convergent neutral evolution would be unlikely. Thus, such features are particularly suitable for phylogenetic reconstructions (Kondrashov, 1999).

Phylogenetic relationships

Based on comparative analyses of the nucleotide sequences of genes encoding ribosomal RNAs and several proteins, molecular phylogeneticists have constructed a 'universal tree of life,' taking it as the basis for a 'natural' hierarchical classification of all living things, providing an objective view of evolutionary relationships and the course of evolution (Kondrashov, 1999). Phylogenetic trees provide a useful means of assessing the evolutionary relationships between putative paralogous and orthologous. Phylogenetic analyses have revealed that gene duplications occurred throughout species history. Interestingly, the analyses of specific families revealed probable rapid proliferation of paralogous at specific times during the evolution of a species, most likely during periods of internal or environmental upheaval. In recent years, detecting signals of natural selection has also become a powerful approach for molecular biologists, biochemists, and virologists to understand the functions of new genes. Positive selection, that is, fixation of advantageous mutations driven by natural selection, has been an exciting topic to evolutionary biologists, because adaptive changes in genes and genomes are ultimately responsible for evolutionary innovations and species differences (Yang, 2005).

Discovering ancestors and origin of novelties

Determining the trait states of the last common ancestor on a rooted tree is necessary for understanding past evolution. This is not easy, because some extra information about the states of the trait in modern species must be available (Kondrashov, 1999). While identifying some of these species as outgroups will help, this cannot be done with absolute confidence. The most direct approach is to polarize individual traits, that is, using some a priori knowledge about their evolution to determine which of their states are ancestral and which are derived. If enough traits are polarized, the tree can be rooted exactly, making it possible to reconstruct the ancestral phenotype (assuming no homoplasy). Unfortunately, our ability to directly polarize morphological traits is rather limited (Kondrashov, 1999).

Comparative genomics have discovered many molecular traits that can be polarized unambiguously because we can confidently infer how they have evolved. At the level of individual nucleotides, small changes which inactivate a pseudogene, as well as Ts, Gs, and Cs in polyA tails of processed pseudogenes, are usually derived (Petrov *et al.*, 1998). With an even greater confidence, we can claim that many individual insertions of transposable elements are irreversible and, thus, derived (Buchanan *et al.*, 1999; Hamdi *et al.*, 1999; Kramerov *et al.*, 1999; Shimamura *et al.*, 1997; Verneau *et al.*, 1998; Waddell *et al.*, 1999). As a result, the origin of novelties will probably be first understood at the molecular level because this process becomes increasingly more complex at the cellular and organismal levels. Origin of new genes is particularly interesting as it is an inherently abrupt process in contrast to cellular and organismal features (Kondrashov, 1999).

Selection versus neutrality

Darwin (1859) realized that traits with no fitness effects would not be subjected to natural selection. Much later, this notion was further developed by Kimura (1968). Kimura's initial reasoning was incorrect that all genes evolve in a neutral fashion as recent studies have identified that a significant proportion of intrapopulation variability can be easily explained by selection. However, the discovery of pseudogenes showed that Darwin and Kimura were both right: evolution of some traits (sites in truly functionless DNA sequences) is not effectively controlled by selection (Li *et al.*, 1981). Kimura's neutral theory of evolution is often used as the null hypothesis for testing modes of molecular evolution for genes and genomes.

Comparative genomics has made it possible to address this issue on a large scale. One approach is to consider a pair of genomes of moderately close species and to calculate the fraction of selectively constrained nucleotides from the pattern of similarity, assuming that all above-random similarity is caused by stabilizing selection (Shabalina and Kondrashov, 1999). Alternatively, one can take a pair of more closely related genomes and compare the levels of divergence between certainly neutral sequences, such as pseudogenes, 'D_n' and between some other sequences, such as protein-coding exons, 'D_f' (Eyre-Walker and Keightley, 1999). If both D_n and D_f were significantly $<< 1$, the fraction of selectively constrained nucleotides within these other sequences can be estimated as $F = 1 - D_f/D_n$. The value F allows the estimate of genomic deleterious mutation rate 'U' (Kondrashov, 1998,1999).

Positive versus stabilizing selection

Positive natural selection causes adaptive allele replacements and, thus, is the driving force of evolution (Darwin 1859). In contrast, negative selection is a conservative force, maintaining status quo (Schmalhausen, 1949). To date, comparative genomic approaches have mostly been used to study negative selection because its impact, DNA-sequence conservation, is relatively easy to detect at the molecular level. In contrast, evolution of sequences as a result of positive selection is harder to distinguish from neutral evolution. Thus, very little is known at present regarding relative importance of positive and negative selection at different time scales. However, very strong positive selection can be detected when the selection causes a significant region of DNA to evolve much faster than neutral sequences. Cases of such selection have been discovered recently. Not surprisingly, strong positive selection (i.e. the classic Darwinian 'natural selection') typically occurred during the acquisition of a new function after gene duplication (Zhang *et al.*, 1998) or as a response to a rapidly changing environment (Duda and Palumbi, 1999; Kordis *et al.*, 1998; Ohno *et al.*, 1998; Papadopoulos *et al.*, 1999).

The standard neutral model of biodiversity

Many ecologists believe that natural ecological communities are inherently complex. The dynamics of each species within a community must be modelled uniquely. An alternative view is provided by neutral theory. The field of neutral theory (Hubbell, 2001) is concerned with stochastic models of community dynamics for which all species in the community share the same conditional distribution of per capita demographic rates (i.e. species are modelled identically rather than uniquely). Neutral models have been used to study biodiversity since Kendall (1948) showed that Fisher's (Fisher *et al.*, 1943) log-series distribution of species abundances could be derived from a neutral birth-death process (as did Volkov *et al.*, 2003) using a more modern theory of birth-death processes). Another well known neutral model is May's (May, 1975) derivation of Preston's (Preston, 1962) log-normal distribution. Other examples of neutral models include Chesson and Warner's (Chesson and Warner, 1981) lottery model, Diserud and Engen's (Deserud and Engen, 2000) u-logistic diffusion model and Bell's (2000) model of dispersal and establishment from a fixed species pool.

In these models, diversity is described by species number S and the variance of frequency or log abundance q~. The frequency distribution of abundance is generally log-normal, skewed to the left by immigration and resembling descriptions of natural communities. Increased immigration and community size always cause S to increase. Their effect on q~ is more complicated, but given biologically reasonable assumptions, S and q~ will be positively correlated in most circumstances. At equilibrium, communities maintain roughly constant levels of S and q~ but change in composition through time; however, for many generations, composition may remain similar. Many prominent patterns observed in natural communities can therefore be generated by a strictly neutral model. This does not show that community structure is determined exclusively by demographic stochasticity, but rather demonstrates the necessity for an appropriate null model when functional hypotheses are being tested. Whether the neutral models of ecology apply to microbial populations have not been tested empirically. Microbial community genomics should help in addressing this issue.

Keeping pace with microevolution

The current availability of more than 700 completed bacterial genome sequences from 18 different phyla has not only led to a reassessment of bacterial systematics, but also to an increasing appreciation of the complexities and dynamism of the genomic environment (Cowan, 2000; Markowitz, 2007). In particular, multiple completed genomes are now available for single species or genera and these data are providing evidence for short-term microevolution in bacteria (Jordan *et al.*, 2002; Whittam and Bumbaugh, 2002). For example, genes found in members of the sequenced strains of *Staphylococcus genus* are probably a subset of the genes in all strains of the named species *S. aureus*. Extending the principle of the core genome to the genes that are present in all bacterial genomes has led to attempts to reconstruct the oldest branches of the bacterial evolutionary tree, and examination of very small genomes from endosymbiotic species has fuelled speculation concerning the minimal gene-set (Klasson and Andersson, 2004).

Patterns of microevolutionary descent examined by MLST data

There are two approaches for examining the levels of relatedness between strains using MLST data: clustering procedures based on pairwise differences in allelic profiles; and phylogenetic analyses based on allele sequences (Feil, 2004). Patterns of descent in clonal complexes have been traditionally examined with MLST or MLEE data using the UPGMA clustering algorithm, which is based on allelic mismatches (Sneath and Sokal, 1973). This technique has been useful for the identification of the main clonal complexes in a population but provides little evidence for the precise pattern of events that occurred in each clonal complex. The combination of a limited number of loci and a large number of very closely related genotypes means that many equally parsimonious pathways can be reconstructed for a single clonal complex (Feil, 2004).

Microarrays and variation in gene content

The completion of a single genome sequence for a species means that it is possible to examine the distribution of each gene in the natural population through the use of microarrays. Microarray data can provide valuable evidence about the evolution and population structure of a given species, and can be used to help identify new virulence determinants (Feil, 2004).

The application of this approach to 15 strains of *H. pylori* has led to a revision of the estimate of the proportion of strain-specific genes from 6–7% (as estimated from a comparison of two genomes) to 12–18% (Salama *et al.*, 2000). Fitzgerald *et al.* (2001) noted from microarray analyses of 36 staphylococcal strains that approximately 20% of the ORFs are strain-specific, and a similar figure was also reported from an analysis of 11 strains of *Campylobacter jejuni* (Dorrell *et al.*, 2001). Microarray analyses can also be used to identify genomic regions that have extremely variable gene content, and highlight the importance of large-scale deletions in the evolution of many species. These regions are often preferred sites for the insertion of genomic islands, or other large mobile elements, and have been referred to as 'plasticity zones' in *H. pylori* (Alm and Trust, 1999) or 'regions of difference' in *S. aureus* (Fitzgerald and Musser, 2001).

Although one limitation of microarray analysis is that it is only possible to identify known genes that are present in the strains being tested by comparison with a reference genome, subtractive hybridization can be used to detect unknown genes, or islands, that are present in the test strains (Guo *et al.*, 2005; Winstanley, 2002).

The range of genetic diversity between species

Although clearly distinct from comparisons between very distant taxa, a tendency to categorize all intra-species or intra-genus comparisons as 'closely related' in the evolutionary literature detracts from the fact that some named species are much more variable than others (Feil, 2004). At one extreme, comparative genomics of *Bacillus* and *Mycobacterium* species confirm a high degree of uniformity in these genera. Sreevatsan and coworkers (Sreevatsan *et al.*, 1997) reported that, for many genes of the *Mycobacterium tuberculosis* complex (which encompasses several known species), the average number of synonymous substitutions per 100 synonymous sites is <0.01, whereas the figure for *E. coli* strains is >11. Indeed, the level of nucleotide diversity in the *Mycobacterium tuberculosis* complex is even lower than that observed for *Shigella sonnei*, which is widely considered to be a specialized clone of *E. coli* (Pupo *et al.*, 2000). The difference in variability between the *Mycobacterium tuberculosis* complex and *E. coli* is reflected both in terms of the allelic variation for genes that belong to the species core genome and in terms of the gene content of the accessory genome (Feil, 2004).

The wide range of intra-species variation reflects the current lack of a universal and meaningful species definition in bacteria (Smith *et al.*, 2000), but it is also a reflection of the period of time since divergence from a common ancestor and the range of ecological niches that different named species have been able to exploit. *E. coli* is estimated to have diverged from *Salmonella* 100–150 million years ago (Ochman and Wilson, 1987), whereas the divergence of *M. bovis* and *M. tuberculosis* might have coincided with the domestication of cattle approximately 15,000–20,000 years ago (Kapur *et al.*, 1994) – although this hypothesis remains controversial (Brosch *et al.*, 2002). As *E. coli* exploits numerous host-associated and environmental niches, it might have greater opportunity to import DNA from unrelated lineages than, for example, *Helicobacter pylori*, which is restricted to the human stomach. *H. pylori* has a highly variable core genome, which reflects an increased rate of mutation and homologous recombination (Achtman and Suerbaum, 2000).

However, Alm and co-worker noted that strain-specific genes tended to be restricted to a limited number of loci and propose a moderate frequency, or selective tolerance, of gene acquisition in *H. pylori* (Alm and Trust, 1999). The example of *H. pylori* illustrates that the rapid diversification of the core genome may not always be mirrored by a fluidity of gene content. Another marked example is the evolution of obligate endosymbionts, such as *Buchnera aphidicola*, a member of the gamma-proteobacteria and a primary endosymbiont of aphids. Obligate endosymbionts have much smaller genomes than their free-living relatives due to the deletion of several large genomic regions. A small population size means that deleterious mutations are more frequently fixed by genetic drift. Indeed, the rate at which the core genome of *B. aphidicola* is believed to have evolved about fourfold faster than that of *E. coli*. However, the absence of phage or other mobile genetic elements, together with strict vertical transmission (resulting in ecological isolation) of the endosymbiont by the aphid host means that *B. aphidicola* rarely import genes from unrelated taxa.

In contrast to the core genome, the gene content and genome architecture of *B. aphidicola* accessory genes suggest an increased stability compared with the *E. coli* genome (Tamas *et al.*, 2002; Wernegreen, 2002). A common characteristic of accessory genes is their propensity for horizontal transfer among cells, so the predominance of phage, plasmids and other mobile genetic elements in the population must also be considered. In the case of *Yersinia pestis* the presence of numerous insertion sequences is thought to have destabilized regions of the genome (Hinchliffe *et al.*, 2003). Other genetic factors, such as natural competence for transformation, mutation rate and the presence of repeat sequences, also contribute towards the differences in variation between species (Aras *et al.*, 2003).

Evolutionary systems biology

The myriad of technological breakthroughs in biology that are leading to the unification of broad scientific fields such as molecular biology, biochemistry, physics, mathematics, and computer science are now known as systems biology (Medina, 2005). Systems biology is a rapidly expanding field that integrates widely diverse areas of science, toward the goal of elucidating the hierarchy of metabolic and regulatory systems in the cell, and ultimately leading to a predictive understanding of the cellular response to perturbations (Ideker *et al.*, 2001; Kitano, 2002). As the theoretical and experimental tools of systems biology rapidly advance, multiple fields are embracing systems biology approaches as a mainstream method of research.

With whole-genome data allowing reconstruction of more robust phylogenies for prokaryotic groups, new biological questions can now be addressed. Genomic and postgenomic data offer a new 'global' view of the function of living systems across the tree of life. These new data suggest that biological systems (e.g. a cell) are composed of discrete 'modules' of interacting components with different functions, and in turn these modules form biological networks that carry out the myriad functions of living systems (Hartwell *et al.*, 1999). Because postgenomics research is taking place throughout the tree of life, comparative approaches are a way to combine data from many organisms to understand the evolution and function of biological systems from the gene to the organismal level. Therefore, systems biology can build on decades of theoretical work in evolutionary biology, and at the same time evolutionary biology can use systems approaches to develop new uncharted directions (Medina, 2005).

Comparative studies can give insight into even the most fundamental principles of life (Medina, 2005). For example, revolutionary findings in network theory have in part come from genomic data from a wide range of organisms, leading scientists to propose laws that seem to govern biological networks (Jeong *et al.*, 2000; Ravasz *et al.*, 2002). Different types of cellular networks (e.g. protein interaction and metabolic networks) seem to share properties with other complex abiotic networks such as their 'scale-free' nature and 'small world' organization. Additional levels in the hierarchy of biological networks and the interactions between them are now being characterized that will allow for integration of data and new theoretical predictions (Ge *et al.*, 2003). Processes widely studied by evolutionary biologists such as selection, gene duplication, and neutral evolution are being examined in the context of network models as opposed to at the level of individual genes or molecules (Hahn *et al.*, 2004; van Noort *et al.*, 2004; Wagner, 2003; Wuchty, 2004).

Microbial phylogenomics – a case study of overlapping genes

As was discussed earlier, with more prokaryotic genomes being sequenced, an increasing interest is to use the genomic sequences in assessing evolutionary relationships among prokaryotic taxa because prokaryotes often lack morphological and behavioural characters amenable to phylogenetic analysis. The approaches based on whole-genome sequences, such as gene content, gene order, and rare genomic changes, capture more phylogenetic signatures and therefore generate more reliable species phylogenies. Currently, due to limitations in taxonomic resolution and the computational time constraints, large-scale systematic comparisons of a section of genomes have rarely been carried out. Luo and coworkers have designed a novel method on the basis of overlapping genes for reliable and time-saving inference of phylogenies among prokaryotic genomes (Luo *et al.*, 2006).

In a further study, they demonstrated the method in determining the phylogenetic relationships for more than 360 completely sequenced prokaryotic genomes (Fig. 2.2). Interestingly, compared with the existing trees constructed using different phylogenetic markers, the topologies of the phylogenies based on overlapping genes for different sets of genomes are remarkably consistent; and their trees offer a higher resolution on the broad relationships among prokaryotic species, especially for deep branches. These results suggest that the method on the basis of overlapping genes as rare genomic changes is appropriate to infer large-scale phylogenies and may provide a framework towards a genome-based taxonomy for prokaryotes.

Power and limitations of comparative genomics

The DNA sequences of entire genomes are being determined at a rapid rate. The explosion in genomic sequence available in public databases has resulted in an unprecedented opportunity for computational whole genome analyses. Whereas initial genome sequencing efforts were for organisms chosen to be widely spaced in the tree of life, there is a growing emphasis on projects to sequence a species that is sufficiently similar to an already-sequenced species to allow direct comparison of those two DNA sequences. This and other changes in genome sequencing strategies have created a strong need for new methods to compare genomic sequences. One of the key computational contributions to the field of biology is arguably the BLAST programs, which is one of the most widely used computational tools in the biological community. Perhaps it should come as no surprise, then, that some of the most widely used tools in bioinformatics today are comparative based. Indeed, with the recent explosions of publicly available genomic sequences, both the opportunities and challenges for computational comparative analyses are greater than ever before.

Comparison of genomic DNA sequences: solved and unsolved problems

The art and science of comparing genomic DNA sequences have changed dramatically in the last two or three years, with respect to the rate, nature, and utilization of the data. It is essential for the bioinformatics community to understand these changes and to respond appropriately. To date, according to GOLD and other databases, we have genomic sequences for 780 bacteria, and other 1790 prokaryotic projects are currently under way. Many of these genomes exist as working 'draft' sequence data that consists of pieces whose relative order and orientation are difficult to determine. When these genomes are completely assembled, more effective comparisons will be used to locate signals that regulate gene transcription, to understand the mechanisms and tempo of genome evolution, and to identify hitherto unimagined segments that modulate the structure and function of the genome.

In 2001, Miller listed the and the following desirable directions of software development for comparing genomic DNA sequences (Miller, 2001): (i) improved software that aligns two genomic sequences with a rigorous statistical basis; (ii) an industrial-strength gene prediction system that effectively combines genomic sequence comparisons, intrinsic sequence properties, and results from searching databases of proteins sequences and ESTs; (iii) reliable and automatic software for aligning three or more genomic sequences; (iv) better methods for displaying and browsing genomic sequence alignments; (v) improved datasets and protocols for evaluating the correctness and performance of genomic alignment software. While some of these goals have been achieved, much remains to be done to allow biologists readily deal with large amount of genomic data in a portable way.

The challenges of applying comparative-based computational methods to whole genomes

A number of promising comparative-based approaches have been developed for gene finding, regulatory element discovery and other purposes, and it is clear that these tools will play

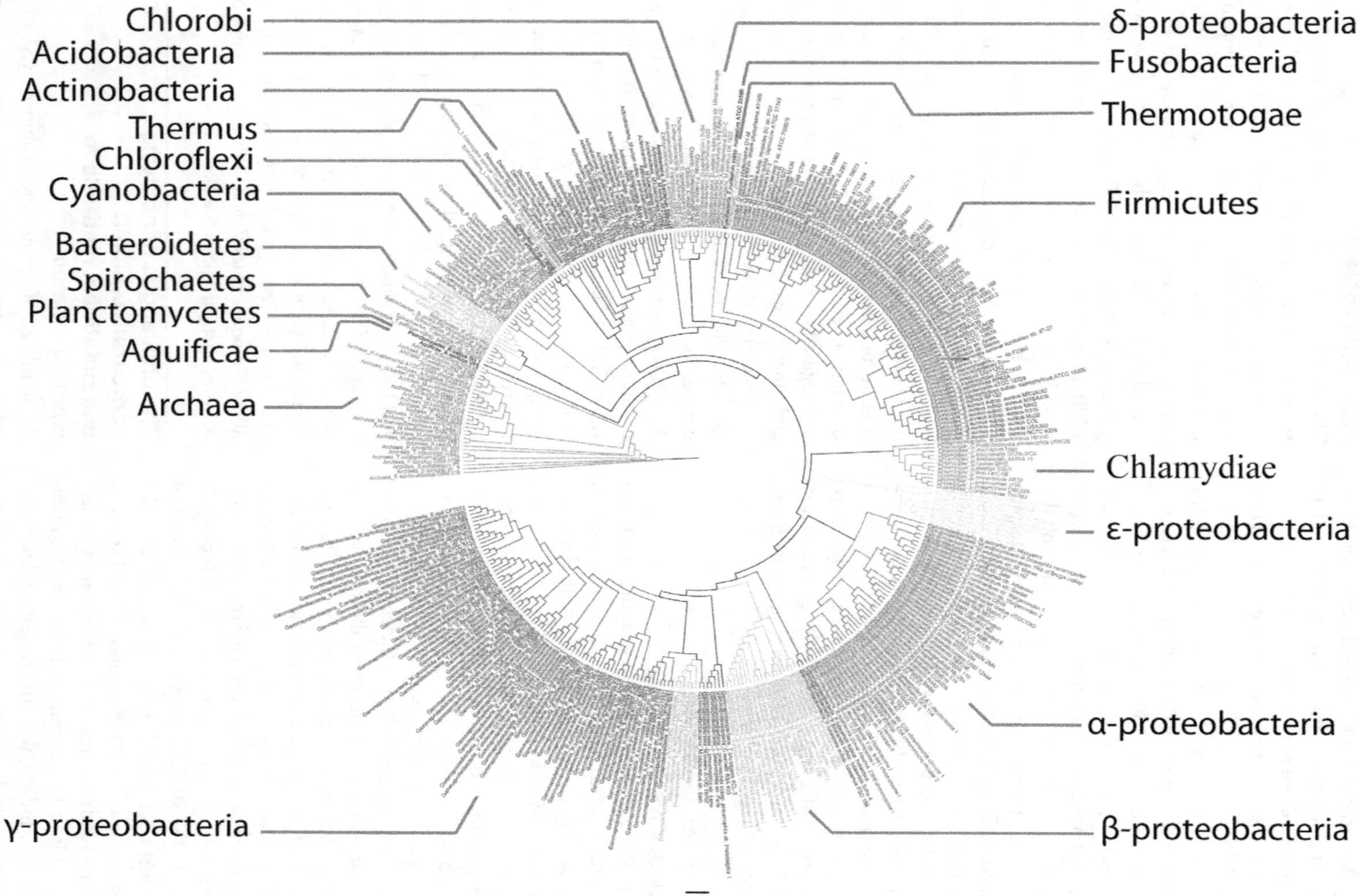

Figure 2.2 The phylogeny derived from overlapping genes inferred by the maximum-evolution method for 366 taxa.

a fundamental role in analysing the enormous amount of new data that is currently being generated. The synthesis of computationally intensive comparative approaches with the requirement for whole genome analysis represents both an unprecedented challenge and an opportunity for computational scientists (Dubchak and Pachter, 2002). A number of recent articles have summarized and reviewed the computational tools and methods being used for comparative genomics (Miller, 2001; Wiehe *et al.*, 2000). The focus here will be three issues: alignment, gene finding and regulatory element discovery. A good, general, and comprehensive reference on genome annotation can be found in Brent (2008).

Alignment

The problem of whole genome alignment has not been extensively researched, perhaps because the computational challenges arising in this context render most classical approaches to alignment infeasible or intractable. The main problem in performing whole genome alignment is that the classical algorithm that forms the basis of most alignment tools, the Smith-Waterman algorithm (Smith and Waterman, 1981), is quadratic in time complexity and is not practical for sequences of size greater than 1000kb, much less for whole genomes in the sizes of billions of bases!

Some of the difficulties in whole genome alignment are (i) algorithms need to be linear so that they are fast; (ii) algorithms need to be robust, i.e. they cannot be too parameter dependent; (iii) algorithms need to be memory efficient; and (iv) algorithms capable of handling rearrangements, gene duplications and repetitive elements. Different solutions have been proposed for these different requirements. A major challenge that remains is the development of a tool that addresses all of the above problems (Dubchak and Pachter, 2002).

A more serious difficulty that is not easy to overcome using basic heuristics is that of rearrangement, repetition and duplication in whole genomes. These features of genomes become important and difficult to address in the context of whole genome comparison. The seeding methods described above are difficult to overcome these problems by being inherently 'local alignment' algorithms. The problem with local alignment is that the added robustness of local alignment comes at the price of less specific matches. On the other hand, anchor-based alignments that avoid the false positive problem described above suffer in their inability to correctly align sequences that have been subjected to rearrangements or inversions, and extra pre- or post-processing is necessary to identify and compensate for such occurrences. In any case, it remains an open problem and a significant challenge to build a single program that will correctly identify duplications and rearrangements, and integrate this process seamlessly into the alignment (Dubchak and Pachter, 2002).

Gene finding

The observation that coding regions are highly conserved in contrast to non-coding regions of the genome has been successfully applied in a case-by-case basis to find genes for the last decade (Hardison *et al.*, 1997). However, the idea of automated comparative-based gene finding is relatively new, originating with solving the ROSTTA problem (Batzoglou *et al.*, 2000; Dubchak and Pachter, 2002). One of the key problems in whole genome gene finding is the draft quality of the sequences, and some programs (e.g. Twinscan; Wang *et al.*, 2003) have been specifically developed to account for this difficulty. In addition, the focus of the program on gene finding in one organism (and not both) avoids the problems of rearrangements in the same way that local alignment avoids these problems. This robustness comes at a price, namely that not all of the extra information available from comparison is taken advantage of in the gene finding stage.

Comparative-based gene finding is related to alignment in a fundamental way. This is because gene finding is enhanced by alignment, and a natural problem that arises in building a program is whether to align before the gene finding (e.g. Twinscan, SGP Wiehe *et al.*, 2001) or to integrate alignment during gene finding (e.g. SLAM Pachter *et al.*, 2002). It remains an open problem to establish the best way of integrating alignment and gene finding on a whole genome scale (Dubchak and Pachter, 2002).

Regulatory elements discovery

The majority of regulatory elements controlling gene expressions are located in non-coding regions. Identifying these elements presents a great challenge because these are no accurate methods to computationally identify regulatory sequences based on scanning the DNA of a single organism. The result of trying to predict transcription factor binding sites in whole genomes, even with known motifs, is that virtually entire genomes will be predicted to be full of binding sites. Furthermore, finding non-coding regions with regulatory functions is not feasible using computational techniques, similar for discovering coding sequences (Dubchak and Pachter, 2002).

The wide distribution and combinatorial nature of these regulatory elements combined with the increasing amounts of available sequence data require faster and more reliable computational tools for predicting and identifying potential regulatory elements in a high-throughput manner. Although several groups have identified evolutionarily conserved sequences using cross-species comparisons and consequently located putative transcription factor binding sites within these regions (Gottgens *et al.*, 2001; Onyango *et al.*, 2000), no study to date has focused on quantitatively and qualitatively evaluating the significance of comparatively predicted transcription factor binding sites. One direction towards high-throughput annotation of regulatory sites is the method (Loots, 2008) that combines transcription factor binding sites database searches with global sequence alignment of syntenic regions and clustering to detect transcription factor binding sites in locally highly conserved regions. This fast procedure reduces the number of predicted transcription factor binding sites by several orders of magnitude.

In summary, the current explosion in the amount of genomic information is leading us into a new era of computational genomics, in which new algorithmic and computational challenges have become the keys to significant biological discoveries. Below we outline a few challenges that lie ahead. First, we need to develop alignment programs that are robust with respect to rearrangements, gene duplications and repetitive alignments. Second, we need to design a whole-genome gene finder that integrates the problems of alignment and gene finding in such a way to allow predicting genes that are not conserved but at the same time takes full advantage of gene conservation in making predictions. Third, there is a pressing need to determine optimal evolutionary distances among genomes that allow critically addressing different problems through comparative genomics (e.g. gene finding or regulatory element detection). And lastly, we need to develop new comparative techniques to take advantage of the availability of an increasing number of whole genome sequences in order to elucidate the structure and organization of transcription factor binding sites.

Acknowledgements

This work was supported by an NSFC Grant (30571037) and Beijing Normal University.

References

Achtman, M. and Suerbaum, S. (2000). Sequence variation in *Helicobacter pylori*. Trends Microbiol. *8*, 57–58.

Allen, E.E. and Banfield, J.F. (2005). Community genomics in microbial ecology and evolution. Nat. Rev. Microbiol. 3, 489–498.

Alm, R.A. and Trust, T.J. (1999). Analysis of the genetic diversity of Helicobacter pylori: the tale of two genomes. J. Mol. Med. 77, 834–846.

Amann, R.I., Ludwig, W., and Schleifer, K.H. (1995). Phylogenetic identification and in situ detection of individual microbial cells without cultivation. Microbiol. Rev. *59*, 143–169.

Andersson, D.I. and Hughes, D. (1996). Muller's ratchet decreases fitness of a DNA-based microbe. Proc. Natl. Acad. Sci. U.S.A. *93*, 906–907.

Andersson, J.O. and Andersson, S.G. (1999). Insights into the evolutionary process of genome degradation. Curr. Opin. Genet. Dev. *9*, 664–671.

Andersson, S.G., Zomorodipour, A., Andersson, J.O., Sicheritz-Ponten, T., Alsmark, U.C., Podowski, R.M., Naslund, A.K., Eriksson, A.S., Winkler, H.H., and Kurland, C.G. (1998). The genome sequence of Rickettsia prowazekii and the origin of mitochondria. Nature *396*, 133–140.

Aras, R.A., Kang, J., Tschumi, A.I., Harasaki, Y., and Blaser, M.J. (2003). Extensive repetitive DNA facilitates prokaryotic genome plasticity. Proc. Natl. Acad. Sci. U.S.A. *100*, 13579–13584.

Bacon, J., James, B.W., Wernisch, L., Williams, A., Morley, K.A., Hatch, G.J., Mangan, J.A., Hinds, J., Stoker, N.G., Butcher, P.D., and Marsh, P.D. (2004). The influence of reduced oxygen availability on pathogenicity and gene expression in *Mycobacterium tuberculosis*. Tuberculosis *84*, 205–217.

Batzoglou, S., Pachter, L., Mesirov, J.P., Berger, B., and Lander, E.S. (2000). Human and mouse gene structure: comparative analysis and application to exon prediction. Genome Res. *10*, 950–958.

Beja, O., Koonin, E.V., Aravind, L., Taylor, L.T., Seitz, H., Stein, J.L., Bensen, D.C., Feldman, R.A., Swanson, R.V., and DeLong, E.F. (2002). Comparative genomic analysis of archaeal genotypic variants in a single population and in two different oceanic provinces. Appl. Environ. Microbiol. *68*, 335–345.

Belda, E., Moya, A., and Silva, F.J. (2005). Genome rearrangement distances and gene order phylogeny in gamma-Proteobacteria. Mol. Biol. Evol. *22*, 1456–1467.

Bell, G. (2000). The distribution of abundance in neutral communities. Am. Nat. *155*, 606–617.

Bernal, A., Ear, U., and Kyrpides, N. (2001). Genomes OnLine Database (GOLD): a monitor of genome projects world-wide. Nucleic Acids Res. *29*, 126–127.

Black, W.C.t., Baer, C.F., Antolin, M.F., and DuTeau, N.M. (2001). Population genomics: genome-wide sampling of insect populations. Annu. Rev. Entomol. *46*, 441–469.

Boucher, Y., Douady, C.J., Papke, R.T., Walsh, D.A., Boudreau, M.E., Nesbo, C.L., Case, R.J., and Doolittle, W.F. (2003). Lateral gene transfer and the origins of prokaryotic groups. Annu. Rev. Genet. *37*, 283–328.

Boucher, Y., Douady, C.J., Sharma, A.K., Kamekura, M., and Doolittle, W.F. (2004). Intragenomic heterogeneity and intergenomic recombination among haloarchaeal rRNA genes. J. Bacteriol. *186*, 3980–3990.

Breitbart, M., Salamon, P., Andresen, B., Mahaffy, J.M., Segall, A.M., Mead, D., Azam, F., and Rohwer, F. (2002). Genomic analysis of uncultured marine viral communities. Proc. Natl. Acad. Sci. U.S.A. *99*, 14250–14255.

Brent, M.R. (2008). Steady progress and recent breakthroughs in the accuracy of automated genome annotation. Nat. Rev. Genet. *9*, 62–73.

Brodin, P., Majlessi, L., Marsollier, L., M.I. de Jonge, Bottai, D., Demangel, C., Hinds, J., Neyrolles, O., Butcher, P.D., Leclerc, C., Cole, S.T., and Brosch, R. (2006). Dissection of ESAT-6 system 1 of *Mycobacterium tuberculosis* and impact on immunogenicity and virulence. Infect. Immun. *74*, 88–98.

Brosch, R., Gordon, S.V., Marmiesse, M., Brodin, P., Buchrieser, C., Eiglmeier, K., Garnier, T., Gutierrez, C., Hewinson, G., Kremer, K., Parsons, L.M., Pym, A.S., Samper, S., D. van Soolingen, and Cole, S.T. (2002). A new evolutionary scenario for the *Mycobacterium tuberculosis* complex. Proc. Natl. Acad. Sci. U.S.A. *99*, 3684–3689.

Brown, P.O. and Botstein, D. (1999). Exploring the new world of the genome with DNA microarrays. Nat. Genet. *21*, 33–37.

Buchanan, F., Crawford, A., Strobeck, C., Palsboll, P., and Plante, Y. (1999). Evolutionary applications of MIRs and SINEs. Anim. Genet. *30*, 47–50.

Butcher, P.D. (2004). Microarrays for *Mycobacterium tuberculosis*. Tuberculosis *84*, 131–137.

Carver, T.J., Rutherford, K.M., Berriman, M., Rajandream, M.A., Barrell, B.G., and Parkhill, J. (2005). ACT: the Artemis Comparison Tool. Bioinformatics *21*, 3422–3423.

Caspi, R., Foerster, H., Fulcher, C.A., Kaipa, P., Krummenacker, M., Latendresse, M., Paley, S., Rhee, S.Y., Shearer, A.G., Tissier, C., *et al.* (2008). The MetaCyc Database of metabolic pathways and enzymes and the BioCyc collection of Pathway/Genome Databases. Nucleic Acids Res. 36: D623–631.

Chesson, P.L. and Warner, R.R. (1981). Environmental variability promotes coexistence in lottery competitive systems. Am. Nat. 117.

Ciccarelli, F.D., Doerks, T., C. von Mering, Creevey, C.J., Snel, B., and Bork, P. (2006). Toward automatic reconstruction of a highly resolved tree of life. Science *311*, 1283–1287.

Cliften, P.F., Hillier, L.W., Fulton, L., Graves, T., Miner, T., Gish, W.R., Waterston, R.H., and Johnston, M. (2001). Surveying Saccharomyces genomes to identify functional elements by comparative DNA sequence analysis. Genome Res. *11*, 1175–1186.

Coenye, T. (2003). A golden age for microbial genomics. Genome Biol. *4*, 347.

Cole, S.T., Eiglmeier, K., Parkhill, J., James, K.D., Thomson, N.R., Wheeler, P.R., Honore, N., Garnier, T., Churcher, C., Harris, D., *et al.* (2001). Massive gene decay in the leprosy bacillus. Nature *409*, 1007–1011.

Cooper, J.E. and Feil, E.J. (2004). Multilocus sequence typing--what is resolved? Trends Microbiol. *12*, 373–377.

Cooper, V.S. and Lenski, R.E. (2000). The population genetics of ecological specialization in evolving *Escherichia coli* populations. Nature *407*, 736–739.

Cowan, D.A. (2000). Microbial genomes--the untapped resource. Trends Biotechnol. *18*, 14–16.

Darwin, C. (1859). The Origin of Species by Means of Natural Selection. (London: Murray).

de Queiroz, K. (1998). The general lineage concept of species, species criteria, and the process of speciation: A conceptual unification and terminological recommendations. In D. J. Howard and S. H. Berlocher (eds), Endless Forms: Species and Speciation (Oxford: Oxford University Press), pp. 57–75.

Deeds, E.J., Hennessey, H., and Shakhnovich, E.I. (2005). Prokaryotic phylogenies inferred from protein structural domains. Genome Res. *15*, 393–402.

Delcher, A.L., Kasif, S., Fleischmann, R.D., Peterson, J., White, O., and Salzberg, S.L. (1999). Alignment of whole genomes. Nucleic Acids Res. *27*, 2369–2376.

Delcher, A.L., Phillippy, A., Carlton, J., and Salzberg, S.L. (2002). Fast algorithms for large-scale genome alignment and comparison. Nucleic Acids Res. *30*, 2478–2483.

DeLong, E.F. (2002). Microbial population genomics and ecology. Curr. Opin. Microbiol. *5*, 520–524.

DeLong, E.F. (2004). Microbial population genomics and ecology: the road ahead. Environ. Microbiol. *6*, 875–878.

DeLong, E.F., Preston, C.M., Mincer, T., Rich, V., Hallam, S.J., Frigaard, N.U., Martinez, A., Sullivan, M.B., Edwards, R., Brito, B.R., Chisholm, S.W., and Karl, D.M. (2006). Community genomics among stratified microbial assemblages in the ocean's interior. Science *311*, 496–503.

Deserud, O.H. and Engen, S. (2000). A general and dynamic species abundance model, embracing the lognormal and he gamma models.. Am. Nat. *155*, 497–511.

Dorrell, N., Mangan, J.A., Laing, K.G., Hinds, J., Linton, D., Al-Ghusein, H., Barrell, B.G., Parkhill, J., Stoker, N.G., Karlyshev, A.V., Butcher, P.D., and Wren, B.W. (2001). Whole genome comparison of Campylobacter jejuni human isolates using a low-cost microarray reveals extensive genetic diversity. Genome Res. *11*, 1706–1715.

Dubchak, I. and Pachter, L. (2002). The computational challenges of applying comparative-based computational methods to whole genomes. Brief. Bioinform. *3*, 18–22.

Duda, T.F., Jr. and Palumbi, S.R. (1999). Molecular genetics of ecological diversification: duplication and rapid evolution of toxin genes of the venomous gastropod Conus. Proc. Natl. Acad. Sci. U.S.A. *96*, 6820–6823.

Eanes, W. (1999). Analysis of selection on enzyme polymorphisms. Annu. Rev. Ecol. Syst. *30*, 301–326.

Eyre-Walker, A. and Keightley, P.D. (1999). High genomic deleterious mutation rates in hominids. Nature *397*, 344–347.

Feil, E.J. (2004). Small change: keeping pace with microevolution. Nat. Rev. Microbiol. *2*, 483–495.

Fisher, R., Coenye, T., and Williams, C. (1943). The relation between the number of species and the number of individuals in a random sample of an animal population. J. Anim. Ecol. *12*, 42–58.

Fitzgerald, J.R. and Musser, J.M. (2001). Evolutionary genomics of pathogenic bacteria. Trends Microbiol. *9*, 547–553.

Fitzgerald, J.R., Reid, S.D., Ruotsalainen, E., Tripp, T.J., Liu, M., Cole, R., Kuusela, P., Schlievert, P.M., Jarvinen, A., and Musser, J.M. (2003). Genome diversification in *Staphylococcus aureus*: Molecular evolution of a highly variable chromosomal region encoding the sstaphylococcal exotoxin-like family of proteins. Infect. Immun. *71*, 2827–2838.

Fitzgerald, J.R., Sturdevant, D.E., Mackie, S.M., Gill, S.R., and Musser, J.M. (2001). Evolutionary genomics of *Staphylococcus aureus*: insights into the origin of methicillin-resistant strains and the toxic shock syndrome epidemic. Proc. Natl. Acad. Sci. U.S.A. *98*, 8821–8826.

Fraser-Liggett, C.M. (2005). Insights on biology and evolution from microbial genome sequencing. Genome Res. *15*, 1603–1610.

Galperin, M.Y., Walker, D.R., and Koonin, E.V. (1998). Analogous enzymes: independent inventions in enzyme evolution. Genome Res. *8*, 779–790.

Ge, H., Walhout, A.J., and Vidal, M. (2003). Integrating 'omic' information: a bridge between genomics and systems biology. Trends Genet. *19*, 551–560.

Gevers, D., Cohan, F.M., Lawrence, J.G., Spratt, B.G., Coenye, T., Feil, E.J., Stackebrandt, E., Van, Y. de Peer, Vandamme, P., Thompson, F.L., and Swings, J. (2005). Opinion: Re-evaluating prokaryotic species. Nat. Rev. Microbiol. *3*, 733–739.

Gill, S.R., Fouts, D.E., Archer, G.L., Mongodin, E.F., Deboy, R.T., Ravel, J., Paulsen, I.T., Kolonay, J.F., Brinkac, L., Beanan, M., Dodson, R.J., Daugherty, S.C., Madupu, R., Angiuoli, S.V., Durkin, A.S., Haft, D.H., Vamathevan, J., Khouri, H., Utterback, T., Lee, C., Dimitrov, G., Jiang, L., Qin, H., Weidman, J., Tran, K., Kang, K., Hance, I.R., Nelson, K.E., and Fraser, C.M. (2005). Insights on evolution of virulence and resistance from the complete genome analysis of an early methicillin-resistant *Staphylococcus aureus* strain and a biofilm-producing methicillin-resistant Staphylococcus epidermidis strain. J. Bacteriol. *187*, 2426–2438.

Gillespie, J. (1994). Non-neutral Evolution (New York: Chapman & Hall).

Gillespie, J.H. (1984). The status of the neutral theory: the neutral theory of molecular evolution. Science *224*, 732–733.

Gilligan, P. and Venkatesh, B. (2004). Application of comparative genomics to the analysis of vertebrate regulatory elements. Brief. Funct. Genomic Proteomic *3*, 7–11.

Glasner, J.D., Liss, P., Plunkett, G., 3rd, Darling, A., Prasad, T., Rusch, M., Byrnes, A., Gilson, M., Biehl, B., Blattner, F.R., and Perna, N.T. (2003). ASAP, a systematic annotation package for community analysis of genomes. Nucleic Acids Res. *31*, 147–151.

Gogarten, J.P., Doolittle, W.F., and Lawrence, J.G. (2002). Prokaryotic evolution in light of gene transfer. Mol. Biol. Evol. *19*, 2226–2238.

Gottgens, B., Gilbert, J.G., Barton, L.M., Grafham, D., Rogers, J., Bentley, D.R., and Green, A.R. (2001). Long-range comparison of human and mouse SCL loci: localized regions of sensitivity to restriction endonucleases correspond precisely with peaks of conserved noncoding sequences. Genome Res. *11*, 87–97.

Guo, H., Sun, S., Finan, T.M., and Xu, J. (2005). Novel DNA sequences from natural strains of the nitrogen-fixing symbiotic bacterium Sinorhizobium meliloti. Appl. Environ. Microbiol. *71*, 7130–7138.

Gupta, R.S. (1998). Protein phylogenies and signature sequences: A reappraisal of evolutionary relationships among archaebacteria, eubacteria, and eukaryotes. Microbiol. Mol. Biol. Rev. *62*, 1435–1491.

Hahn, M.W., Conant, G.C., and Wagner, A. (2004). Molecular evolution in large genetic networks: does connectivity equal constraint? J. Mol. Evol. *58*, 203–211.

Hamdi, H., Nishio, H., Zielinski, R., and Dugaiczyk, A. (1999). Origin and phylogenetic distribution of Alu DNA repeats: irreversible events in the evolution of primates. J. Mol. Biol. *289*, 861–871.

Hardison, R.C. (2003). Comparative genomics. PLoS Biol. 1: E58.

Hardison, R.C., Oeltjen, J., and Miller, W. (1997). Long human-mouse sequence alignments reveal novel regulatory elements: a reason to sequence the mouse genome. Genome Res. *7*, 959–966.

Hartwell, L.H., Hopfield, J.J., Leibler, S., and Murray, A.W. (1999). From molecular to modular cell biology. Nature 402: C47–52.

Hinchliffe, S.J., Isherwood, K.E., Stabler, R.A., Prentice, M.B., Rakin, A., Nichols, R.A., Oyston, P.C., Hinds, J., Titball, R.W., and Wren, B.W. (2003). Application of DNA microarrays to study the evolutionary genomics of *Yersinia pestis* and *Yersinia pseudotuberculosis*. Genome Res. *13*, 2018–2029.

Holden, M.T., Titball, R.W., Peacock, S.J., Cerdeno-Tarraga, A.M., Atkins, T., Crossman, L.C., Pitt, T., Churcher, C., Mungall, K., Bentley, S.D., *et al.* (2004). Genomic plasticity of the causative agent of melioidosis, Burkholderia pseudomallei. Proc. Natl. Acad. Sci. U.S.A. *101,* 14240–14245.

Hubbell, S.P. (2001). The Unified Neutral Theory of Biodiversity and Biogeography (Princeton: Princeton Univ. Press).

Huber, H., Hohn, M.J., Rachel, R., Fuchs, T., Wimmer, V.C., and Stetter, K.O. (2002). A new phylum of Archaea represented by a nanosized hyperthermophilic symbiont. Nature *417,* 63–67.

Hunt, D.M., Dulai, K.S., Cowing, J.A., Julliot, C., Mollon, J.D., Bowmaker, J.K., Li, W.H., and Hewett-Emmett, D. (1998). Molecular evolution of trichromacy in primates. Vision Res. *38,* 3299–3306.

Huynen, M.A., Snel, B., and V. van Noort. (2004). Comparative genomics for reliable protein-function prediction from genomic data. Trends Genet. *20,* 340–344.

Ideker, T., Galitski, T., and Hood, L. (2001). A new approach to decoding life: systems biology. Annu. Rev. Genomics Hum. Genet. *2,* 343–372.

Jeong, H., Tombor, B., Albert, R., Oltvai, Z.N., and Barabasi, A.L. (2000). The large-scale organization of metabolic networks. Nature *407,* 651–654.

Jin, Q., Yuan, Z., Xu, J., Wang, Y., Shen, Y., Lu, W., Wang, J., Liu, H., Yang, J., Yang, F., *et al.* (2002). Genome sequence of Shigella flexneri 2a: insights into pathogenicity through comparison with genomes of *Escherichia coli* K12 and O157. Nucleic Acids Res. *30,* 4432–4441.

Jordan, I.K., Rogozin, I.B., Wolf, Y.I., and Koonin, E.V. (2002). Microevolutionary genomics of bacteria. Theor. Popul. Biol. *61,* 435–447.

Jorde, L.B., Watkins, W.S., and Bamshad, M.J. (2001). Population genomics: a bridge from evolutionary history to genetic medicine. Hum. Mol. Genet. *10,* 2199–2207.

Jukes, T.H. and Kimura, M. (1984). Evolutionary constraints and the neutral theory. J. Mol. Evol. *21,* 90–92.

Kaeberlein, T., Lewis, K., and Epstein, S.S. (2002). Isolating 'uncultivable' microorganisms in pure culture in a simulated natural environment. Science *296,* 1127–1129.

Kanehisa, M., Araki, M., Goto, S., Hattori, M., Hirakawa, M., Itoh, M., Katayama, T., Kawashima, S., Okuda, S., Tokimatsu, T., and Yamanishi, Y. (2008). KEGG for linking genomes to life and the environment. Nucleic Acids Res. 36: D480–484.

Kanehisa, M. and Goto, S. (2000). KEGG: kyoto encyclopedia of genes and genomes. Nucleic Acids Res. *28,* 27–30.

Kanehisa, M., Goto, S., Hattori, M., Aoki-Kinoshita, K.F., Itoh, M., Kawashima, S., Katayama, T., Araki, M., and Hirakawa, M. (2006). From genomics to chemical genomics: new developments in KEGG. Nucleic Acids Res. 34: D354–357.

Kapur, V., Whittam, T.S., and Musser, J.M. (1994). Is *Mycobacterium tuberculosis* 15, 000 years old? J. Infect. Dis. *170,* 1348–1349.

Karp, P.D., Keseler, I.M., Shearer, A., Latendresse, M., Krummenacker, M., Paley, S.M., Paulsen, I., Collado-Vides, J., Gama-Castro, S., Peralta-Gil, M., Santos-Zavaleta, A., Penaloza-Spinola, M.I., Bonavides-Martinez, C., and Ingraham, J. (2007). Multidimensional annotation of the *Escherichia coli* K-12 genome. Nucleic Acids Res. *35,* 7577–7590.

Karp, P.D., Ouzounis, C.A., Moore-Kochlacs, C., Goldovsky, L., Kaipa, P., Ahren, D., Tsoka, S., Darzentas, N., Kunin, V., and Lopez-Bigas, N. (2005). Expansion of the BioCyc collection of pathway/genome databases to 160 genomes. Nucleic Acids Res. *33,* 6083–6089.

Kellis, M., Patterson, N., Endrizzi, M., Birren, B., and Lander, E.S. (2003). Sequencing and comparison of yeast species to identify genes and regulatory elements. Nature *423,* 241–254.

Kendall, M.G. (1948). The Advanced Theory of Statistics (London Griffin).

Kent, W.J. (2002). BLAT--the BLAST-like alignment tool. Genome Res. *12,* 656–664.

Kimura, M. (1968). Evolutionary rate at the molecular level. Nature *217,* 624–626.

Kimura, M. (1979). The neutral theory of molecular evolution. Sci. Am. 241, 98–100, 102, 108.

Kitano, H. (2002). Systems biology: a brief overview. Science *295,* 1662–1664.

Klasson, L. and Andersson, S.G. (2004). Evolution of minimal-gene-sets in host-dependent bacteria. Trends Microbiol. *12,* 37–43.

Kondrashov, A.S. (1998). Measuring spontaneous deleterious mutation process. Genetica 102–103, 183–197.

Kondrashov, A.S. (1999). Comparative genomics and evolutionary biology. Curr. Opin. Genet. Dev. *9,* 624–629.

Konstantinidis, K.T. and Tiedje, J.M. (2004). Trends between gene content and genome size in prokaryotic species with larger genomes. Proc. Natl. Acad. Sci. U.S.A. *101,* 3160–3165.

Koonin, V.E. and Galperin, Y.M. (2003). Sequence–Evolution–Function: Computational Approaches in Comparative (Boston: Kluwer Academic Publishers).

Kordis, D., Bdolah, A., and Gubensek, F. (1998). Positive Darwinian selection in Vipera palaestinae phospholipase A2 genes is unexpectedly limited to the third exon. Biochem. Biophys. Res. Commun. *251,* 613–619.

Kramerov, D., Vassetzky, N., and Serdobova, I. (1999). The evolutionary position of dormice (Gliridae) in Rodentia determined by a novel short retroposon. Mol. Biol. Evol. *16,* 715–717.

Krummenacker, M., Paley, S., Mueller, L., Yan, T., and Karp, P.D. (2005). Querying and computing with BioCyc databases. Bioinformatics *21,* 3454–3455.

Kurtz, S., Phillippy, A., Delcher, A.L., Smoot, M., Shumway, M., Antonescu, C., and Salzberg, S.L. (2004). Versatile and open software for comparing large genomes. Genome Biol. 5: R12.

Kyrpides, N.C. (1999). Genomes OnLine Database (GOLD 1.0): a monitor of complete and ongoing genome projects world-wide. Bioinformatics *15,* 773–774.

Lawrence, J.G. and Hendrickson, H. (2005). Genome evolution in bacteria: order beneath chaos. Curr. Opin. Microbiol. *8*, 572–578.

Leipe, D.D., Aravind, L., and Koonin, E.V. (1999). Did DNA replication evolve twice independently? Nucleic Acids Res. *27*, 3389–3401.

Li, W.H., Gojobori, T., and Nei, M. (1981). Pseudogenes as a paradigm of neutral evolution. Nature *292*, 237–239.

Liolios, K., Mavromatis, K., Tavernarakis, N., and Kyrpides, N.C. (2008). The Genomes On Line Database (GOLD) in 2007: status of genomic and metagenomic projects and their associated metadata. Nucleic Acids Res. 36: D475–479.

Liolios, K., Tavernarakis, N., Hugenholtz, P., and Kyrpides, N.C. (2006). The Genomes On Line Database (GOLD) v.2: a monitor of genome projects worldwide. Nucleic Acids Res. 34: D332–334.

Loots, G.G. (2008). Chapter 10 genomic identification of regulatory elements by evolutionary sequence comparison and functional analysis. Adv. Genet. *61*, 269–293.

Luo, Y., Fu, C., Zhang, D.Y., and Lin, K. (2006). Overlapping genes as rare genomic markers: the phylogeny of gamma-Proteobacteria as a case study. Trends Genet. *22*, 593–596.

Luo, Y., Fu, C., Zhang, D.Y., and Lin, K. (2007). BPhyOG: an interactive server for genome-wide inference of bacterial phylogenies based on overlapping genes. BMC Bioinf. *8*, 266.

Lynch, M. and Walsh, B. (1998). Genetics and Analysis of Quantitative Traits (Sunderland, MA: Sinauer).

Mahan, M.J., Slauch, J.M., and Mekalanos, J.J. (1993). Selection of bacterial virulence genes that are specifically induced in host tissues. Science *259*, 686–688.

Maiden, M.C., Bygraves, J.A., Feil, E., Morelli, G., Russell, J.E., Urwin, R., Zhang, Q., Zhou, J., Zurth, K., Caugant, D.A., *et al.* (1998). Multilocus sequence typing: a portable approach to the identification of clones within populations of pathogenic microorganisms. Proc. Natl. Acad. Sci. U.S.A. *95*, 3140–3145.

Markowitz, V.M. (2007). Microbial genome data resources. Curr. Opin. Biotechnol. *18*, 267–272.

Markowitz, V.M., Ivanova, N.N., Szeto, E., Palaniappan, K., Chu, K., Dalevi, D., Chen, I.M., Grechkin, Y., Dubchak, I., Anderson, I., *et al.* (2008a). IMG/M: a data management and analysis system for metagenomes. Nucleic Acids Res. 36: D534–538.

Markowitz, V.M., Korzeniewski, F., Palaniappan, K., Szeto, E., Werner, G., Padki, A., Zhao, X., Dubchak, I., Hugenholtz, P., Anderson, I., *et al.* (2006). The integrated microbial genomes (IMG) system. Nucleic Acids Res. 34: D344–348.

Markowitz, V.M. and Kyrpides, N.C. (2007). Comparative genome analysis in the integrated microbial genomes (IMG) system. Methods Mol. Biol. *395*, 35–56.

Markowitz, V.M., Szeto, E., Palaniappan, K., Grechkin, Y., Chu, K., Chen, I.M., Dubchak, I., Anderson, I., Lykidis, A., Mavromatis, K., Ivanova, N.N., and Kyrpides, N.C. (2008b). The integrated microbial genomes (IMG) system in 2007: data content and analysis tool extensions. Nucleic Acids Res. 36: D528–533.

Maruyama, K., Sato, N., and Ohta, N. (1999). Conservation of structure and cold-regulation of RNA-binding proteins in cyanobacteria: probable convergent evolution with eukaryotic glycine-rich RNA-binding proteins. Nucleic Acids Res. *27*, 2029–2036.

May, R.M. (1975). Patterns of Species Abundance and Diversity. (Harvard, MA: Belknap Press).

McClelland, M., Florea, L., Sanderson, K., Clifton, S.W., Parkhill, J., Churcher, C., Dougan, G., Wilson, R.K., and Miller, W. (2000). Comparison of the *Escherichia coli* K-12 genome with sampled genomes of a Klebsiella pneumoniae and three Salmonella enterica serovars, Typhimurium, Typhi and Paratyphi. Nucleic Acids Res. *28*, 4974–4986.

Medina, M. (2005). Genomes, phylogeny, and evolutionary systems biology. Proc. Natl. Acad. Sci. U.S.A. 102 Suppl *1*, 6630–6635.

Medini, D., Donati, C., Tettelin, H., Masignani, V., and Rappuoli, R. (2005). The microbial pan-genome. Curr. Opin. Genet. Dev. *15*, 589–594.

Miller, W. (2001). Comparison of genomic DNA sequences: solved and unsolved problems. Bioinformatics *17*, 391–397.

Mira, A., Ochman, H., and Moran, N.A. (2001). Deletional bias and the evolution of bacterial genomes. Trends Genet. *17*, 589–596.

Mongodin, E.F., Hance, I.R., Deboy, R.T., Gill, S.R., Daugherty, S., Huber, R., Fraser, C.M., Stetter, K., and Nelson, K.E. (2005). Gene transfer and genome plasticity in Thermotoga maritima, a model hyperthermophilic species. J. Bacteriol. *187*, 4935–4944.

Moran, N.A. and Mira, A. (2001). The process of genome shrinkage in the obligate symbiont *Buchnera aphidicola*. Genome Biol. 2: RESEARCH0054.

Moran, N.A. and Plague, G.R. (2004). Genomic changes following host restriction in bacteria. Curr. Opin. Genet. Dev. *14*, 627–633.

Nakata, N., Tobe, T., Fukuda, I., Suzuki, T., Komatsu, K., Yoshikawa, M., and Sasakawa, C. (1993). The absence of a surface protease, OmpT, determines the intercellular spreading ability of Shigella: the relationship between the ompT and kcpA loci. Mol. Microbiol. *9*, 459–468.

Nelson, K.E., Clayton, R.A., Gill, S.R., Gwinn, M.L., Dodson, R.J., Haft, D.H., Hickey, E.K., Peterson, J.D., Nelson, W.C., Ketchum, K.A., *et al.* (1999). Evidence for lateral gene transfer between Archaea and bacteria from genome sequence of Thermotoga maritima. Nature *399*, 323–329.

Nesbo, C.L., Nelson, K.E., and Doolittle, W.F. (2002). Suppressive subtractive hybridization detects extensive genomic diversity in Thermotoga maritima. J. Bacteriol. *184*, 4475–4488.

Nierman, W.C., DeShazer, D., Kim, H.S., Tettelin, H., Nelson, K.E., Feldblyum, T., Ulrich, R.L., Ronning, C.M., Brinkac, L.M., Daugherty, S.C., *et al.* (2004). Structural flexibility in the Burkholderia mallei genome. Proc. Natl. Acad. Sci. U.S.A. *101*, 14246–14251.

Ochman, H. and Davalos, L.M. (2006). The nature and dynamics of bacterial genomes. Science *311*, 1730–1733.

Ochman, H. and Wilson, A.C. (1987). Evolution in bacteria: evidence for a universal substitution rate in cellular genomes. J. Mol. Evol. *26*, 74–86.

Ohno, M., Menez, R., Ogawa, T., Danse, J.M., Shimohigashi, Y., Fromen, C., Ducancel, F., Zinn-Justin, S., Le, M.H., Du Boulain, J.C., *et al.* (1998). Molecular evolution of snake toxins: is the functional diversity of snake toxins associated with a mechanism of accelerated evolution? Prog. Nucleic Acid Res. Mol. Biol. *59*, 307–364.

Onyango, P., Miller, W., Lehoczky, J., Leung, C.T., Birren, B., Wheelan, S., Dewar, K., and Feinberg, A.P. (2000). Sequence and comparative analysis of the mouse 1-megabase region orthologous to the human 11p15 imprinted domain. Genome Res. *10*, 1697–1710.

Pace, N.R. (1997). A molecular view of microbial diversity and the biosphere. Science *276*, 734–740.

Pachter, L., Alexandersson, M., and Cawley, S. (2002). Applications of generalized pair hidden Markov models to alignment and gene finding problems. J. Comput. Biol. *9*, 389–399.

Papadopoulos, D., Schneider, D., Meier-Eiss, J., Arber, W., Lenski, R.E., and Blot, M. (1999). Genomic evolution during a 10, 000-generation experiment with bacteria. Proc. Natl. Acad. Sci. U.S.A. *96*, 3807–3812.

Park, J., Park, B., Jung, K., Jang, S., Yu, K., Choi, J., Kong, S., Park, J., Kim, S., Kim, H., Kim, S., Kim, J.F., Blair, J.E., Lee, K., Kang, S., and Lee, Y.H. (2008). CFGP: a web-based, comparative fungal genomics platform. Nucleic Acids Res. 36: D562–571.

Parkhill, J., Sebaihia, M., Preston, A., Murphy, L.D., Thomson, N., Harris, D.E., Holden, M.T., Churcher, C.M., Bentley, S.D., Mungall, K.L., *et al.* (2003). Comparative analysis of the genome sequences of *Bordetella pertussis, Bordetella parapertussis* and *Bordetella bronchiseptica*. Nat. Genet. 35, 32–40.

Parkhill, J., Wren, B.W., Thomson, N.R., Titball, R.W., Holden, M.T., Prentice, M.B., Sebaihia, M., James, K.D., Churcher, C., Mungall, K.L., *et al.* (2001). Genome sequence of *Yersinia pestis*, the causative agent of plague. Nature *413*, 523–527.

Peterson, J.D., Umayam, L.A., Dickinson, T., Hickey, E.K., and White, O. (2001). The comprehensive microbial resource. Nucleic Acids Res. *29*, 123–125.

Petrov, D.A., Chao, Y.C., Stephenson, E.C., and Hartl, D.L. (1998). Pseudogene evolution in Drosophila suggests a high rate of DNA loss. Mol. Biol. Evol. *15*, 1562–1567.

Preston, F.W. (1962). The canonical distribution of commonness and rarity. Ecology *43*, 254–283.

Pupo, G.M., Lan, R., and Reeves, P.R. (2000). Multiple independent origins of Shigella clones of *Escherichia coli* and convergent evolution of many of their characteristics. Proc. Natl. Acad. Sci. U.S.A. *97*, 10567–10572.

Raoult, D., Fournier, P.E., and Drancourt, M. (2004). What does the future hold for clinical microbiology? Nat. Rev. Microbiol. *2*, 151–159.

Ravasz, E., Somera, A.L., Mongru, D.A., Oltvai, Z.N., and Barabasi, A.L. (2002). Hierarchical organization of modularity in metabolic networks. Science *297*, 1551–1555.

Riesenfeld, C.S., Schloss, P.D., and Handelsman, J. (2004). Metagenomics: genomic analysis of microbial communities. Annu. Rev. Genet. *38*, 525–552.

Roff, D.A. (1997). Evolutionary Quantitative Genetics (New York: Chapman & Hall).

Rokas, A. (2006). Genomics. Genomics and the tree of life. Science *313*, 1897–1899.

Rokas, A. and Holland, P.W. (2000). Rare genomic changes as a tool for phylogenetics. Trends Ecol. Evol. *15*, 454–459.

Salama, N., Guillemin, K., McDaniel, T.K., Sherlock, G., Tompkins, L., and Falkow, S. (2000). A whole-genome microarray reveals genetic diversity among Helicobacter pylori strains. Proc. Natl. Acad. Sci. U.S.A. *97*, 14668–14673.

Schmalhausen, I. (1949). Factors of Evolution: the Theory of Stabilizing Selection (Philadelphia: Blakiston).

Schroeder, B.G., Peterson, L.M., and Fleischmann, R.D. (2002). Improved quantitation and reproducibility in *Mycobacterium tuberculosis* DNA microarrays. J Mol. Microbiol. Biotechnol. *4*, 123–126.

Schwartz, S., Elnitski, L., Li, M., Weirauch, M., Riemer, C., Smit, A., Green, E.D., Hardison, R.C., and Miller, W. (2003). MultiPipMaker and supporting tools: Alignments and analysis of multiple genomic DNA sequences. Nucleic Acids Res. *31*, 3518–3524.

Schwartz, S., Zhang, Z., Frazer, K.A., Smit, A., Riemer, C., Bouck, J., Gibbs, R., Hardison, R., and Miller, W. (2000). PipMaker--a web server for aligning two genomic DNA sequences. Genome Res. *10*, 577–586.

Shabalina, S.A. and Kondrashov, A.S. (1999). Pattern of selective constraint in *C. elegans* and *C. briggsae* genomes. Genet. Res. *74*, 23–30.

Shen, Z. and Jacobs-Lorena, M. (1999). Evolution of chitin-binding proteins in invertebrates. J. Mol. Evol. *48*, 341–347.

Shimamura, M., Yasue, H., Ohshima, K., Abe, H., Kato, H., Kishiro, T., Goto, M., Munechika, I., and Okada, N. (1997). Molecular evidence from retroposons that whales form a clade within even-toed ungulates. Nature *388*, 666–670.

Smith, J.M., Feil, E.J., and Smith, N.H. (2000). Population structure and evolutionary dynamics of pathogenic bacteria. Bioessays 22, 1115–1122.

Smith, T.F. and Waterman, M.S. (1981). Identification of common molecular subsequences. J. Mol. Biol. *147*, 195–197.

Sneath, P.H. and Sokal, R.R. (1973). Numerical Taxonomy (San Francisco, CA: W.H. Freeman).

Sobral, R.G., Jones, A.E., Des, S.G., Etages, Dougherty, T.J., Peitzsch, R.M., Gaasterland, T., Ludovice, A.M., de Lencastre, H., and Tomasz, A. (2007). Extensive and genome-wide changes in the transcription profile of *Staphylococcus aureus* induced by modulating the transcription of the cell wall synthesis gene murF. J. Bacteriol. *189*, 2376–2391.

Spratt, B.G., Hanage, W.P., and Feil, E.J. (2001). The relative contributions of recombination and point mutation to the diversification of bacterial clones. Curr. Opin. Microbiol. *4*, 602–606.

Sreevatsan, S., Pan, X., Stockbauer, K.E., Connell, N.D., Kreiswirth, B.N., Whittam, T.S., and Musser, J.M.

(1997). Restricted structural gene polymorphism in the *Mycobacterium tuberculosis* complex indicates evolutionarily recent global dissemination. Proc. Natl. Acad. Sci. U.S.A. *94*, 9869–9874.

Stewart, G.R., Wernisch, L., Stabler, R., Mangan, J.A., Hinds, J., Laing, K.G., Young, D.B., and Butcher, P.D. (2002). Dissection of the heat-shock response in *Mycobacterium tuberculosis* using mutants and microarrays. Microbiology *148*, 3129–3138.

Strauss, E.J. and Falkow, S. (1997). Microbial pathogenesis: genomics and beyond. Science *276*, 707–712.

Strous, M., Pelletier, E., Mangenot, S., Rattei, T., Lehner, A., Taylor, M.W., Horn, M., Daims, H., Bartol-Mavel, D., Wincker, P., *et al.* (2006). Deciphering the evolution and metabolism of an anammox bacterium from a community genome. Nature *440*, 790–794.

Tamas, I., Klasson, L., Canback, B., Naslund, A.K., Eriksson, A.S., Wernegreen, J.J., Sandstrom, J.P., Moran, N.A., and Andersson, S.G. (2002). 50 million years of genomic stasis in endosymbiotic bacteria. Science *296*, 2376–2379.

Tatusov, R.L., Fedorova, N.D., Jackson, J.D., Jacobs, A.R., Kiryutin, B., Koonin, E.V., Krylov, D.M., Mazumder, R., Mekhedov, S.L., Nikolskaya, A.N., *et al.* (2003). The COG database: an updated version includes eukaryotes. BMC Bioinf. *4*, 41.

Teichmann, S.A. and Mitchison, G. (1999). Is there a phylogenetic signal in prokaryote proteins? J. Mol. Evol. *49*, 98–107.

Tringe, S.G., C. von Mering, Kobayashi, A., Salamov, A.A., Chen, K., Chang, H.W., Podar, M., Short, J.M., Mathur, E.J., Detter, J.C., Bork, P., Hugenholtz, P., and Rubin, E.M. (2005). Comparative metagenomics of microbial communities. Science *308*, 554–557.

Tsolaki, A.G., Hirsh, A.E., DeRiemer, K., Enciso, J.A., Wong, M.Z., Hannan, M., Goguet, Y.O. de la Salmoniere, Aman, K., Kato-Maeda, M., and Small, P.M. (2004). Functional and evolutionary genomics of *Mycobacterium tuberculosis*: insights from genomic deletions in 100 strains. Proc. Natl. Acad. Sci. U.S.A. *101*, 4865–4870.

Tyson, G.W., Chapman, J., Hugenholtz, P., Allen, E.E., Ram, R.J., Richardson, P.M., Solovyev, V.V., Rubin, E.M., Rokhsar, D.S., and Banfield, J.F. (2004). Community structure and metabolism through reconstruction of microbial genomes from the environment. Nature *428*, 37–43.

Uchiyama, I. (2007). MBGD: a platform for microbial comparative genomics based on the automated construction of orthologous groups. Nucleic Acids Res. 35: D343–346.

Utaida, S., Dunman, P.M., Macapagal, D., Murphy, E., Projan, S.J., Singh, V.K., Jayaswal, R.K., and Wilkinson, B.J. (2003). Genome-wide transcriptional profiling of the response of *Staphylococcus aureus* to cell-wall-active antibiotics reveals a cell-wall-stress stimulon. Microbiology *149*, 2719–2732.

Valdivia, R.H. and Falkow, S. (1997). Fluorescence-based isolation of bacterial genes expressed within host cells. Science *277*, 2007–2011.

van Noort, V., Snel, B., and Huynen, M.A. (2004). The yeast coexpression network has a small-world, scale-free architecture and can be explained by a simple model. EMBO Rep. *5*, 280–284.

Venter, J.C., Remington, K., Heidelberg, J.F., Halpern, A.L., Rusch, D., Eisen, J.A., Wu, D., Paulsen, I., Nelson, K.E., Nelson, W., *et al.* (2004). Environmental genome shotgun sequencing of the Sargasso Sea. Science *304*, 66–74.

Verneau, O., Catzeflis, F., and Furano, A.V. (1998). Determining and dating recent rodent speciation events by using L1 (LINE-1) retrotransposons. Proc. Natl. Acad. Sci. U.S.A. *95*, 11284–11289.

Volkov, I., Banavar, J.R., Hubbell, S.P., and Maritan, A. (2003). Neutral theory and relative species abundance in ecology. Nature *424*, 1035–1037.

Waddell, P.J., Okada, N., and Hasegawa, M. (1999). Towards resolving the interordinal relationships of placental mammals. Syst. Biol. *48*, 1–5.

Wagner, A. (2003). How the global structure of protein interaction networks evolves. Proc. Biol. Sci. *270*, 457–466.

Wang, M., Buhler, J., and Brent, M.R. (2003). The effects of evolutionary distance on TWINSCAN, an algorithm for pair-wise comparative gene prediction. Cold Spring Harb. Symp. Quant. Biol. *68*, 125–130.

Welch, R.A., Burland, V., Plunkett, G., 3rd, Redford, P., Roesch, P., Rasko, D., Buckles, E.L., Liou, S.R., Boutin, A., Hackett, J., *et al.* (2002). Extensive mosaic structure revealed by the complete genome sequence of uropathogenic *Escherichia coli*. Proc. Natl. Acad. Sci. U.S.A. *99*, 17020–17024.

Wernegreen, J.J. (2002). Genome evolution in bacterial endosymbionts of insects. Nat. Rev. Genet. *3*, 850–861.

Whitaker, R.J. and Banfield, J.F. (2006). Population genomics in natural microbial communities. Trends Ecol. Evol. *21*, 508–516.

Whittam, T.S. and Bumbaugh, A.C. (2002). Inferences from whole-genome sequences of bacterial pathogens. Curr. Opin. Genet. Dev. *12*, 719–725.

Wiehe, T., Gebauer-Jung, S., Mitchell-Olds, T., and Guigo, R. (2001). SGP-1: prediction and validation of homologous genes based on sequence alignments. Genome Res. *11*, 1574–1583.

Wiehe, T., Guigo, R., and Miller, W. (2000). Genome sequence comparisons: hurdles in the fast lane to functional genomics. Brief Bioinform. *1*, 381–388.

Winstanley, C. (2002). Spot the difference: applications of subtractive hybridisation to the study of bacterial pathogens. J. Med. Microbiol. *51*, 459–467.

Woese, C.R. (1987). Bacterial evolution. Microbiol. Rev. *51*, 221–271.

Woese, C.R. and Fox, G.E. (1977). Phylogenetic structure of the prokaryotic domain: the primary kingdoms. Proc. Natl. Acad. Sci. U.S.A. *74*, 5088–5090.

Woese, C.R., Kandler, O., and Wheelis, M.L. (1990). Towards a natural system of organisms: proposal for the domains Archaea, Bacteria, and Eucarya. Proc. Natl. Acad. Sci. U.S.A. *87*, 4576–4579.

Wren, B.W. (2000). Microbial genome analysis: insights into virulence, host adaptation and evolution. Nat. Rev. Genet. *1*, 30–39.

Wuchty, S. (2004). Evolution and topology in the yeast protein interaction network. Genome Res. *14*, 1310–1314.

Yang, Z. (2005). The power of phylogenetic comparison in revealing protein function. Proc. Natl. Acad. Sci. U.S.A. *102*, 3179–3180.

Zhang, J., Rosenberg, H.F., and Nei, M. (1998). Positive Darwinian selection after gene duplication in primate ribonuclease genes. Proc. Natl. Acad. Sci. U.S.A. *95*, 3708–3713.

Zhang, Z., Schwartz, S., Wagner, L., and Miller, W. (2000). A greedy algorithm for aligning DNA sequences. J. Comput. Biol. *7*, 203–214.

Patterns of Horizontal Gene Transfer in Bacteria

3

Weilong Hao and G. Brian Golding

Abstract

Horizontal gene transfer, as a major force in shaping bacterial gene content, has gained incredible attention over the last decade. Along with the fast growing bacterial genome sequence data, there have been an increasingly large number of studies focused on horizontal gene transfer. The studies have been gradually transformed from identifying individual genes that have been horizontally transferred to assessing the general patterns of horizontal gene transfer and evaluating the systematic consequences of massive gene transfers. The rates of gene transfers have been measured by various methods such as parsimony and maximum likelihood methods. Different phylogenetic methods were applied to a variety of data sets to assess whether there exists a congruent and meaningful bacterial tree. Even though some consensus has been reached, many contradictions have emerged and need to be solved in future studies. Our goal here is to review recent studies on this subject with an emphasis on the emerging patterns in horizontal gene transfer research.

Introduction

Horizontal gene transfer (HGT, also known as lateral gene transfer) is a process that transfers genetic material between species (Doolittle, 1999). Horizontally transferred genes are commonly observed in free-living bacteria (Garcia-Vallvé *et al.*, 2000). A well-known example is the comparison of the first three sequenced genomes of *Escherichia coli* where about 60% of the genes are not shared among all three strains (Welch *et al.*, 2002). Evidence of HGT has also been observed in obligate intracellular bacteria (Blanc *et al.*, 2007), which were traditionally thought to be rarely affected by HGT due to their isolated life styles (Silva *et al.*, 2003). While some genes are known to have a higher probability to be transferred than others (Jain *et al.*, 1999), HGT has been found to affect almost all functional categories of genes (Zhaxybayeva *et al.*, 2006). For instance, genes coding for ribosomal RNAs and ribosomal proteins are commonly inherited vertically and have been used extensively for the reconstruction of evolutionary relationships among species (Fox *et al.*, 1980; Maidak *et al.*, 1997; Brown *et al.*, 2001). However, several studies have shown that some ribosomal RNA and ribosomal protein genes are subject to HGT (Yap *et al.*, 1999; Brochier *et al.*, 2000). As more horizontally transferred genes are uncovered, HGT has been gradually accepted as an important force in bacterial genome evolution. However, the frequency and adaptive roles of HGT in bacterial evolution are still heavily debated (Gogarten *et al.*, 2002; Daubin *et al.*, 2003; Kurland *et al.*, 2003; Bapteste *et al.*, 2005; Doolittle and Bapteste, 2007).

The architecture of transferred genes

Genomic islands are clusters of genes of potential horizontal origin in a prokaryotic genome. Genomic islands harbour a large number of genes that might not have any known homologues or only have homologues in distantly related species. Many of these genes are related to virulence in pathogens and play a significant role in pathogenicity and expansion of the host bacteria into new niches (Hacker *et al.*, 1997; Hacker and Kaper, 2000; Hacker and Carniel, 2001; Gophna *et al.*, 2003; Waterman and Holden, 2003; Hsiao *et al.*, 2005; Lawrence, 2005; Coleman *et al.*, 2006). Many genomic islands are large and have anomalous base compositions different from the rest of the genomes (Hsiao *et al.*, 2003; Zhang and Zhang, 2003). On the other hand, a large proportion of horizontally transferred genes are not in distinct genomic islands (Lawrence and Ochman, 1998), and these genes can be difficult to trace and confirm. One hypothesis for the origins of these small fragments is that they have been transferred in small fragments. Alternatively, a large number of genes could have been transferred in a single or few events but that extensive genome rearrangements could have scattered the horizontally transferred genes across the chromosome. Indeed, in *E. coli*, after successful transfer of large pieces of DNA sequences, randomized distribution of the acquired sequences can be detected during short-term experimental evolution (Babic *et al.*, 2008). Even though there have been studies attempting to measure the size of insertions and deletions in the face of gene movement (Mira *et al.*, 2001), the general pattern of the size of initial gene transfers and gene losses is still not well understood.

Many horizontally transferred genes form operon clusters. However, the role of HGT in operon formation has been hotly debated. Lawrence (1999) proposed HGT as a primary force in operon formation and this proposal is also known as the selfish operon theory. Several studies have criticized this theory based on the fact that recently transferred genes are less likely to be a member of operons (Pal and Hurst, 2004; Price *et al.*, 2005, 2008). It is noteworthy that recently transferred genes have high rates of gene turn-over and many transferred genes in a genome are possibly on the way to being eliminated from the host cells (Hao and Golding, 2004, 2006). Genes that are going to be deleted are possibly not functional or of significant adaptive value to the host cells. It's therefore not surprising that these genes are randomly distributed and not in operon clusters. If only the genes that are likely to be retained are considered in the analysis, one should expect a higher fraction of recently transferred genes in operon clusters. There have been a large number of case studies showing HGT of gene clusters and metabolic pathways (Top and Springael, 2003; Diaz, 2004; Fux *et al.*, 2005; Cunneen and Reeves, 2007; Homma *et al.*, 2007; Templeton, 2007; Chen *et al.*, 2008; Lima and Menck, 2008), suggesting that horizontal transfer or replacement of a metabolic pathway may occur under specific environmental conditions. These studies provide evidence for the role of HGT in gene clustering. Broader-scale investigations are needed to address the relative roles of HGT plays in operon formation.

As was mentioned above, some genomic islands can be easily and successfully detected by their atypical base compositions (Hsiao *et al.*, 2003; Zhang and Zhang, 2003). However, the base composition approach is often not reliable for identifying HGTs in general (Koski *et al.*, 2001). It has been clearly demonstrated that genes, after horizontal transfer, will ameliorate to reflect the base composition of the new genome (Lawrence and Ochman, 1997; Marri and Golding, 2008), and successfully transferred genes are probably from organisms with a similar codon usage as the new host (Medrano-Soto *et al.*, 2004). Therefore, ancient transfers may have undergone complete amelioration and such genes can not be undetected as horizontally transferred by the base composition approach. Instead, other approaches are needed. For example, HGTs can be identified by their uneven presence or absence pattern across genomes (Koonin *et al.*, 2001). A patchy distribution where a gene is present in one species and absent in all close relatives can be considered a signature for HGT(s). However, in cases like these, the possibility of multiple gene losses cannot be ruled out either. Under such circumstances, phylogenetic information can be of great help in screening for HGTs. HGT can be identified by examining potential phylogenetic conflicts among homologous genes. This is because transferred genes have phylogenetic information from the donor rather than the new host. Significant incongruence has been widely accepted as a reliable indicator for HGT. However, a lack of incongruent signals does not necessarily suggest that there have been no HGTs.

Pattern of gene insertions/deletions

Parsimony methods have been widely used to estimate gene insertions/deletions (gene gains/losses) based on the pattern of gene presence/absence (phyletic pattern). The essential idea of parsimony methods is to map phyletic patterns of orthologous genes to a 'species tree'. In previous studies, an HGT penalty, or 'relative frequency of HGT versus gene loss' was introduced in order to achieve more reasonable interpretations such as constant genome size (Snel *et al.*, 2002; Kunin and Ouzounis, 2003). When the penalty was set to 1 (equal weights assigned to a gain and a loss), there were always more HGTs than gene losses in the observed results. The 'optimal' HGT penalty was suggested to be between 2 and 3 in (Kunin and Ouzounis, 2003), suggesting that in order to keep roughly constant genome size, two or three gene losses per HGT event are needed to explain the observed pattern in the data. Koonin and co-workers analysed a broader spectrum of bacterial genomes together with archaeal and eukaryotic genomes and reached a conclusion that the number of predicted HGT events that occurred during the evolution of prokaryotes might be approximately the same as the number of gene losses (Mirkin *et al.*, 2003). Clearly, there is no a priori best penalty value in general, since different studies suggested different 'optimal' penalties. Focusing on the tips of phylogenies for closely related bacterial species, Hao and Golding (2004) illustrated that there are more insertions on the recent branches and that many recently transferred genes are deleted rapidly during evolution. This finding provided a plausible explanation for the conflicting results on HGT penalty values in previous studies. Since many transferred genes are rapidly deleted, there are always more inferred insertions over deletions especially for recently evolved branches. This is because the retained transferred genes on ancient branches are not deleted rapidly and are likely to be better uncovered in the parsimony estimation. Hence, one should expect the 'optimal' penalties in distantly related species to be low and the penalties in closely related species to be high. In fact, a low penalty value (about 1) was observed by Mirkin *et al.* (2003) in a deep phylogeny that covers bacteria, archaea and eukaryote, while a high penalty value (between 2 and 3) was observed in a group of more closely related bacterial species (Kunin and Ouzounis, 2003).

Recently, Hao and Golding (2006) introduced a maximum likelihood method to study HGT and assumed an equal rate of gene insertions and deletions. The results revealed that the rates of both gene gain and loss on recent branches are much higher than those on ancient branches. Furthermore, the rates of whole gene insertions/deletions are comparable to the rate of nucleotide substitution. Such a result supports the view that recently transferred genes are deleted rapidly by assuming roughly stable genome sizes over evolution (as shown in Fig. 3.1). The fast rate of gene turnover further suggests that recently transferred and retained genes are likely to be species specific and some genes might play an important role in host's adaptation to new niches (Marri *et al.*, 2006, 2007). The quantification of HGTs has also been tackled from different angles. Dagan and Martin (2007) estimated the rate of gene transfers with a restriction to the size of the ancestral genome. Given a phyletic pattern. If the ancestral genome size is small, there should be more transfers. Conversely, if the ancestral genome size is large, there might be fewer transfers but more deletions. The results suggest that by assuming reasonable ancestral genome sizes, at least two-thirds and probably all genes, have been affected by HGT at some time in their evolutionary past (Dagan and Martin, 2007).

Among horizontally transferred genes, many do not have any existing homologues in the current database and these genes are often called ORFans (Fischer and Eisenberg, 1999). ORFans are known to be shorter in length and have faster evolutionary rates (Daubin and Ochman, 2004). Some consider ORFans as due to over-annotation of bacterial genomes. However, there is experimental evidence suggesting that some ORFans are functional and several have available structural information (Siew and Fischer, 2003; Siew and Fischer, 2004). Most of the ORFans show K_a/K_s values significantly smaller than 1, indicating they are under purifying selection, further supporting their potential functional significance (Daubin and Ochman, 2004). Compared with genes that have distant homologues, ORFans are much more likely to persist (van Passel *et al.*,

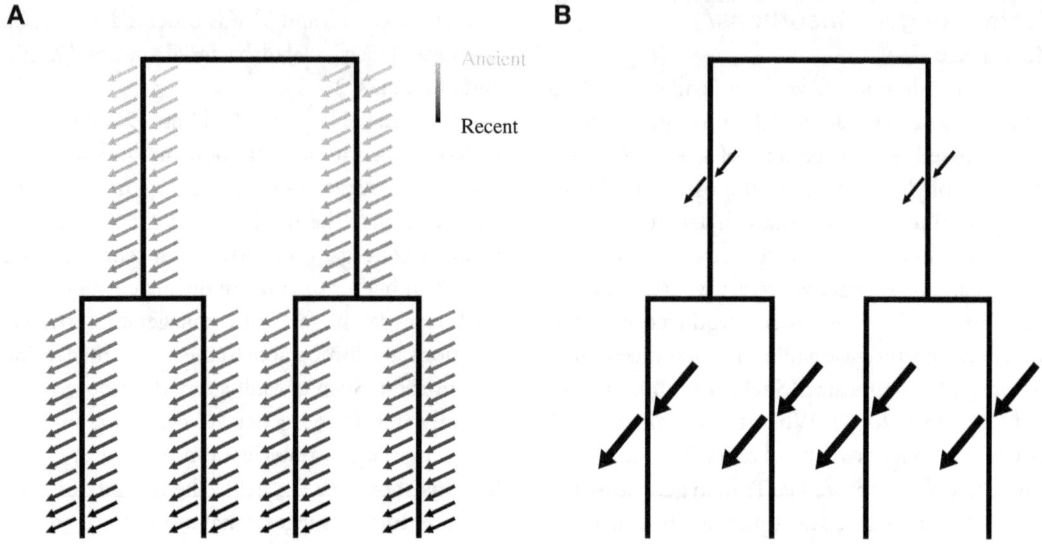

Figure 3.1 A cartoon demonstrating the actual and observed patterns of gene insertions/deletions. (A) A probably true pattern of gene insertions/deletions. Recently transferred genes are deleted rapidly and many transferred genes are transient. (B) Inferred rates (numbers) of gene insertions/deletions. High rates (numbers) of gene insertions/deletions are observed on recent branches.

2008). The different potential of persistence in a genome suggests that many adaptive traits may be conferred by novel genes originated in currently unknown bacterial genomes (van Passel *et al.*, 2008). However, the adaptive significance of these novel genes is still being debated. For example, most horizontally transferred genes in *E. coli* have shown to be less likely to be translated into proteins (Taoka *et al.*, 2004). It is possible that the purifying selection on ORFans was the consequence of selection against novel deleterious function instead of a need to retain a selective advantage that these genes provide to their host (Gogarten and Townsend, 2005). Therefore, any adaptive role of these novel genes needs to be further evaluated quantitatively through experimentation.

Rate variation of gene insertions/deletions

Given the massive amounts of horizontally transferred genes in closely related species, if all genes are equally mobile, one would expect little retention of vertically transferred genetic materials between distantly related species. However, not all genes are affected by HGT in the same manner and at similar rates. It has been shown that informational genes (i.e. genes involved in transcription and translation) have more interacting partners than operational genes (i.e. those responsible for metabolic processes) and that informational genes are less likely to be horizontally transferred than operational genes (Rivera *et al.*, 1998; Jain *et al.*, 1999). Such observations have led to the formulation of the 'complexity hypothesis'.

To test the complexity hypothesis and further understand the rate variation in gene insertions/deletions, we have estimated the degree of rate variation from genomic data using a maximum likelihood method. Consistent with the 'complexity hypothesis', informational genes showed lower rates of gene insertions/deletions compared with non-informational genes. However, when informational genes were removed from the analysis, the amount of rate variation did not decrease significantly and still remained high (Hao and Golding, 2008). That is, the difference between information genes and operational genes is only a small part of the rate variation in gene insertions/deletions. Results also revealed a substantial amount of rate variation within both informational genes and non-informational genes. Taken together, the data indicate that the rate variation of gene transfers is much more complex than simply implied by the 'complexity hypothesis' alone.

The number of horizontally transferred genes also varies substantially among different bacterial lineages. For instance, horizontally transferred genes in some endosymbiotic bacteria are comparatively rare (Silva *et al.*, 2003), while approximately 24% of genes in the free-living thermophile *Thermotoga maritima* might have undergone HGT (Garcia-Vallvé *et al.*, 2000). The possibility of HGTs into a genome could be affected by many internal and external environmental factors, such as genome size, codon usage, carbon utilization, isolated niches, and biochemical properties (Jain *et al.*, 2003; Silva *et al.*, 2003; Medrano-Soto *et al.*, 2004; Ringemann *et al.*, 2006). Highly varied internal and external environmental conditions will result in different amounts and different types of genes successfully transferred and retained. In fact, the rates of insertions/deletions for genes in different functional classification tend to be strongly influenced by host habitat (Marri *et al.*, 2006). The number of gene deletions can be altered by selection pressure (Hershberg *et al.*, 2007) and extensive HGTs have been found in the microflora-rich environment (e.g. gastrointestinal tract; Nicolas *et al.*, 2007). Furthermore, H-NS has been reported as a machinery responsible for the silencing of horizontally transferred genes in bacteria (Lucchini *et al.*, 2006; Navarre *et al.*, 2006). Variations of the silencing machinery in different species could also result in different patterns of HGTs.

Most recent studies of HGT have focused on single-copy genes to avoid the confounding effects of duplication. For example, copies of duplicated genes quite often undergo asymmetric evolution (Conant and Wagner, 2003; Zhang *et al.*, 2003), with one copy experiencing accelerated evolution that could lead to incorrect phylogenetic inference. Secondly, gene conversion might happen after duplication (Hughes, 2000; Lathe and Bork, 2001), and gene conversion along separate lineages has caused much concern in phylogenomic studies (Gao and Innan, 2004; Lin *et al.*, 2006). It has been shown that gene conversion occurs between paralogues during bacterial evolution (Santoyo and Romero, 2005), and the rate of gene conversion is about 10-fold less than the fixation of point mutations (Hashimoto *et al.*, 2003). Finally, duplication followed by differential loss (Fig. 3.2) could be invoked as an alternative to horizontal gene transfer and vice versa (Gogarten and Townsend, 2005). Several recent studies attempted analysing HGT in duplicated genes and genes that have a high duplicability, which describes the likelihood of a gene having one or more paralogues (Yang *et al.*, 2003). The results suggest that duplicated genes or genes that have a high duplicability might be more likely undergoing HGT than the single-copy genes (Hooper and Berg, 2003; Wellner *et al.*, 2007). When analysing duplicated genes, the picture of rate variation for horizontal gene transfers becomes more complex and methods that incorporate gene duplication information will be desirable for future studies.

Is there a tree of life?

While the importance of HGT during bacterial genome evolution has been widely acknowledged, the amount of HGTs and the consequence of HGT are still hotly debated. HGTs, to some people who study phylogenomics, are noises that obscure phylogenetic signals. Even if only a small fraction of HGTs are retained during any given time period, there is no doubt that the effect will be accumulated over evolutionary time. Given sufficient time, the phylogenetic signals of the host genomes could be wiped out in the course of evolution by the noises of HGT. A fundamental question then rises of whether it is possible to obtain a bacterial phylogeny. It has been suggested that a tree-like phylogeny exists at the tips of prokaryotes, but a tree-like phylogeny might be difficult to obtain for deeply divergent groups of prokaryotes because of HGT (Creevey *et al.*, 2004). Phylogenetic conflict in a substantial portion of genes has also been observed in other studies. For example, in a divergent cyanobacteria group, most of the orthologues exhibit their own unique topologies (Shi and Falkowski, 2008). Similarly, as much as 40% of the 'core' genome in gastrointestinal strains of bacteria *Lactobacillus acidophilus* and *L. johnsonii* may be horizontally transferred (Nicolas *et al.*, 2007).

Phylogenetic conflict among genes is a commonly used signal for identifying HGT among genome sequences. Current studies attempt to address whether trees inferred from core sets of protein-coding genes significantly disagree with a reference tree (e.g. a rRNA tree). It has been suggested that a select set of orthologous genes

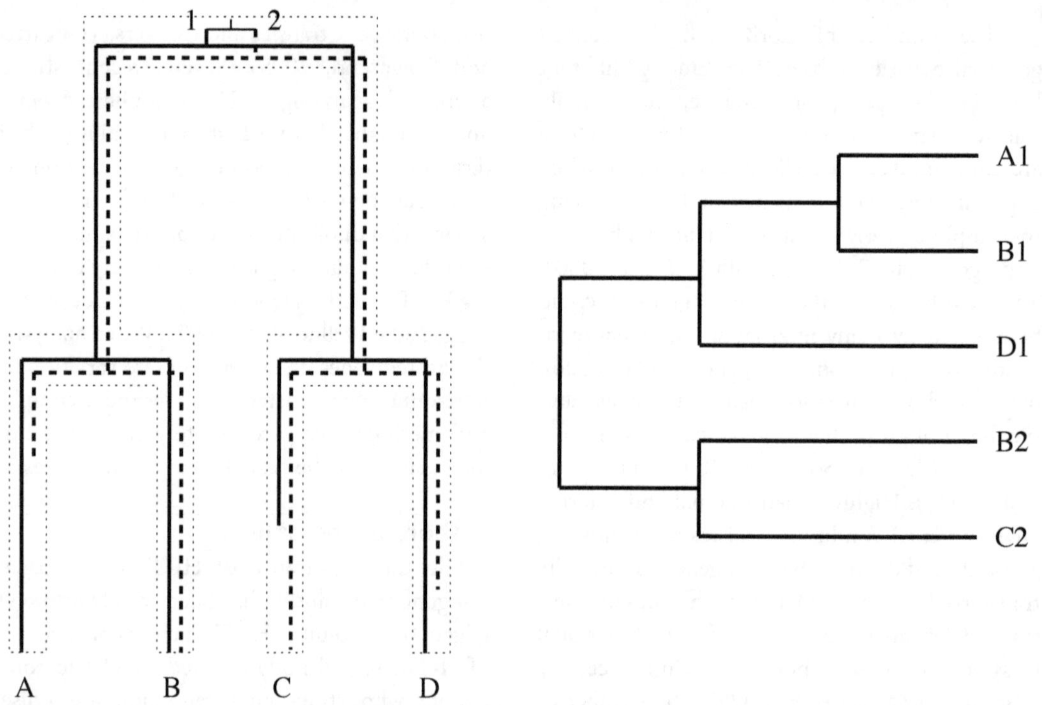

Figure 3.2 An HGT-like evolutionary relationship due to duplication. Differential losses after an ancient duplication occur in A, C, D species (on the left), and the gene phylogeny on the right yields an HGT-like evolutionary relationship.

fail to reject congruence with a single tree (Lerat *et al.*, 2003; Daubin *et al.*, 2003; Lerat *et al.*, 2005). By employing a different statistical method (heat maps), the opposite conclusion was reached: that is, orthologues do not always support a tree-like evolutionary history (Bapteste *et al.*, 2005; Susko *et al.*, 2006; Bapteste *et al.*, 2008). It is noteworthy that these conflicting conclusions (Lerat *et al.*, 2003 and Susko *et al.*, 2006) were drawn from the same set of data. The conflict might be due to a variety of reasons. First, the data might lack phylogenetic signals. Orthologues used in the studies were identified upon sequence conservation, and a highly conserved set of sequence might contain less phylogenetic information. It is plausible that widely conserved genes used in the studies might not have much statistical power to significantly reject (or significantly support) any single tree. This would be particularly true in studies of highly divergent groups. For sequences with little phylogenetic signals, different results might accrue from using different methods in HGT detection. Some statistical methods, such as the SH test (Shimodaira and Hasegawa, 1999) are known to be conservative to infer incongruence among phylogenies (Shimodaira, 2002). The SH test used in (Lerat *et al.*, 2003) might not be able to successfully identify all incongruent signals, whereas the 'heat maps' method used in Susko *et al.* (2006) might be more sensitive to incongruent phylogenetic signals.

However, the debate does not seem to end here. A recent study (Galtier, 2007) argues that gene tree incongruence *per se* is not indicative of HGT but can be false signals due to variation in sequence lengths and the stochastic features of the substitution process. The study also suggests that super-tree methods making use of more available data are robust to HGT (Galtier, 2007). A similar conclusion was reached earlier, i.e. many genes are vertically inherited and a relatively small number of genes are horizontally transferred (Beiko *et al.*, 2005; Ge *et al.*, 2005). To quantitatively evaluate the effect of HGT, more powerful methods are desirable. In the phylogenetic studies discussed above, ORFan genes (which are results of HGT but do not have homologues) were typically not considered. Furthermore, duplicated genes that have been shown to have a higher probability of being horizontally transferred (Hooper and Berg, 2003; Wellner *et al.*, 2007) were also excluded from these studies. As new statistical methods are developed, more precise quantitative assessments of the impact of HGT on phylogeny reconstruction could be obtained. As more HGTs are discovered, the significance of the core gene-based tree for bacteria and archaea becomes much weaker than it was before (McInerney *et al.*, 2008). A phylogenetic tree is supposed to represent the evolutionary relationships among different organisms. However, here, the bacterial tree (or the archaeal tree), if there is one, is a topology of an ever-shrinking fraction of genes in cells (Charlebois and Doolittle, 2004), which Dagan and Martin (2006) called 'the tree of one percent' and might not be representative of the tree of cells/organisms (Doolittle and Bapteste, 2007).

Are there barriers to HGT?

Successful gene transfer not only requires successful integration of the foreign gene into the host genomes but also needs fixation or long-term persistence in the population. Earlier population genetic studies suggest that horizontally transferred sequences can be fixed in a population only when they confer a substantial selective advantage upon the recipient (Berg and Kurland, 2002; Kurland *et al.*, 2003). After introducing an additional parameter for 'invasion', a recent study suggests that horizontally transferred sequences can be fixed in a population or at least persist for a long time in a substantial fraction of individuals in the population even when they are neutral or slightly deleterious (Novozhilov *et al.*, 2005). The argument (Berg and Kurland, 2002; Kurland *et al.*, 2003) relies on an assumption of clear species boundaries. However, whether there is a bacterial species boundary is still an open question (Woese, 2000; Gogarten *et al.*, 2002; Lawrence, 2002; Gevers *et al.*, 2005; Konstantinidis *et al.*, 2006; Fraser *et al.*, 2007). For instance, gene transfer via homologous recombination has played an important role in blurring prokaryotic speciation (Lawrence, 2002; Nesbo *et al.*, 2006; Fraser *et al.*, 2007; Papke *et al.*, 2007) and continuous homologous recombination can destroy an existing species boundary and lead to species convergence (Sheppard *et al.*, 2008). Given the prevalence of HGTs during evolution, the bacterial species boundary, if exists, must be very fuzzy.

The advantage that a foreign gene confers on the host cell is not the only factor that can influence successful gene transfer. HGT tends to take place among organisms that share similar internal and external environmental factors (Jain *et al.*, 2003). For instance, compatible codon usage is thought to increase the chance of successful HGT into the host (Medrano-Soto *et al.*, 2004). Environmental factors, while different from those of the host genome, could potentially act as barriers to HGT. Duplicability has shown to be another barrier to HGT, since genes that exhibit low duplicability are less likely to be horizontally transferred (Wellner *et al.*, 2007). In addition, foreign genes might be toxic to the new host. As a result, successful transfer requires low toxicity of the foreign genes to the host (Kurland *et al.*, 2003). Integration of horizontally transferred genes into regulatory interaction networks is also essential for their retention, since most recent arrivals have low or no interaction within the network (Lercher and Pal, 2008). After HGT, genes not able to integrate into the network will probably be eliminated from the genome over time.

One should expect generally low barriers to HGT in most organisms since so many genes have apparently been transferred. Laboratory approaches have demonstrated that barriers to HGT are indeed very low and no single gene family is completely non-transferable in *E. coli* (Sorek *et al.*, 2007). Other evidence is also consistent with this finding. Successfully horizontally transferred genes are more likely to be connected to core genes than expected by chance (Ochman *et al.*, 2007; Davids and Zhang, 2008), and they are more likely than other genes to be regulated by multiple regulators (Price *et al.*, 2008). This suggests that many transferred genes indeed can integrate into the network of the new host.

Conclusion

Despite recent progresses, several major challenges remain in understanding the pattern of HGT. These challenges include identifying all horizontally transferred genes, addressing their influences on the adaptation of their hosts to new niches, examining the factors that influence the rate of gene gain and gene loss, and determining the time of specific transfer events. The fast turnover of recently transferred genes suggests that some transferred genes play crucial roles in adaptation. With the growing amount of sequence data, we should be able to identify the genes of adaptive significance and examine how these genes differ in their rates of gene transfer, including their intragenomic dynamics during evolution. The answers to these questions will further improve our understanding of bacterial genome evolution.

Acknowledgements

G.B.G. is supported by a grant from the Natural Sciences and Engineering Council of Canada.

References

Babic, A., Lindner, A.B., Vulic, M., Stewart, E.J., and Radman, M. (2008). Direct visualization of horizontal gene transfer. Science. *319*, 1533–1536.

Bapteste, E., Susko, E., Leigh, J., MacLeod, D., Charlebois, R.L., and Doolittle, W.F. (2005). Do orthologous gene phylogenies really support tree-thinking? BMC Evol Biol. *5*, 33.

Bapteste, E., Susko, E., Leigh, J., Ruiz-Trillo, I., Bucknam, J., and Doolittle, W.F. (2008). Alternative methods for concatenation of core genes indicate a lack of resolution in deep nodes of the prokaryotic phylogeny. Mol. Biol. Evol. *25*, 83–91.

Beiko, R.G., Harlow, T.J., and Ragan, M.A. (2005). Highways of gene sharing in prokaryotes. Proc. Natl. Acad. Sci. U.S.A. *102*, 14332–14337.

Berg, O.G. and Kurland, C.G. (2002). Evolution of microbial genomes: sequence acquisition and loss. Mol. Biol. Evol. *19*, 2265–2276.

Blanc, G., Ogata, H., Robert, C., Audic, S., Claverie, J.M., and Raoult, D. (2007). Lateral gene transfer between obligate intracellular bacteria: Evidence from the *Rickettsia massiliae* genome. Genome Res. *17*, 1657–1664.

Brochier, C., Philippe, H., and Moreira, D. (2000). The evolutionary history of ribosomal protein RpS14: horizontal gene transfer at the heart of the ribosome. Trends Genet. *16*, 529–533.

Brown, J.R., Douady, C.J., Italia, M.J., Marshall, W.E., and Stanhope, M.J. (2001). Universal trees based on large combined protein sequence data sets. Nat Genet. *28*, 281–285.

Charlebois, R.L. and Doolittle, W.F. (2004). Computing prokaryotic gene ubiquity: rescuing the core from extinction. Genome Res. *14*, 2469–2477.

Chen, Y.Y., Feng, C.W., Chiu, C.F., and Burne, R.A. (2008). *cadDX* operon of *Streptococcus salivarius* 57.I. Appl. Environ. Microbiol. *74*, 1642–1645.

Coleman, M.L., Sullivan, M.B., Martiny, A.C., Steglich, C., Barry, K., Delong, E.F., and Chisholm, S.W. (2006). Genomic islands and the ecology and evolution of *Prochlorococcus*. Science. *311*, 1768–1770.

Conant, G.C. and Wagner, A. (2003). Asymmetric sequence divergence of duplicate genes. Genome Res. *13*, 2052–2058.

Creevey, C.J., Fitzpatrick, D.A., Philip, G.K., Kinsella, R.J., O'Connell, M.J., Pentony, M.M., Travers, S.A., Wilkinson, M., and McInerney, J.O. (2004). Does a tree-like phylogeny only exist at the tips in the prokaryotes? Proc. Biol. Sci. *271*, 2551–2558.

Cunneen, M.M. and Reeves, P.R. (2007). The *Yersinia kristensenii* O11 O-antigen gene cluster was acquired by lateral gene transfer and incorporated at a novel chromosomal locus. Mol. Biol. Evol. *24*, 1355–1365.

Dagan, T. and Martin, W. (2006). The tree of one percent. Genome Biol. *7*, 118.

Dagan, T. and Martin, W. (2007). Ancestral genome sizes specify the minimum rate of lateral gene transfer during prokaryote evolution. Proc. Natl. Acad. Sci. U.S.A. *104*, 870–875.

Daubin, V., Moran, N.A., and Ochman, H. (2003). Phylogenetics and the cohesion of bacterial genomes. Science. *301*, 829–832.

Daubin, V. and Ochman, H. (2004). Bacterial genomes as new gene homes: the genealogy of ORFans in *E. coli*. Genome Res. *14*, 1036–1042.

Davids, W. and Zhang, Z. (2008). The impact of horizontal gene transfer in shaping operons and protein interaction networks-direct evidence of preferential attachment. BMC Evol Biol. *8*, 23.

Diaz, E. (2004). Bacterial degradation of aromatic pollutants: a paradigm of metabolic versatility. Int. Microbiol. *7*, 173–180.

Doolittle, W.F. (1999). Phylogenetic classification and the universal tree. Science *284*, 2124–2129.

Doolittle, W.F. and Bapteste, E. (2007). Pattern pluralism and the Tree of Life hypothesis. Proc. Natl. Acad. Sci. U.S.A. *104*, 2043–2049.

Fischer, D. and Eisenberg, D. (1999). Finding families for genomic ORFans. Bioinformatics *15*, 759–762.

Fox, G.E., Stackebrandt, E., Hespell, R.B., Gibson, J., Maniloff, J., Dyer, T.A., Wolfe, R.S., Balch, W.E., Tanner, R.S., Magrum, L.J., *et al.* (1980). The phylogeny of prokaryotes. Science *209*, 457–463.

Fraser, C., Hanage, W.P., and Spratt, B.G. (2007). Recombination and the nature of bacterial speciation. Science. *315*, 476–480.

Fux, C.A., Costerton, J.W., Stewart, P.S., and Stoodley, P. (2005). Survival strategies of infectious biofilms. Trends Microbiol. *13*, 34–40.

Galtier, N. (2007). A model of horizontal gene transfer and the bacterial phylogeny problem. Syst Biol. *56*, 633–642.

Gao, L.Z. and Innan, H. (2004). Very low gene duplication rate in the yeast genome. Science. *306*, 1367–1370.

Garcia-Vallvé, S., Romeu, A., and Palau, J. (2000). Horizontal gene transfer in bacterial and archaeal complete genomes. Genome Res. *10*, 1719–1725.

Ge, F., Wang, L.S., and Kim, J. (2005). The cobweb of life revealed by genome-scale estimates of horizontal gene transfer. PLoS Biol. 3, e316.

Gevers, D., Cohan, F.M., Lawrence, J.G., Spratt, B.G., Coenye, T., Feil, E.J., Stackebrandt, E., Van de Peer, Y., Vandamme, P., Thompson, F.L., *et al.* (2005). Opinion: Re-evaluating prokaryotic species. Nat. Rev. Microbiol. 3, 733–739.

Gogarten, J.P., Doolittle, W.F., and Lawrence, J.G. (2002). Prokaryotic evolution in light of gene transfer. Mol. Biol. Evol. *19*, 2226–2238.

Gogarten, J.P. and Townsend, J.P. (2005). Horizontal gene transfer, genome innovation and evolution. Nat. Rev. Microbiol. 3, 679–687.

Gophna, U., Ron, E.Z., and Graur, D. (2003). Bacterial type III secretion systems are ancient and evolved by multiple horizontal-transfer events. Gene *312*, 151–163.

Hacker, J., Blum-Oehler, G., Muhldorfer, I., and Tschape, H. (1997). Pathogenicity islands of virulent bacteria: structure, function and impact on microbial evolution. Mol Microbiol. *23*, 1089–1097.

Hacker, J. and Carniel, E. (2001). Ecological fitness, genomic islands and bacterial pathogenicity. A Darwinian view of the evolution of microbes. EMBO Rep. 2, 376–381.

Hacker, J. and Kaper, J.B. (2000). Pathogenicity islands and the evolution of microbes. Annu Rev Microbiol. *54*, 641–679.

Hao, W. and Golding, G.B. (2004). Patterns of bacterial gene movement. Mol. Biol. Evol. 21, 1294- 1307.

Hao, W. and Golding, G.B. (2006). The fate of laterally transferred genes: Life in the fast lane to adaptation or death. Genome Res. *16*, 636–643.

Hao, W. and Golding, G.B. (2008). Uncovering rate variation of lateral gene transfer during bacterial genome evolution. BMC Genomics 9, 235.

Hashimoto, J.G., Stevenson, B.S., and Schmidt, T.M. (2003). Rates and consequences of recombination between rRNA operons. J Bacteriol. *185*, 966–972.

Hershberg, R., Tang, H., and Petrov, D.A. (2007). Reduced selection leads to accelerated gene loss in Shigella. Genome Biol. *8*, R164.

Homma, K., Fukuchi, S., Nakamura, Y., Gojobori, T., and Nishikawa, K. (2007). Gene cluster analysis method identifies horizontally transferred genes with high reliability and indicates that they provide the main mechanism of operon gain in 8 species of gamma-Proteobacteria. Mol. Biol. Evol. *24*, 805–813.

Hooper, S.D. and Berg, O.G. (2003). Duplication is more common among laterally transferred genes than among indigenous genes. Genome Biol. *4*, R48.

Hsiao, W., Wan, I., Jones, S.J., and Brinkman, F.S. (2003). IslandPath: aiding detection of genomic islands in prokaryotes. Bioinformatics *19*, 418–420.

Hsiao, W.W., Ung, K., Aeschliman, D., Bryan, J., Finlay, B.B., and Brinkman, F.S. (2005). Evidence of a large novel gene pool associated with prokaryotic genomic islands. PLoS Genet. *1*, e62.

Hughes, D. (2000). Co-evolution of the tuf genes links gene conversion with the generation of chromosomal inversions. J Mol Biol. *297*, 355–364.

Jain, R., Rivera, M.C., and Lake, J.A. (1999). Horizontal gene transfer among genomes: the complexity hypothesis. Proc. Natl. Acad. Sci. U.S.A. *96*, 3801–3806.

Jain, R., Rivera, M.C., Moore, J.E., and Lake, J.A. (2003). Horizontal gene transfer accelerates genome innovation and evolution. Mol. Biol. Evol. *20*, 1598–1602.

Konstantinidis, K.T., Ramette, A., and Tiedje, J.M. (2006). The bacterial species definition in the genomic era. Phil. Trans. R. Soc. Lond. B Biol. Sci. *361*, 1929–1940.

Koonin, E.V., Makarova, K.S., and Aravind, L. (2001). Horizontal gene transfer in prokaryotes: Quantification and classification. Annu. Rev. Microbiol. *55*, 709–742.

Koski, L.B., Morton, R.A., and Golding, G.B. (2001). Codon bias and base composition are poor indicators of horizontally transferred genes. Mol. Biol. Evol. *18*, 404–412.

Kunin, V. and Ouzounis, C.A. (2003). The balance of driving forces during genome evolution in prokaryotes. Genome Res. *13*, 1589–1594.

Kurland, C.G., Canback, B., and Berg, O.G. (2003). Horizontal gene transfer: a critical view. Proc. Natl. Acad. Sci. U.S.A. *100*, 9658–9662.

Lathe, 3rd, W.C. and Bork, P. (2001). Evolution of *tuf* genes: ancient duplication, differential loss and gene conversion. FEBS Lett. *502*, 113–116.

Lawrence, J. (1999). Selfish operons: the evolutionary impact of gene clustering in prokaryotes and eukaryotes. Curr. Opin. Genet Dev. *9*, 642–648.

Lawrence, J.G. (2002). Gene transfer in bacteria: speciation without species? Theor. Popul. Biol. *61*, 449–460.

Lawrence, J.G. (2005). Horizontal and vertical gene transfer: the life history of pathogens. Contrib. Microbiol. *12*, 255–271.

Lawrence, J.G. and Ochman, H. (1997). Amelioration of bacterial genomes: rates of change and exchange. J. Mol. Evol. *44*, 383–397.

Lawrence, J.G. and Ochman, H. (1998). Molecular archaeology of the *Escherichia coli* genome. Proc. Natl. Acad. Sci. U.S.A. *95*, 9413–9417.

Lerat, E., Daubin, V., and Moran, N.A. (2003). From gene trees to organismal phylogeny in prokaryotes: The case of the gamma-Proteobacteria. PLoS Biol. *1*, E19.

Lerat, E., Daubin, V., Ochman, H., and Moran, N.A. (2005). Evolutionary origins of genomic repertoires in Bacteria. PLoS Biol. 3, e130.

Lercher, M.J. and Pal, C. (2008). Integration of horizontally transferred genes into regulatory interaction networks takes many million years. Mol. Biol. Evol. *25*, 559–567.

Lima, W.C. and Menck, C.F. (2008). Replacement of the arginine biosynthesis operon in xanthomonadales by lateral gene transfer. J. Mol. Evol. *66*, 266–275.

Lin, Y.S., Byrnes, J.K., Hwang, J.K., and Li, W.H. (2006). Codon-usage bias versus gene conversion in the evolution of yeast duplicate genes. Proc. Natl. Acad. Sci. U.S.A. *103*, 14412–14416.

Lucchini, S., Rowley, G., Goldberg, M.D., Hurd, D., Harrison, M., and Hinton, J.C. (2006). H-NS mediates the silencing of laterally acquired genes in bacteria. PLoS Pathog. 2, e81.

Maidak, B.L., Olsen, G.J., Larsen, N., Overbeek, R., McCaughey, M.J., and Woese, C.R. (1997). The RDP (Ribosomal Database Project). Nucleic Acids Res. *25*, 109–111.

Marri, P.R. and Golding, G.B. (2008). Gene amelioration demonstrated: the journey of nascent genes in bacteria. Genome *51*, 164–168.

Marri, P.R., Hao, W., and Golding, G.B. (2006). Gene gain and gene loss in *Streptococcus*: Is it driven by habitat? Mol. Biol. Evol. *23*, 2379–2391.

Marri, P.R., Hao, W., and Golding, G.B. (2007). The role of laterally transferred genes in adaptive evolution. BMC Evol. Biol. 7(Suppl. *1*), S8.

McInerney, J.O., Cotton, J.A., and Pisani, D. (2008). The prokaryotic tree of life: past, present...and future? Trends Ecol. Evol. *23*, 276–281.

Medrano-Soto, A., Moreno-Hagelsieb, G., Vinuesa, P., Christen, J.A., and Collado-Vides, J. (2004). Successful lateral transfer requires codon usage compatibility between foreign genes and recipient genomes. Mol. Biol. Evol. *21*, 1884–1894.

Mira, A., Ochman, H., and Moran, N.A. (2001). Deletional bias and the evolution of bacterial genomes. Trends Genet. *17*, 589–596.

Mirkin, B.G., Fenner, T.I., Galperin, M.Y., and Koonin, E.V. (2003). Algorithms for computing parsimonious evolutionary scenarios for genome evolution, the last universal common ancestor and dominance of horizontal gene transfer in the evolution of prokaryotes. BMC Evol. Biol. 3, 2.

Navarre, W.W., Porwollik, S., Wang, Y., McClelland, M., Rosen, H., Libby, S.J., and Fang, F.C. (2006). Selective silencing of foreign DNA with low GC content by the H-NS protein in *Salmonella*. Science *313*, 236–238.

Nesbo, C.L., Dlutek, M., and Doolittle, W.F. (2006). Recombination in thermotoga: implications for species concepts and biogeography. Genetics *172*, 759–769.

Nicolas, P., Bessieres, P., Ehrlich, S.D., Maguin, E., and van de Guchte, M. (2007). Extensive horizontal transfer of core genome genes between two *Lactobacillus* species found in the gastrointestinal tract. BMC Evol. Biol. 7, 141.

Novozhilov, A.S., Karev, G.P., and Koonin, E.V. (2005). Mathematical modeling of evolution of horizontally transferred genes. Mol. Biol. Evol. *22*, 1721–1732.

Ochman, H., Liu, R., and Rocha, E.P. (2007). Erosion of interaction networks in reduced and degraded genomes. J. Exp. Zool. B Mol. Dev. Evol. *308*, 97–103.

Pal, C. and Hurst, L.D. (2004). Evidence against the selfish operon theory. Trends Genet. *20*, 232–234.

Papke, R.T., Zhaxybayeva, O., Feil, E.J., Sommerfeld, K., Muise, D., and Doolittle, W.F. (2007). Searching for species in haloarchaea. Proc. Natl. Acad. Sci. U.S.A. *104*, 14092–14097.

Price, M.N., Dehal, P.S., and Arkin, A.P. (2008). Horizontal gene transfer and the evolution of transcriptional regulation in *Escherichia coli*. Genome Biol. *9*, R4.

Price, M.N., Huang, K.H., Arkin, A.P., and Alm, E.J. (2005). Operon formation is driven by co-regulation and not by horizontal gene transfer. Genome Res. *15*, 809–819.

Ringemann, C., Ebenhoh, O., Heinrich, R., and Ginsburg, H. (2006). Can biochemical properties serve as selective pressure for gene selection during inter-species and endosymbiotic lateral gene transfer? Syst Biol (Stevenage) 153, 212–222.

Rivera, M.C., Jain, R., Moore, J.E., and Lake, J.A. (1998). Genomic evidence for two functionally distinct gene classes. Proc. Natl. Acad. Sci. U.S.A. *95*, 6239–6244.

Santoyo, G. and Romero, D. (2005). Gene conversion and concerted evolution in bacterial genomes. FEMS Microbiol. Rev. *29*, 169–183.

Sheppard, S.K., McCarthy, N.D., Falush, D., and Maiden, M.C.J. (2008). Convergence of *Campylobacter* species: Implications for bacterial evolution. Science. *320*, 237–239.

Shi, T. and Falkowski, P.G. (2008). Genome evolution in cyanobacteria: the stable core and the variable shell. Proc. Natl. Acad. Sci. U.S.A. *105*, 2510–2515.

Shimodaira, H. (2002). An approximately unbiased test of phylogenetic tree selection. Syst Biol. *51*, 492–508.

Shimodaira, H. and Hasegawa, M. (1999). Multiple comparison of log-likelihoods with applications to phylogenetic inference. Mol. Biol. Evol. *16*, 1114–1116.

Siew, N. and Fischer, D. (2003). Analysis of singleton ORFans in fully sequenced microbial genomes. Proteins. *53*, 241–251.

Siew, N. and Fischer, D. (2004). Structural biology sheds light on the puzzle of genomic ORFans. J Mol Biol. *342*, 369–373.

Silva, F.J., Latorre, A., and Moya, A. (2003). Why are the genomes of endosymbiotic bacteria so stable? Trends Genet. *19*, 176–180.

Snel, B., Bork, P., and Huynen, M.A. (2002). Genomes in flux: the evolution of archaeal and proteobacterial gene content. Genome Res. *12*, 17–25.

Sorek, R., Zhu, Y., Creevey, C.J., Francino, M.P., Bork, P., and Rubin, E.M. (2007). Genome-wide experimental determination of barriers to horizontal gene transfer. Science *318*, 1449–1452.

Susko, E., Leigh, J., Doolittle, W.F., and Bapteste, E. (2006). Visualizing and assessing phylogenetic congruence of core gene sets: a case study of the gamma-proteobacteria. Mol. Biol. Evol. 23, 1019–1030.

Taoka, M., Yamauchi, Y., Shinkawa, T., Kaji, H., Motohashi, W., Nakayama, H., Takahashi, N., and Isobe, T. (2004). Only a small subset of the horizontally transferred chromosomal genes in *Escherichia coli* are translated into proteins. Mol. Cell Proteomics 3, 780–787.

Templeton, T.J. (2007). Whole-genome natural histories of apicomplexan surface proteins. Trends Parasitol. 23, 205–212.

Top, E.M. and Springael, D. (2003). The role of mobile genetic elements in bacterial adaptation to xenobiotic organic compounds. Curr. Opin. Biotechnol. *14*, 262–269.

van Passel, M.W., Marri, P.R., and Ochman, H. (2008). The emergence and fate of horizontally acquired genes in *Escherichia coli*. PLoS Comput. Biol. *4*, e1000059.

Waterman, S.R. and Holden, D.W. (2003). Functions and effectors of the *Salmonella* pathogenicity island 2 type III secretion system. Cell Microbiol. *5*, 501–511.

Welch, R.A., Burland, V., Plunkett, 3rd, G., Redford, P., Roesch, P., Rasko, D., Buckles, E.L., Liou, S.R., Boutin, A., Hackett, J., *et al.* (2002). Extensive mosaic structure revealed by the complete genome sequence of uropathogenic *Escherichia coli*. Proc. Natl. Acad. Sci. U.S.A. *99*, 17020–17024.

Wellner, A., Lurie, M.N., and Gophna, U. (2007). Complexity, connectivity, and duplicability as barriers to lateral gene transfer. Genome Biol. *8*, R156.

Woese, C.R. (2000). Interpreting the universal phylogenetic tree. Proc. Natl. Acad. Sci. U.S.A. *97*, 8392–8396.

Yang, J., Lusk, R., and Li, W.H. (2003). Organismal complexity, protein complexity, and gene duplicability. Proc. Natl. Acad. Sci. U.S.A. *100*, 15661–15665.

Yap, W.H., Zhang, Z., and Wang, Y. (1999). Distinct types of rRNA operons exist in the genome of the actinomycete *Thermomonospora chromogena* and evidence for horizontal transfer of an entire rRNA operon. J. Bacteriol. *181*, 5201–5209.

Zhang, P., Gu, Z., and Li, W.H. (2003). Different evolutionary patterns between young duplicate genes in the human genome. Genome Biol. *4*, R56.

Zhang, R. and Zhang, C.T. (2003). Identification of genomic islands in the genome of *Bacillus cereus* by comparative analysis with *Bacillus anthracis*. Physiol Genomics. *16*, 19–23.

Zhaxybayeva, O., Gogarten, J.P., Charlebois, R.L., Doolittle, W.F., and Papke, R.T. (2006). Phylogenetic analyses of cyanobacterial genomes: Quantification of horizontal gene transfer events. Genome Res. *16*, 1099–1108.

Population Genetics of Human Pathogenic Bacteria: Implications for Source Tracking and Rapid Identification

4

Ruifu Yang, Yujun Cui, Yanjun Li and Yanfeng Yan

Abstract

Population genetics examine variation in genes among a group of strains of a particular species. Its major theme is to look at how different environmental factors and selective pressures can affect the distribution of genes and alleles. In this chapter *Yersinia pestis* was employed as an example to illustrate how the techniques are used for population genetic studies and how the achievements of these kinds of studies can be used for rapid identification and tracing the origin of pathogenic bacteria. New emerging techniques, including high throughput sequencing technologies, will give us unprecedented opportunities to understand microevolution and pathogenesis of bacterial pathogens.

Introduction

Most genetic studies of bacteria examine the genetic material of one strain at a time, often testing how one gene influences the phenotypic features of a specific strain. In contrast, population genetic studies examine variation in genes among a group of strains of a particular species. One major theme of population genetics is to look at how different environmental factors and selective pressures can affect the distribution of genes and alleles. Recent research activities in this area have centred on studying the patterns of and the potential mechanisms for genetic variations of target microorganisms at various geographical scales, from those within micro-niches to among populations from different continents. Such studies are important for understanding the evolutionary history, epidemiology, and population dynamics of pathogens, for revealing their potential for and mode of the evolution of antibiotic resistance, and ultimately for designing better strategies for controlling infectious disease and maintaining public health (Perez-Losada *et al.*, 2006). Indeed, recent interests in population genetic studies of human pathogenic bacteria have been largely driven by the rapid and worldwide emergence of drug resistance pathogens, the impact of global warming on worldwide pathogen distributions, the increasing threat of bioterrorism, and the continuing emergence and/or re-emergence of infectious diseases, and the rapid spread of many pathogens.

In this chapter, we first briefly describe the general approaches and methods for studying populations of human bacterial pathogens. We then summarize recent research progresses on the population genetics of these pathogens, using the plague pathogen *Yersinia pestis* as an example.

Methods used for population genetics research

One of the major objectives for studying population genetics of a bacterium is to find ways to discriminate between different strains within a bacterial species. The allelic distributions among strains and populations are then used to define the genetic structure of bacterial populations of interest. We can use the distribution of genetic variations in the target bacterial populations to reveal the possible evolutionary processes that shaped the populations in the past.

Many molecular markers and analytical methods have been employed to characterize bacterial strains that were isolated from different hosts, vectors or clinical specimens in different geographical regions at various times. Most recent markers focused on the differences at the DNA sequence level (Table 4.1). Based on whether the marker loci are under significant selective pressure, the markers used for indexing DNA polymorphisms can be divided into two main groups: neutral markers and contingency markers (Aras *et al.*, 2002). Neutralists consider that the great majority of evolutionary changes at the DNA level did not result from Darwinian natural selection acting on advantageous mutations, but, rather, from random fixation of selectively neutral or very nearly neutral mutations through stochastic genetic drifts and random migrations. These markers may be distributed throughout the genome. However, some of these neutral markers may be influenced by selective pressures because of their tight linkages with loci under strong selection (i.e. genetic hitch-hiking). Many techniques have been used to analyse variation at the neutral genetic loci: (i) the restriction sites-based DNA fingerprint techniques, including restriction fragment length polymorphism (RFLP), pulsed-field gel electrophoresis (PFGE), and amplified fragment length polymorphism (AFLP) etc; (ii) the repetitive sequences-based DNA fingerprint techniques, including arbitrarily-primed PCR (AP-PCR), insertion element (ISs) etc; and (iii) non-mobile interspersed repetitive sequences-based DNA fingerprint techniques, including repetitive element-based PCR (rep-PCR), REP- (repetitive extragenic palindromic), ERIC- (enterobacterial repetitive intergenic consensus), XRIC (*Xanthomonas* repetitive intergenic consensus), and BOX element-based fingerprinting techniques. However, though generally assumed, not all polymorphisms revealed by these markers were neutral. For example, Enright and Spratt reported that extensive variation in the *ddl* gene in penicillin-resistant *Streptococcus pneumoniae* was the result of hitchhiking driven by the penicillin-binding protein 2b gene (Enright and Spratt, 1999).

The techniques for analysing variation at contingency markers include those for analysing variable number of tandem repeats (VNTR), short sequence repeats (SSR), clustered regularly interspaced short palindromic repeat (CRISPR), and genes of functional importance such as the flagellin genes, and microarray-based methods. The sequence-based analysis, including 16S rRNA gene sequencing, single nucleotide polymorphism (SNP), whole genome comparison, and multilocus sequence typing (MLST) etc, could be easily standardized and are convenient for data exchange for population genetics research.

Though many techniques are available, the effectiveness of a procedure for analysing each microbial species must be validated experimentally by using representative strains isolated at different times from various sources (van Belkum *et al.*, 2001). Ideally, this would require laboratories involved in such studies to have the same set of strains for referencing. However, in the age of bioterrorism, it is often difficult to exchange strains or even DNAs of pathogenic organisms. One way to overcome the problem is to produce a binary output (numbers or characters) for easy exchange among researchers and institutions. For example, PulseNet USA developed an extensive database by PFGE for food-borne pathogens for disease surveillance. Similar databases have also been built for other typing methods, including MLVA, CRISPR, SNPs, and MLST. These databases have greatly facilitated international collaborations.

Population genetics of important pathogenic bacteria

Overview of *Yersinia pestis* populations

The population genetics of human pathogenic bacteria is a very large field and impossible to include in a single chapter. Indeed, many specialized books in recent years have dealt with different aspects of this topic. Instead, in this section, we focus on the populations of several closely related species within the genus *Yersinia* to illustrate the power of a variety of molecular markers that have helped understand the history and epidemiological patterns of these species.

One of the most deadly human pathogens is *Yersinia pestis*. It is the causative agent of plague that has killed hundreds of millions of people in the three major historical plague pandemics (Anisimov *et al.*, 2004; Perry and Fetherston, 1997). It belongs to the genus *Yersinia* that includes 15 species (www.bacterio.cict.fr/xz/yersinia.html). Three species in this genus are pathogenic to humans: *Y. pestis*, *Y. pseudotuberculosis* and *Y. enterocolitica*. The last two species

Table 4.1 Techniques used for study on bacterial population genetics

Genetic markers	Principle	Advantages	Disadvantages	References
DNA fingerprinting				
RFLP	DNA is digested completely to different fragments due to variation in sequences at the restriction site. The DNA fragments were separated by gel electrophoresis to generate DNA fingerprints. Ribotyping using rRNA gene as probe and IS -based typing using insertion sequence as probe were developed for reducing numbers of restrict fragments and increasing reproducibility and stability of RFLP	Simple to operate. There is commercially available machine for ribotyping	If the whole genomic DNA is restricted by a frequent cleaving enzyme, fragments appear as smears that are difficult to differentiate, causing low reproducibility and stability of this method. This method needs a large quantity of high-quality DNA	Botstein *et al.* (1980), Liu and Sanderson (1995a,b), Struelens (1998)
PFGE	DNA is digested completely to different large fragments using a rare cutter and these fragments are separated by gel electrophoresis in a special oriented electronic field	High discriminatory power, reproducibility and stability	Difficult to operate, needing special electrophoresis apparatus for fragment separation and live pathogens for sample preparation	
AFLP	DNA is first digested by a frequent cutter and a rare cutter and then the resulting fragments are ligated to the restriction sites-specific double stranded oligonucleotide adapters for using as template for the following PCR based on specific primers that extend beyond the adapter sequences at the 3′ end by one to three arbitrary nucleotides. The amplified fragments are separated by electrophoresis using fluorescent, radioactive or silver nitrate labelling for visualization	Cheap, fast and reliable, small amounts of DNA needed. High discriminatory power, reproducibility and stability	So many fragments obtained can produce much random noise that will obscure informative and phylogenetic signals	Vos *et al.* (1995)
Rep-PCR	PCR based on specific primers for non-mobile interspersed repetitive sequences, including REP, ERIC, XRIC and BOX elements, amplifies	Cheap, fast and reliable, small amounts of DNA needed.	Discriminatory power is limited. Not all bacteria contain these non-mobile interspersed repetitive sequences	Williams *et al.* (1990)

Table 4.1 Continued

Genetic markers	Principle	Advantages	Disadvantages	References
MLVA	Variable number of tandem repeat (VNTR) is minisatellite sequences, sometimes called short sequence repeats (SSR), with a repeat size ranging from 1 to 6 bp and repeat number ranging from 9 to 100 in the genome. VNTR loci may be generated by slipped strand mispairing. Multi-locus VNTR analysis (MLVA) is a technique that analyses multiple VNTR loci at same time for revealing powerful discrimination of bacterial strains within the same species	Very high discriminatory power, reproducibility and stability	Either using conventional gel electophoresis without labels or employing capillary electrophoresis with fluorescent labels to differentiate repeat numbers is laborious and the latter method is expensive. The genome information should be understood before performing MLVA	Al Dahouk *et al.* (2007), Grissa *et al.* (2008)
Functional genes	Different functional genes, including flagellin genes and avirulence genes that have well-conserved N- and C-termini but hypervariable central portions, even for different genera. These hypervariable portions could be employed to study population structure and evolution of pathogenic bacteria	Easily to be standardized with high reproducibility and stability	PCR and sequencing should be employed to obtain the sequences for comparison. The discriminatory power is relatively low. Some evolution-significant locus, such as O antigen gene cluster, is very long. It is difficult to obtain it for analysis	Gomez-Gomez and Boller (2002)
Microarray-based analysis	A microarray is solid substrate immobilized with oligonucleotide or cDNA probes targeting whole genome DNA of a specific organism. Sample DNA labelled with fluorescent dye is used to hybridize these probes for presence of target DNAs. Several species of bacterial pathogens have been investigated by using whole genome microarray hybridization for identification of gene content.	High-throughput method with high reproducibility and stability	It is only used for analysing known genes because the microarray is prepared based on the whole genome sequences of interested	Zhou *et al.* (2004a,b)
DFR analysis	The Different regions (DFRs) represent the dynamic regions of a pathogen's genome in a population, demonstrating its characteristics of acquisition or deletion in the adaptive evolution of the pathogen in a specific population.	Horizontal gene transfer (HGT) could be analysed using this method with high reproducibility and stability	Genome sequence information should be understood before applying this technique	Li *et al.* (2008), Zhou *et al.* (2004a,b)

Sequence-based comparison				
Single-locus analysis	The ubiquitous and function-constant 16S rRNA gene (rDNA) has been used widely for almost all bacteria studied in order to obtaining phylogenetic and taxonomic results	Sequencing of 16S rDNA has led to infer a universal phylogenetic tree for revealing relationship between bacteria. This method is easy to operate and there is a special database for this molecule	This single locus-based analysis sometimes is not accurate and the universal tree deduced from this information has been found to be different from the trees inferred based on other sequences. Polyphasic taxonomy is needed for identification purpose and multiple loci should be analysed for phylogenetic tree	Struelens (1998)
CRISPR	A family of elements that typically consist of non-contiguous direct repeats (DR, 24 bp-47 bp) separated by stretches of similarly sized non-repetitive sequences, providing the basis of spoligotyping technology	Sequence-based comparison provides potential of high reproducibility and stability	To obtain the sequences for comparison one has to know the basic genomic sequence information, and to do PCR and sequencing, which is time-consuming	Brudey *et al.* (2006), Grissa *et al.* (2008)
MLST	Multilocus sequence typing (MLST) is a technology based on sequencing of multiple housekeeping gene fragments scattered around the genome of interest from diverse strains for reliable recognition of genetically related bacteria within clonal species. This method has been applied to all the major human bacterial pathogens	This method has been standardized for many bacteria by establishing a globally accessible database via internet (http://www.mlst.net)	The number of nucleotide's difference between alleles is not considered by data analysis, therefore, point mutations and HGT are equally weighted	Jolley *et al.* (2004), Perez-Losada *et al.* (2006), Saunders and Holmes (2007)
Whole genome sequence comparison	Whole genome sequences for bacterial pathogens are accumulated rapidly, which will promote comparison of closely related bacteria for panoramic view of genome differences	All the differences at genome level could be obtained for revealing 'natural' relationship between bacteria in a population	Although there are some ultra-low cost sequencing technologies invented, it is still expensive to obtain whole genome sequences for large quantities of bacteria. Sequencing quality is critical for sequence comparison	Bolotin *et al.* (2005), (Monecke *et al.* (2007)

are enteropathogenic, while *Y. pestis* has a very different ecology and epidemiology and causes a markedly different disease. All these three yersiniae target lymph tissues during infection and carry a virulence plasmid, pCD1. This plasmid is required for infection in these tissues and is involved in overcoming the host's defence mechanism (Carniel, 2002).

Based on their abilities to reduce nitrate and utilize glycerol, strains of *Y. pestis* are traditionally classified into three biovars: *Antiqua* (positive for both), *Medievalis* (negative for nitrate reduction and positive for glycerol utilization) and *Orientalis* (positive for nitrate reduction and negative for glycerol utilization). Recently, a new biovar, *Microtus*, was proposed based on the analyses of whole genome sequences (Song *et al.*, 2004; Zhou *et al.*, 2004a,b).

Y. pestis is a multi-host and multi-vector pathogen, involving more than 200 species of wild rodents as hosts and over 80 species of fleas as vectors (Anisimov *et al.*, 2004). Different hosts and vectors have their own specific ecological landscapes, shaping numerous niches for *Y. pestis*. During its expansion and adaptation to new niches, *Y. pestis* genomes have undergone considerable evolution to generate genetic variations in response to natural selective pressures. As has been revealed by the wide variety of molecular studies, there is considerable genome diversity of strains from different plague foci. Detailed examination of the genome diversity of *Y. pestis* will help us better understand the origin and expansion of plague, and provide us with solid data for developing a reliable tracking system for this bacterium.

Molecular techniques used for studying *Y. pestis* evolution

Recent analyses suggest that *Y. pestis* most likely evolved from *Y. pseudotuberculosis* serotype O1:b within the last 20,000 years (Achtman *et al.*, 1999, 2004). The very short evolutionary history of *Y. pestis* helps explain the limited phenotypic and genetic diversities. How did *Y. pestis* evolve so rapidly from a mammalian enteropathogen widely found in the environment into a blood-borne pathogen capable of parasitizing fleas? This topic has attracted significant attentions from the scientific community. Indeed, many techniques have been employed to uncover the mystery behind *Y. pestis* microevolution. These techniques include SNP analysis (Achtman *et al.*, 1999, 2004), PFGE (pulsed-field gel electrophoresis) (Guiyoule *et al.*, 1994; Revazishvili *et al.*, 2008), MLST (multilocus sequence typing) (Achtman *et al.*, 1999), different region analysis (Li *et al.*, 2008), MLVA (multiocus variable number of tandem repeat analysis) (Adair *et al.*, 2000; Klevytska *et al.*, 2001; Lowell *et al.*, 2005; Pourcel *et al.*, 2004), CRISPR (clustered regularly interspaced short palindromic repeat) analysis (Pourcel *et al.*, 2005; Cui *et al.*, 2008), ribotyping (Guiyoule *et al.*, 1994, 1997), RAPD (Huang *et al.*, 2000; Yu, 2003), PCR-based technique (Hai, 2004), IS (insertion sequence)-based typing (Odaert *et al.*, 2008), whole genome microarray-based analysis (Dorrell *et al.*, 2005; Wang *et al.*, 2007; Zhou *et al.*, 2004a,b); and whole genome sequencing (Parkhill *et al.*, 2001; Deng *et al.*, 2002; Chain *et al.*, 2004; Song *et al.*, 2004; Auerbach *et al.*, 2007; Touchman *et al.*, 2007).

Achtman *et al.* (1999) analysed the population genetic structure of *Y. pestis* and compared them with the two other pathogenic *Yersinia* species. Fragments of five housekeeping genes as well as a gene involved in the synthesis of lipopolysaccharide were sequenced from 36 strains representing the global diversity of *Y. pestis* and including 12 to 13 strains from each of the two other species. Very little or no sequence polymorphism was found in any of the six genes in *Y. pestis*. Indeed, the alleles from *Y. pestis* were identical or almost identical to alleles from strains in *Y. pseudotuberculosis*. The sequence data was the basis for the hypothesis that *Y. pestis* is a clone that evolved from *Y. pseudotuberculosis* 1500–20,000 years ago, shortly before the first known pandemics of the human plague. Later, He and his group (Achtman *et al.*, 2004) employed three molecular methods, including genome-wide synonymous SNP analysis, variation in number of tandem repeats, and insertion of IS100 elements, to define eight populations, and proposed an evolutionary tree for these populations, rooted on *Yersinia pseudotuberculosis* and dated to about 6500 years ago. They proposed a new genetic classification system for *Y. pestis*. Specifically, 'ORI', 'MED', 'ANT' and 'PE' were designated to represent *Orientalis, Medievalis, Antiqua,* and *Pestoides* (*Microtus*) genotypes. A number plus a point before the above designations were employed to illustrate the evolutionary branches within the same genotype group. For example, 1.ORI represented *Orientalis* strains within evolutionary branch #1, classical Medievalis isolates are referred to as 2.MED, and *Antiqua* isolates were split into two distinct groups designated 1.ANT and 2.ANT, which were isolated in Africa and East Asia, respectively (Achtman *et al.*, 2004). Branch 0 includes almost all *Pestoides* isolates (groups 0.PE1, 0.PE2, and 0.PE3) as well as the *Microtus* isolate, 91001 (0.PE4). 0.PE4 was recently proposed as biovar *Microtus* according to their CRISPR diversities (Cui *et al.*, 2008).

Our laboratory used whole-genome microarray hybridizations and revealed that 22 genomic regions (termed differential regions, DFRs) were absent in one or more of 36 *Y. pestis* isolates studied (Zhou *et al.*, 2004a,b). In addition, PCR was employed to screen the presence or absence of these DFRs among 260 representative strains. Our analyses identified 14 genomovars, genomovar01 to genomovar14 (Zhou *et al.*, 2004a,b). More recently, Dai *et al.* (2005) found a new DFR, DFR23. Variation within this genomic region allowed the differentiation of some closely related strains. These 23 DFRs were subsequently applied to genotyping 909 *Y. pestis* strains and the analyses identified a total of 32 genomovars (Li *et al.*, 2008). The DFR profiles and neighbour-joining dendrogram of the 32 genomovars were shown in Fig. 4.1. Most of the genomovars were clustered into three groups, namely A, B and C, except for

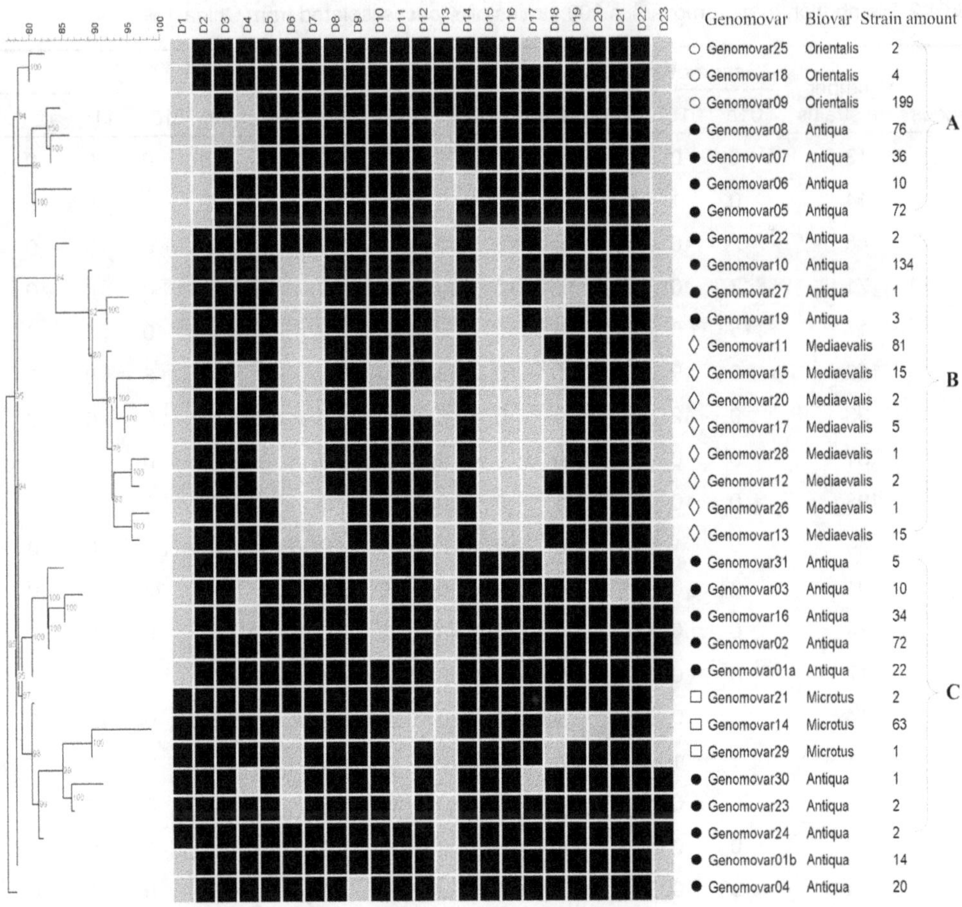

Figure 4.1 Neighbour-joining dendrogram of the 32 genomovars based on DFR profiles. The black and grey squares indicated the presence and absence of the corresponding DFRs, respectively. The white circle stands for genomovars of biovar *Orientalis*, black circle for *Antique*, diamond for *Medievalis* and square for *Microtus*. There are clearly three clusters (A, B and C) for all the genomovars.

genomovar01b and genomovar04. All *Orientalis* strains (205 strains in three genomovars) were grouped in cluster A; all *Medievalis* strains (122 strains in eight genomovars) in cluster B; and all *Pestoides* (plus *Microtus*) strains (66 strains in three genomovars) in cluster C. This result also clearly illustrated the close relationship of strains from the same biovar in China. However, different from the above three genotype groups, strains of the *Antiqua* genotype (516 strains in 18 genomovars) were distributed in all three clusters, revealing considerable genome diversities of the *Antiqua* strains.

SNPs analysis has identified two different molecular groups of *Antiqua* strains on two evolutionary lineages of *Y. pestis* (1.ANT and 2.ANT). These two groups were found belonging to the *Orientalis* and *Medievalis* branches, respectively (Achtman *et al.*, 2004). CRIPSR analysis also identified two clusters of *Antiqua* strains, one from Asia and another from Africa (Pource *et al.*, 2005). The genomovars in China are often plague focus-specific (Table 4.2) (Li *et al.*, 2008). Strains in a specific plague focus almost always fall into a few genomovars, with the majority of strains in one genomovar. Therefore, the new terms major genomovar and minor genomovar were coined to describe the regional specificity of genome plasticity for *Y. pestis* strains. Almost all the strains belonging to the major genomovars in a specific focus were isolated from the main hosts and vectors and were generally distributed throughout the focus (Fig. 4.2). In contrast, strains in the minor genomovar were isolated mainly from

Table 4.2 Distribution of genomovars in 909 strains of *Y. pestis* isolated from China

Focus or subfocus	Number of strains	Genomovar[a]													
		01a	01b	02	03	04	05	06	07	08	09	10	11	12	13
A	13	0	0	0	0	*11*	0	0	0	0	0	0	0	0	0
B1	14	0	1	03	0	*09*	0	0	0	0	0	0	0	0	0
B2	46	2	0	*43*	0	0	0	0	0	0	0	0	0	0	0
B3	71	3	0	*24*	10	0	0	0	0	0	0	0	0	0	0
B4	17	*15*	0	2	0	0	0	0	0	0	0	0	0	0	0
C	152	0	9	0	0	0	*63*	10	10	*48*	0	8	0	0	0
D	33	0	2	0	0	0	3	0	0	*27*	0	0	0	0	0
E	31	0	0	0	0	0	0	0	*23*	0	8	0	0	0	0
F	198	0	0	0	0	0	0	0	1	1	*191*	0	0	0	0
G	44	0	1	0	0	0	0	0	0	0	0	*39*	0	0	0
H	91	0	0	0	0	0	0	0	1	0	0	*87*	0	0	0
I	74	0	0	0	0	0	0	0	0	0	0	0	*69*	2	0
J	17	0	0	0	0	0	0	0	1	0	0	0	0	0	*15*
K1	14	1	0	0	0	0	0	0	0	0	0	0	*12*	0	0
K2	11	1	1	0	0	0	*6*	0	0	0	0	0	0	0	0
L	22	0	0	0	0	0	0	0	0	0	0	0	0	0	0
M	46	0	0	0	0	0	0	0	0	0	0	0	0	0	0
O	15	0	0	0	0	0	0	0	0	0	0	0	0	0	0
Sum	909	22	14	72	10	20	72	10	36	76	199	134	81	2	15

a The corresponding genomovars of italic numbers were focus-specific major genomovars.

minor hosts and were distributed sporadically along the border of neighbouring foci. Major genomovars probably play important roles in conserving the genotypic traits of the pathogen involved in specific individual plague focus. The strains belonging to the minor genomovars were typically distributed sporadically and probably contribute little to the major features of the plague focus. However, these minor genomovars could still play important roles in particular stages during the adaptive microevolution of *Y. pestis* because major and minor genomovars are defined based on their relative frequencies within specific plague foci. The major genomovar in one plague focus might be the minor one in the other. For example, genomovar09 was the major genomovar in focus F, but the minor one in focus E (Fig. 4.2). Each focus has its own characteristic major genomovar(s).

It should be noted that some strains from certain foci might have indistinguishable DFR profiles. For instance, strains from foci G and H shared genomovar10, foci K2 and I shared genomovar11 and foci L and M shared genomovar14. This result suggests the close relationship between the strains among different foci. The strains might be the result of recent spread from one focus to another and there have not been enough time to accumulate DFR variations. Other methods with higher resolution are needed to differentiate such strains from different foci. For example, using CRISPR and MLVA analyses, we were able to differentiate strains from foci L and M, as well as K2 and I (Cui *et al.*, 2008).

14	15	16	17	18	19	20	21	22	23	24	25	26	27	28	29	30	31
0	0	1	0	0	0	0	0	0	0	0	1	0	0	0	0	0	0
0	0	0	0	0	0	0	0	0	0	0	0	0	0	0	0	0	1
0	0	1	0	0	0	0	0	0	0	0	0	0	0	0	0	0	0
0	0	*32*	0	0	0	0	0	0	0	0	0	0	0	0	0	0	2
0	0	0	0	0	0	0	0	0	0	0	0	0	0	0	0	0	0
0	0	0	0	0	1	0	0	0	1	2	0	0	0	0	0	0	0
0	0	0	0	0	0	0	0	0	1	0	0	0	0	0	0	0	0
0	0	0	0	0	0	0	0	0	0	0	0	0	0	0	0	0	0
0	0	0	0	3	0	0	0	0	0	0	1	1	0	0	0	0	0
0	0	0	1	0	1	0	0	2	0	0	0	0	0	0	0	0	0
0	0	0	1	1	0	0	0	0	0	0	0	0	1	0	0	0	0
0	0	0	0	0	1	2	0	0	0	0	0	0	0	0	0	0	0
0	0	0	1	0	0	0	0	0	0	0	0	0	0	0	0	0	0
0	0	0	0	0	0	0	0	0	0	0	0	0	0	1	0	0	0
0	0	0	1	0	0	0	0	0	0	0	0	0	0	0	0	0	2
20	0	0	1	0	0	0	0	0	0	0	0	0	0	0	1	0	0
43	0	0	0	0	0	0	2	0	0	0	0	0	0	0	0	1	0
0	*15*	0	0	0	0	0	0	0	0	0	0	0	0	0	0	0	0
63	15	34	5	4	3	2	2	2	2	2	2	1	1	1	1	1	5

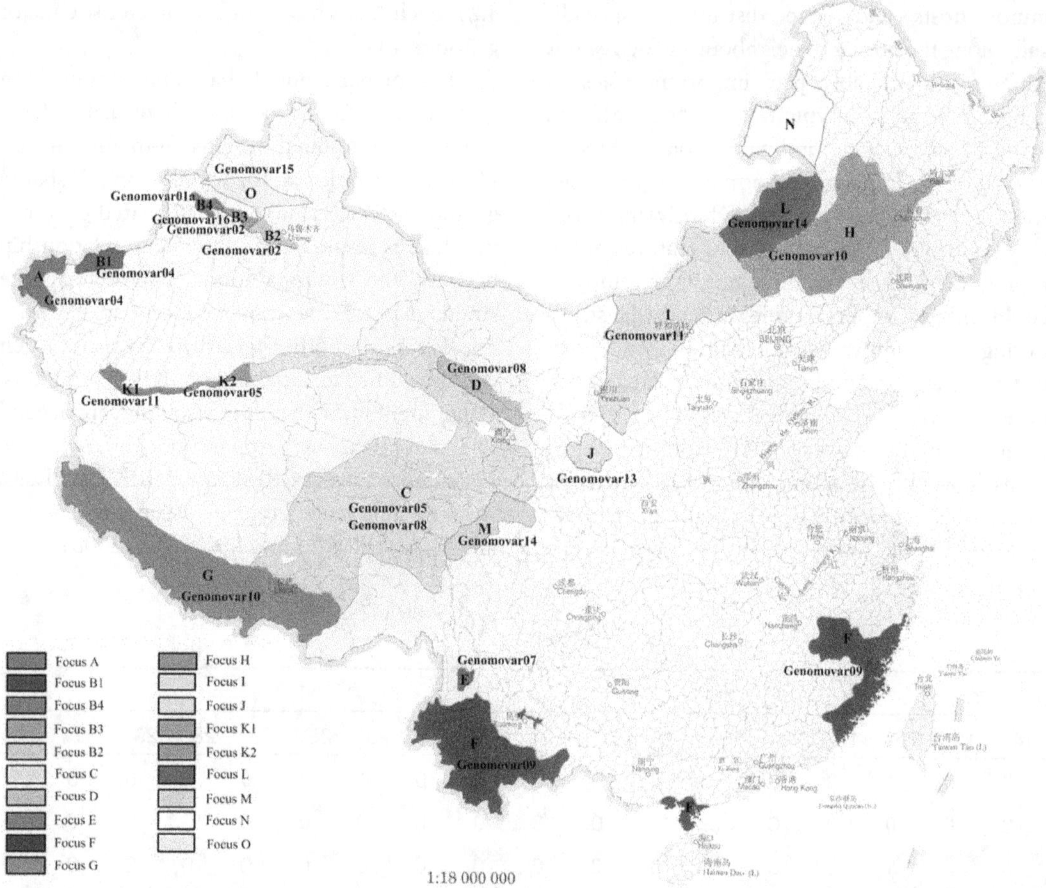

Figure 4.2 The distribution of major genomovars in natural plague foci of China. There are 15 plague foci in China. Focus A: *Marmota caudate* plague focus of the Pamirs Plateau; focus B: *Marmota baibacina-Spermophilus undulates* plague focus of the Tianshan Mountains; focus C: *Marmota himalayana* plague focus of the Qinghai–Gansu–Tibet Grassland; focus D: *Marmota himalayana* plague focus of the Qilian Mountain; focus E: *Apodemus chevrieri–Eothenomys miletus* plague focus of the highland of Northwestern Yunnan Province; focus F: *Rattus flavipectus* plague focus of the Yunnan–Guangdong–Fujian provinces; focus G: *Marmota himalayana* plague focus of the Gangdisi Mountains; focus H: *Spermophilus dauricus* plague focus of the Song-Liao Plain; focus I: *Meriones unguiculatus* plague focus of the Inner Mogolian Plateau; focus J: *Spermophilus dauricus alaschanicus* plague focus of the Loess Plateau in Gansu and Ningxia provinces; focus K: *Marmota himalayana* plague focus of the Kunlun Mountains; focus L: *Microtus brandti* plague focus of the Xilin Gol Grassland; focus M: *Microtus fuscus* plague focus of the Qinghai–Tibet Plateau; focus N: *Marmota sibirica* plague focus of the Hulun Buir Plateau of Inner Mongonia; focus O: *Rhombomys opimus* plague focus of the Junggar Basin of Xinjiang. B1, B2, B3 and B4 are subfoci of focus B, K1 and K2 are subfoci of focus K. Notably, focus N is a silent plague focus without strains available for this study. A colour version of this figure is available at http://www.caister.com/supplementary/population-genetics.

By comparing DFR profiles of the six sequenced *Orientalis* strains (Auerbach *et al.*, 2007; Holt *et al.*, 2008; Touchman *et al.*, 2007; Welch *et al.*, 2007) with the tested *Orientalis* strains in China, a microevolutionary pattern of *Orientalis* strains based on the reductionist principle was deduced (Fig. 4.3) and a virtual genomovarX was proposed as the missing link between genomovar09 to MG05-1020 and IP275 (Welch *et al.*, 2007). Interestingly, MG05-1020 and IP275 were both isolated in Madagascar. If this genomovar in Madagascar or elsewhere could be found, it will be of great help for better understanding the spreading of the third plague pandemics, which was believed to be originated from Yunnan province of China about 1500 years ago.

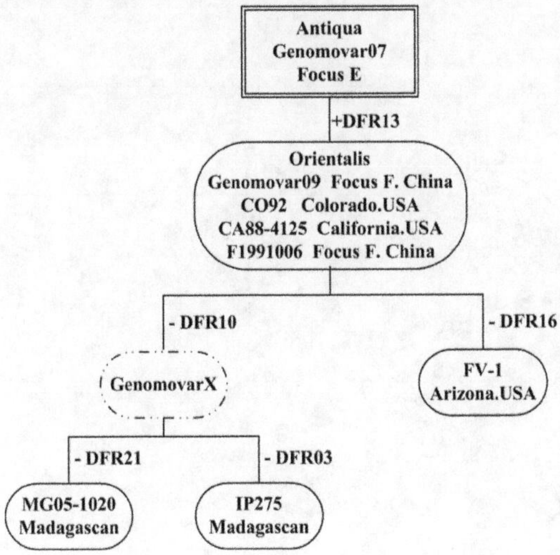

Figure 4.3 The microevolution scenario of *Orientalis* strains based on the gain-and-loss of DFRs. Orientalis strains evolved from genomovar07 Antiqua strains of focus E, by acquiring DFR13, and then evolved as different genomovars by losing certain DFR. A virtual genomovarX was included to illustrate the step-by-step reductionism evolution.

The *Orientalis* group is believed to be a 'young' biovar of *Y. pestis*, and the timing for its spread throughout the world is probably less than 120 years. *Orientalis* strains in China are very homologous in their DFR pattern (only three genomovars in 205 strains with 199 strains as genomovar09) whereas strains from outside China show considerable heterogeneity (four genomovars in five strains). The heterogeneity outside of China might be due to parallel and differential loss of DFRs during their adaptation to various geographic regions and ecological niches. The adaptive microevolution might lead to the discrete segregation between the progenitor and offspring strain genotypes. This genome reduction would gradually cause the offspring strains to inhabit a more specific host niche, with little overlap with their progenitors.

CRISPR analyses have been successfully used for studying *Y. pestis* microevolution (Pourcel *et al.*, 2005; Vergnaud *et al.*, 2007). CRISPRs are a family of elements which typically consist of non-contiguous direct repeats (DR, 24–47 bp) separated by stretches of similarly sized unique sequences (Jansen *et al.*, 2002; Mojica *et al.*, 2005). One or more CRISPR loci are found in 40% of the sequenced bacterial genomes as well as in most sequenced archaea (Grissa *et al.*, 2008). CRISPR loci, Cas (CRISPR-associated proteins), and leader sequences (the non-coding sequences flanking the CRISPR loci that acts as a promoter) have been suggested to constitute a prokaryotic immune system against bacteriophage attack (Barrangou *et al.*, 2007). This was demonstrated in *Streptococcus thermophilus* (Barrangou *et al.*, 2002; Deveau *et al.*, 2008). The unique sequences in CRISPR loci, the so-called 'spacers', show considerable divergence within a given species. A fraction of spacers are observed to be homologous with pre-existing sequences such as bacteriophage and conjugative plasmids (Bolotin *et al.*, 2005; Lillestol *et al.*, 2006; Pourcel *et al.*, 2005). Diversity within a CRISPR locus was first used for typing *Mycobacterium tuberculosis* strains (Brudey *et al.*, 2006). In this study, 43 spacers were screened by hybridization using a nylon membrane (Goyal *et al.*, 1997) in an assay called 'spoligotyping' (spacer oligonucleotide typing). An international database for the spacers comprising more than 2000 patterns from almost 40,000 isolates has been created and is accessible through the internet for *M. tuberculosis* (Brudey *et al.*, 2006). More recently, a similar assay was applied for genotyping *Corynebacterium diphtheriae* (Mokrousov *et al.*, 2007).

To better understand the adaptive microevolution of *Y. pestis* in Asia, a representative collection of 125 strains of *Y. pestis* from 26 natural plague foci of China, the former Soviet Union and Mongolia were analysed by our laboratory, in collaboration with French and Russian scientists. The distribution of spacers among *Y. pestis* strains was found largely geographic and plague focus specific. The data allowed the construction of a hypothetic evolutionary model of *Y. pestis* (Fig. 4.4). This model suggests a transmission route of *Microtus* strains encircling Takla Makan Desert and ZhunGer Basin in north-western China. Starting from Tajikistan, one branch passed through the Kunlun Mountains, and moved to the Qinghai-Tibet Plateau. Another branch went north via the Pamirs Plateau, the Tianshan Mountains, the Altai Mountains and the Inner Mongolian Plateau. Other *Y. pestis* lineages might have evolved during their spread along those routes for certain specific areas.

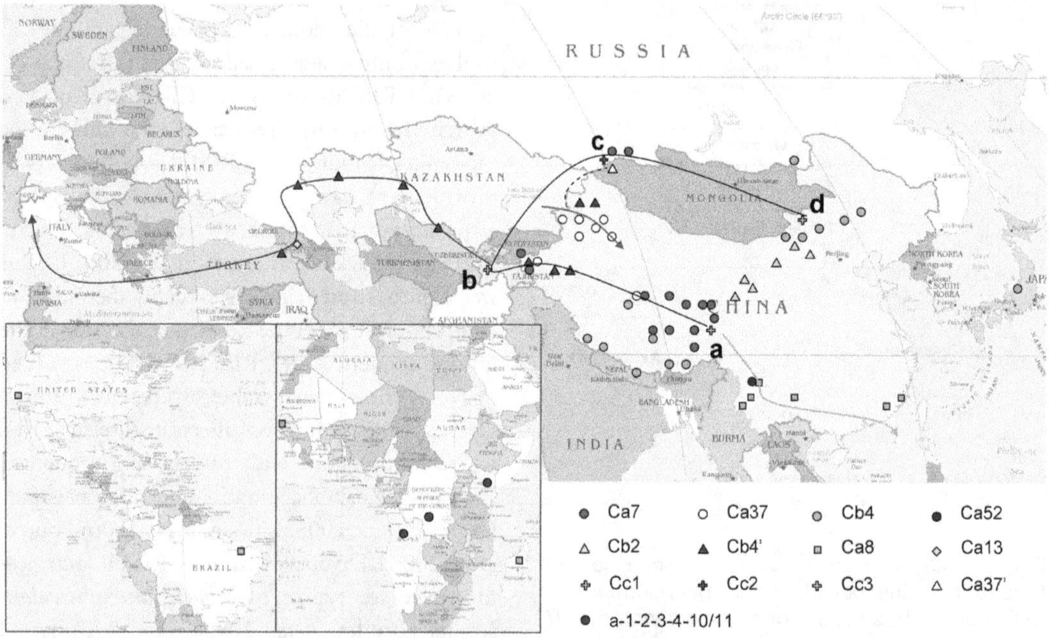

Figure 4.4 Hypothetic transmission route of *Y. pestis*. Black line shows the main transmission route of *Y. pestis* bv. microtus, red line bv. mediaevalis, pink line bv. orientalis, and green line Ca37. Black dot line shows the potential relationship between Ca37 and Ca37′. Various symbols represent different biovars (see legend of Fig. 4.2 for detail, one more symbol, diamond, is used to represent subsp. *caucasica* in this figure). The spacers array information of isolates from South America and Africa comes from previous research (Deveau *et al.*, 2007*). Red circles represent isolates from Africa, with spacers array 'a-1-2-3-4-10/11' in YPa (data from Pourcel et. al*, 2005). All the known African isolates contain 'a10' spacer. Therefore, a10 is possibly a characteristic spacer of isolates from Africa. Nevertheless, we do not consider these isolates as one cluster due to the limited data available from African strains. A colour version of this figure is available at http://www.caister.com/supplementary/population-genetics

The typing method MLVA has been used mainly by two groups for studying *Y. pestis* (Girard *et al.*, 2004; Pourcel *et al.*, 2004; Lista *et al.*, 2006), one group was from the US and the second from France. The American group employed 43 VNTR loci (Girard *et al.*, 2004) and the French group 25 markers (Pourcel *et al.*, 2004). Only six loci were shared by these two groups. Most strains used in their studies belonged to the *Orientalis* group, indicating that the results might be biased. We are currently studying a very diverse collection of strains of *Y. pestis*, including strains of *Orientalis*, *Medievalis*, *Antiqua* and *Microtus* groups, to evaluate the ability of these two sets of VNTR markers to differentiate *Y. pestis* strains. Our preliminary results demonstrated that some of the markers in these two sets are very informative while others are not informative at all (unpublished data). We hope to propose a robust set of VNTR markers for genotyping *Y. pestis* in the near future.

Rapid identification of pathogenic bacteria based on the analysis of DNA diversities

The genotyping results established through population genetic studies can help develop rapid methods to accurately detect and identify the target pathogens. For example, based on a single SNP, Vogler *et al.* (2008) developed a TaqMan-minor groove binding (MGB) assay to separate the *Yersinia pestis* strain CO92 from all other strains. Another SNP allowed the discrimination of North American strains from strains in other parts of the world (Touchman *et al.*, 2007). We showed that an array of 23 PCRs based on the DFR typing scheme enabled us to rapidly identify the source of *Y. pestis* (Zhou *et al.*, 2004a,b; Li *et al.*, 2008). Lista *et al.* (2006) developed a genotyping method for *Y. pestis* and *Bacillus anthracis* in the field. Their method was based on their previously described multiple locus VNTR assays (MLVA), in combination with the Agilent 2100 BioAnalyzer. This platform allowed on-site MLVA genotyping of bioterrorism agents as well as other bacterial pathogens. It is an alternative to the more expensive and technically demanding capillary electrophoresis methods, and to the less expensive but more time-consuming classical gel electrophoresis approach.

Many databases have been developed based on genotyping results of populations of microbial pathogens for rapid identification and source-tracing of target pathogens. Some of the well-known ones include PulseNet USA for food-borne disease surveillance (Gerner-Smidt *et al.*, 2006); USDA VetNet (Jackson *et al.*, 2007) for subtyping zoonotic pathogens; MLVA databases and CRISPR databases for several pathogens (Pourcel *et al.*, 2005; Garcia-Yoldi *et al.*, 2006; Vergnaud *et al.*, 2007); MLST databases for many bacteria, including important human pathogens (Aanensen and Spratt, 2005); an international spoligotyping database (SpolDB4) for the classification, population genetics and epidemiology of *Mycobacterium tuberculosis* (Brudey, Driscoll *et al.*, 2006); a genoBASE pylori for *Helicobacter pylori* (Ahmed *et al.*, 2007), including DNA fingerprinting data in the form of fluorescent amplified fragment length polymorphism (FAFLP); several enterobacterial repetitive intergenic consensus (ERIC) and repetitive extragenic palindromic (REP) signature profiles (http://210.212.212.4/HP/index.html); a knowledgebase AmpliBASE MT for *Mycobacterium tuberculosis* diversity (Majeed *et al.*, 2004), including high-resolution DNA fingerprints representing fluorescent amplified fragment length polymorphism (FAFLP) profiles or amplitypes developed for the *Mycobacterium tuberculosis* complex strains from 48 different countries (Majeed *et al.*, 2004). These preliminary databases are critical for development of microbial forensics (McEwen *et al.*, 2006).

Perspectives

With so many techniques used for typing *Y. pestis*, it is reasonable to ask which one(s) are better. The answer to this question depends on our objective and different methods will be best for addressing different issues. For microevolutionary studies, such as for determining the spread of strains during the third pandemic of *Y. pestis* infection and for how the strains for this pandemic evolved, the application of SNPs, especially synonymous SNPs, will be the best marker. If we want to trace the origin of *Y. pestis*, MLVA might be the most highly discriminative marker. For ordinary genotyping, we can use all the different techniques depending on the technical platform we have already established in different labs. DFR is simple because it only depends on ordinary PCR and the results are judged by the presence or absence of the amplicons.

However, we would like to point out that the applications of techniques and analytical methods suitable for other pathogenic bacteria might be different because different bacterial populations have different characteristics. *Y. pestis* is a recently evolved, homogeneous species with only one serogroup and is closely related to *Y. pseudotuberculosis*. For other species with greater genetic variation, such as species in *Salmonella*, we could employ markers with a low-mutation rate such as SNPs to address specific questions such as the evolution of antibiotic resistant. While MLST provides no information for strain discrimination in *Y. pestis*, it can provide us abundant information for genetically diversified bacteria, including *Neisseria meningitidis* (Shao *et al.*, 2006), erocytotoxin-producing *Escherichia coli* (Ziebell *et al.*, 2008), *Chlamydia trachomatis* (Pannekoek *et al.*, 2008), and *Staphylococcus epidermidis* (Miragaia *et al.*, 2008), etc. Indeed, the MLST technique developed for genetic population studies of these bacteria have played important roles in characterizing emerging pathogenic strains in these species. For example, using MLST, a new clonal complex of *Neisseria meningitidis* was identified as responsible for causing outbreaks of the serogroup C meningococcal disease from 2003 to 2005 in China (Shao *et al.*, 2006). Similarly, a clone of methicillin-resistant *Staphylococcus aureus* strains was identified responsible for an epidemic in Finland (Vainio *et al.*, 2008).

Aside from the above techniques, the advantages of high throughput sequencing technologies, represented by 454, Solexa and SoLiD technology (Holt *et al.*, 2008), will probably play increasingly important roles in population studies and identifications of human pathogenic bacteria. It is now possible to sequence multiple bacterial genomes of one species at a reasonable cost. Such data will provide abundant genetic information for a better understanding of their phenotypes. Such whole genome-based studies will revolutionize our understanding of bacterial microevolution, pathogenesis and life cycle.

Acknowledgements

The research work on *Yersinia pestis* was supported by The National Basic Research Program of China (Program 973, No. 2009CB522600), the National High Technology Research and Development Program of China (863 program) (contract no. 2006AA02Z438), the National Natural Science Foundation of China for Distinguished Young Scholars (No. 30525025), and the National Natural Science Foundation of China (no. 30430620). We thank Dr Jianping Xu for his help in improving this manuscript.

References

Aanensen, D.M., and Spratt, B.G. (2005). The multilocus sequence typing network: mlst.net. Nucleic Acids Res. *33*, W728–33.

Achtman, M., Zurth, K., Morelli, G., Torrea, G., Guiyoule, A., and Carniel, E. (1999). *Yersinia pestis*, the cause of plague, is a recently emerged clone of *Yersinia pseudotuberculosis*. Proc. Natl. Acad. Sci. U.S.A. *96*, 14043–14048.

Achtman, M., Morelli, G., Zhu, P., Wirth, T., Diehl, I., Kusecek, B., Vogler, A.J., Wagner, D.M., Allender, C.J., Easterday, W.R., *et al.* (2004). Microevolution and history of the plague bacillus, *Yersinia pestis*. Proc. Natl. Acad. Sci. U.S.A. *101*, 17837–17842.

Adair, D.M., Worsham, P.L., Hill, K.K., Klevytska, A.M., Jackson, P.J., Friedlander, A.M., and Keim, P. (2000). Diversity in a variable-number tandem repeat from *Yersinia pestis*. J. Clin. Microbiol. *38*, 1516–1519.

Ahmed, N., Majeed, A.A., Ahmed, I., Hussain, M.A., Alvi, A., Devi, S.M., Rizwan, M., Ranjan, A., Sechi, L.A., and Megraud, F. (2007). genoBASE pylori: a genotype search tool and database of the human gastric pathogen *Helicobacter pylori*. Infect. Genet. Evol. *7*, 463–468.

Al Dahouk, S., Fleche, P.L., Nockler, K., Jacques, I., Grayon, M., Scholz, H.C., Tomaso, H., Vergnaud, G., and Neubauer, H. (2007). Evaluation of Brucella MLVA typing for human brucellosis. J. Microbiol. Methods *69*, 137–145.

Anisimov, A.P., Lindler, L.E., and Pier, G.B. (2004). Intraspecific diversity of *Yersinia pestis*. Clin. Microbiol. Rev. *17*, 434–464.

Aras, R.A., Small, A.J., Ando, T., and Blaser, M.J. (2002). *Helicobacter pylori* interstrain restriction-modification diversity prevents genome subversion by chromosomal DNA from competing strains. Nucleic Acids Res. *30*, 5391–5397.

Auerbach, R.K., Tuanyok, A., Probert, W.S., Kenefic, L., Vogler, A.J., Bruce, D.C., Munk, C., Brettin, T.S., Eppinger, M., Ravel, J., Wagner, D.M., and Keim, P. (2007). *Yersinia pestis* evolution on a small timescale: comparison of whole genome sequences from North America. PLoS ONE 2, e770.

Bannerman, T.L., Hancock, G.A., Tenover, F.C., and Miller, J.M. (1995). Pulsed-field gel electrophoresis as a replacement for bacteriophage typing of *Staphylococcus aureus*. J. Clin. Microbiol. *33*, 551–555.

Barrangou, R., Yoon, S.S., Breidt, F., Jr., Fleming, H.P., and Klaenhammer, T.R. (2002). Identification and characterization of *Leuconostoc fallax* strains isolated from an industrial sauerkraut fermentation. Appl. Environ. Microbiol. *68*, 2877–2884.

Barrangou, R., Freemaux, C., Deveau, H., Richards, M., Boyaval, P., Moineau, S., Romero, D. A, and Horvath, P. (2007). CRISPR provides acquired resistance against viruses in prokaryotes. Science *315*, 1709–1712.

Bolotin, A., Quinquis, B., Sorokin, A., and Ehrlich, S.D. (2005). Clustered regularly interspaced short palindrome repeats (CRISPRs) have spacers of extrachromosomal origin. Microbiology *151*, 2551–2561.

Botstein, D., White, R.L., Skolnick, M., and Davis, R.W. (1980). Construction of a genetic linkage map in man using restriction fragment length polymorphisms. Am. J. Hum. Genet. *32*, 314–331.

Brudey, K., Driscoll, J.R., Rigouts, L., Prodinger, W.M., Gori, A., Al-Hajoj, S.A., Allix, C., Aristimuno, L., Arora, J., Baumanis, V., *et al.* (2006). *Mycobacterium tuberculosis* complex genetic diversity: mining the fourth international spoligotyping database (SpolDB4) for classification, population genetics and epidemiology. BMC Microbiol. *6*, 23.

Carlton, J.M. (2003). Genome sequencing and comparative genomics of tropical disease pathogens. Cell. Microbiol. *5*, 861–873.

Carniel, E. (2002). Plasmids and pathogenicity islands of *Yersinia*. Curr. Top. Microbiol. Immunol. *264*, 89–108.

Chain, P.S., Carniel, E., Larimer, F.W., Lamerdin, J., Stoutland, P.O., Regala, W.M., Georgescu, A.M., Vergez, L.M., Land, M.L., Motin, V.L., *et al.* (2004). Insights into the evolution of *Yersinia pestis* through whole-genome comparison with *Yersinia pseudotuberculosis*. Proc. Natl. Acad. Sci. U.S.A. *101*, 13826-13831.

Cui, Y., Li, Y., Gorge, O., Platonov, M.E., Yan, Y., Guo, Z., Pourcel, C., Dentovskaya, S.V., Balakhonov, S.V., Wang, X., Song, Y., Anisimov, A.P., Vergnaud, G., and Yang, R. (2008). Insight into microevolution of *Yersinia pestis* by clustered regularly interspaced short palindromic repeats. PLoS ONE 3, e2652.

Dai, E., Tong, Z., Wang, X., Li, M., Cui, B., Dai, R., Zhou, D., Pei, D., Song, Y., Zhang, J., *et al.* (2005). Identification of different regions among strains of *Yersinia pestis* by suppression subtractive hybridization. Res. Microbiol. *156*, 785–789.

Deng, W., Burland, V., Plunkett, G. III, Boutin, A., Mayhew, G.F., Liss, P., Perna, N.T., Rose, D.J., Mau, B., Zhou, S., *et al.* (2002). Genome sequence of *Yersinia pestis* KIM. J. Bacteriol. *184*, 4601–4611.

Deveau, H., Barrangou, R., Garneau, J.E., Labonte, J., Fremaux, C., Boyaval, P., Romero, D.A., Horvath, P., and Moineau, S. (2007). Phage response to CRISPR-encoded resistance in *Streptococcus thermophilus*. J. Bacteriol. *190*, 1390–1400.

Dorrell, N., S Hinchliffe, J., and Wren, B.W. (2005). Comparative phylogenomics of pathogenic bacteria by microarray analysis. Curr. Opin. Microbiol. *8*, 620–626.

Enright, M.C., and Spratt, B.G. (1999). Extensive variation in the ddl gene of penicillin-resistant *Streptococcus pneumoniae* results from a hitchhiking effect driven by the penicillin-binding protein 2b gene. Mol. Biol. Evol. *16*, 1687–1695.

Garcia-Yoldi, D., Le Fleche, P., Marin, C.M., De Miguel, M.J., Munoz, P.M., Vergnaud, G., and Lopez-Goni, I. (2006). Assessment of genetic stability of Brucella melitensis Rev 1 vaccine strain by multiple-locus variable-number tandem repeat analysis. Vaccine *25*, 2858–2862.

Gerner-Smidt, P., Hise, K., Kincaid, J., Hunter, S., Rolando, S., Hyytia-Trees, E., Ribot, E.M., and Swaminathan, B. (2006). PulseNet USA: a five-year update. Foodborne Pathog Dis 3, 9–19.

Girard, J.M., Wagner, D.M., Vogler, A.J., Keys, C., Allender, C.J., Drickamer, L.C., and Keim, P. (2004). Differential plague-transmission dynamics determine *Yersinia pestis* population genetic structure on local, regional, and global scales. Proc. Natl. Acad. Sci. U.S.A. *101*, 8408–8413.

Gomez-Gomez, L., and Boller, T. (2002). Flagellin perception: a paradigm for innate immunity. Trends Plant Sci. *7*, 251–256.

Goyal, M., Saunders, N.A., J.D. van Embden, Young, D.B., and Shaw, R.J. (1997). Differentiation of *Mycobacterium tuberculosis* isolates by spoligotyping and IS6110 restriction fragment length polymorphism. J. Clin. Microbiol. *35*, 647–651.

Grissa, I., Bouchon, P., Pourcel, C., and Vergnaud, G. (2008). On-line resources for bacterial micro-evolution studies using MLVA or CRISPR typing. Biochimie *90*, 660–668.

Guillemot, D., and Leclercq, R. (2005). [Impact of population exposure on the risk of bacterial resistance]. Med. Mal. Infect. 35 (Suppl. 3), S212–20.

Guiyoule, A., Grimont, F., Iteman, I., Grimont, P.A., Lefevre, M., and Carniel, E. (1994). Plague pandemics investigated by ribotyping of *Yersinia pestis* strains. J. Clin. Microbiol. *32*, 634–641.

Guiyoule, A., Rasoamanana, B., Buchrieser, C., Michel, P., Chanteau, S., and Carniel, E. (1997). Recent emergence of new variants of *Yersinia pestis* in Madagascar. J. Clin. Microbiol. *35*, 2826–2833.

Hai, R., Yu, D.Z., Wei, J.C., Xia, L.X., Shi, X.M., Zhang, Z.K., and Zhang, E.M. (2004). Molecular biological characteristics and genetic significance of *Yersinia pestis* in China. Zhonghua Liu Xing Bing Xue Za Zhi *25*, 509–513.

Holt, K.E., Parkhill, J., Mazzoni, C.J., Roumagnac, P., Weill, F.X., Goodhead, I., Rance, R., Baker, S., Maskell, D.J., Wain, J., Dolecek, C., Achtman, M., and Dougan, G. (2008). High-throughput sequencing provides insights into genome variation and evolution in Salmonella Typhi. Nat. Genet. *40*, 987–993.

Huang, F., Yu, D., Hai, R., and Cai, H. (2000). [Study on the application of random amplified polymorphic DNA in *Yersinia pestis* genotyping]. Zhonghua Liu Xing Bing Xue Za Zhi *21*, 424–426.

Jackson, C.R., Fedorka-Cray, P.J., Wineland, N., Tankson, J.D., Barrett, J.B., Douris, A., Pearson Gresham, C., Hall, M.C., Mcglinchey, B., and Price, M.V. (2007). Introduction to USDA VetNet: status of *Salmonella* and *Campylobacter* databases from 2004 through 2005. Foodborne Pathog. Dis. *4*, 241–248.

Jansen, R., Embden, J.D., Gaastra, W., and Schouls, L.M. (2002). Identification of genes that are associated with DNA repeats in prokaryotes. Mol. Microbiol. *43*, 1565–1575.

Jolley, K.A., Chan, M.S., and Maiden, M.C. (2004). mlstdbNet – distributed multi-locus sequence typing (MLST) databases. BMC Bioinf. *5*, 86.

Klevytska, A.M., Price, L.B., Schupp, J.M., Worsham, P.L., Wong, J., and Keim, P. (2001). Identification and characterization of variable-number tandem repeats in the *Yersinia pestis* genome. J. Clin. Microbiol. *39*, 3179–3185.

Li, Y., Dai, E., Cui, Y., Li, M., Zhang, Y., Wu, M., Zhou, D., Guo, Z., Dai, X., Cui, B., Qi, Z., Wang, Z., Wang, H., Dong, X., Song, Z., Zhai, J., Song, Y., and Yang, R. (2008). Different region analysis for genotyping *Yersinia pestis* isolates from China. PLoS ONE *3*, e2166.

Lillestol, R.K., Redder, P., Garrett, R.A., and Brugger, K. (2006). A putative viral defence mechanism in archaeal cells. Archaea 2, 59–72.

Lista, F., Faggioni, G., Valjevac, S., Ciammaruconi, A., Vaissaire, J., le Doujet, C., Gorge, O., De Santis, R., Carattoli, A., Ciervo, A., Fasanella, A., Orsini, F., D'Amelio, R., Pourcel, C., Cassone, A., and Vergnaud, G. (2006). Genotyping of *Bacillus anthracis* strains based on automated capillary 25-loci multiple locus variable-number tandem repeats analysis. BMC Microbiol. *6*, 33.

Liu, S.L., and Sanderson, K.E. (1995a). I-CeuI reveals conservation of the genome of independent strains of *Salmonella typhimurium*. J. Bacteriol. *177*, 3355–3357.

Liu, S.L., and Sanderson, K.E. (1995b). Rearrangements in the genome of the bacterium *Salmonella typhi*. Proc. Natl. Acad. Sci. U.S.A. *92*, 1018–1022.

Lowell, J.L., Wagner, D.M., Atshabar, B., Antolin, M.F., Vogler, A.J., Keim, P., Chu, M.C., and Gage, K.L. (2005). Identifying sources of human exposure to plague. J. Clin. Microbiol. *43*, 650–656.

Majeed, A.A., Ahmed, N., Rao, K.R., Ghousunnissa, S., Kauser, F., Bose, B., Nagarajaram, H.A., Katoch, V.M., Cousins, D.V., Sechi, L.A., Gilman, R.H., and Hasnain, S.E. (2004). AmpliBASE MT: a *Mycobacterium tuberculosis* diversity knowledgebase. Bioinformatics *20*, 989–992.

McEwen, S.A., Wilson, T.M., Ashford, D.A., Heegaard, E.D., and Kournikakis, B. (2006). Microbial forensics for natural and intentional incidents of infectious disease involving animals. Rev. Sci. Tech. *25*, 329–339.

Miragaia, M., Carrico, J.A., Thomas, J.C., Couto, I., Enright, M.C., and de Lencastre, H. (2008). Comparison of molecular typing methods for characterization of

Staphylococcus epidermidis: proposal for clone definition. J. Clin. Microbiol. *46*, 118–129.

Mojica, F.J., Diez-Villasenor, C., Garcia-Martinez, J., and Soria, E. (2005). Intervening sequences of regularly spaced prokaryotic repeats derive from foreign genetic elements. J. Mol. Evol. *60*, 174–182.

Mokrousov, I., Limeschenko, E., Vyazovaya, A., and Narvskaya, O. (2007). *Corynebacterium diphtheriae* spoligotyping based on combined use of two CRISPR loci. Biotechnol. J. *2*, 901–906.

Monecke, S., Berger-Bachi, B., Coombs, G., Holmes, A., Kay, I., Kearns, A., Linde, H.J., O'Brien, F., Slickers, P., and Ehricht, R. (2007). Comparative genomics and DNA array-based genotyping of pandemic *Staphylococcus aureus* strains encoding Panton–Valentine leukocidin. Clin. Microbiol. Infect. *13*, 236–249.

Odaert, M., Devalckenaere, A., Trieu-Cuot, P., and Simonet, M. (1998). Molecular characterization of IS1541 insertions in the genome of *Yersinia pestis*. J. Bacteriol. *180*, 178–181.

Pannekoek, Y., Morelli, G., Kusecek, B., Morre, S.A., Ossewaarde, J.M., Langerak, A.A., and van der Ende, A. (2008). Multi locus sequence typing of Chlamydiales: clonal groupings within the obligate intracellular bacteria *Chlamydia trachomatis*. BMC Microbiol. *8*, 42.

Parkhill, J., Wren, B.W., Thomson, N.R., Titball, R.W., Holden, M.T., Prentice, M.B., Sebaihia, M., James, K.D., Churcher, C., Mungall, K.L., *et al.* (2001). Genome sequence of *Yersinia pestis*, the causative agent of plague. Nature *413*, 523–527.

Perez-Losada, M., Browne, E.B., Madsen, A., Wirth, T., Viscidi, R.P., and Crandall, K.A. (2006). Population genetics of microbial pathogens estimated from multilocus sequence typing (MLST) data. Infect. Genet. Evol. *6*, 97–112.

Perry, R.D., and Fetherston, J.D. (1997). *Yersinia pestis* – etiologic agent of plague. Clin. Microbiol. Rev. *10*, 35–66.

Pourcel, C., Andre-Mazeaud, F., Neubauer, H., Ramisse, F., and Vergnaud, G. (2004). Tandem repeats analysis for the high resolution phylogenetic analysis of *Yersinia pestis*. BMC Microbiol. *4*, 22.

Pourcel, C., Salvignol, G., and Vergnaud, G. (2005). CRISPR elements in *Yersinia pestis* acquire new repeats by preferential uptake of bacteriophage DNA, and provide additional tools for evolutionary studies. Microbiology *151*, 653–663.

Revazishvili, T., Rajanna, C., Bakanidze, L., Tsertsvadze, N., Imnadze, P., O'Connell K., Kreger, A., Stine, O.C., Morris, J.G. Jr, and Sulakvelidze, A. (2008). Characterisation of *Yersinia pestis* isolates from natural foci of plague in the Republic of Georgia, and their relationship to *Y. pestis* isolates from other countries. Clin. Microbiol. Infect. *14*, 429-436.

Saunders, N.A., and Holmes, A. (2007). Multilocus sequence typing (MLST) of *Staphylococcus aureus*. Methods Mol. Biol. *391*, 71–85.

Schwartz, D.C., and Cantor, C.R. (1984). Separation of yeast chromosome-sized DNAs by pulsed field gradient gel electrophoresis. Cell 37, 67–75.

Shao, Z., Li, W., Ren, J., Liang, X., Xu, L., Diao, B., Li, M., Lu, M., Ren, H., Cui, Z., Zhu, B., Dai, Z., Zhang, L., Chen, X., Kan, B., and Xu, J. (2006). Identification of a new Neisseria meningitidis serogroup C clone from Anhui province, China. Lancet *367*, 419–423.

Song, Y., Tong, Z., Wang, J., Wang, L., Guo, Z., Han, Y., Zhang, J., Pei, D., Zhou, D., Qin, H., *et al.* (2004). Complete genome sequence of *Yersinia pestis* strain 91001, an isolate avirulent to humans. DNA Res. *11*, 179–197.

Struelens, M.J. (1998a). Molecular epidemiologic typing systems of bacterial pathogens: current issues and perspectives. Mem Inst Oswaldo Cruz *93*, 581–585.

Struelens, M.J. (1998b). Tracking the epidemiology of antimicrobial drug resistance in hospitals: time to deploy molecular typing. J. Med. Microbiol. *47*, 1035–1036.

Touchman, J.W., Wagner, D.M., Hao, J., Mastrian, S.D., Shah, M.K., Vogler, A.J., Allender, C.J., Clark, E.A., Benitez, D.S., Youngkin, D.J., *et al.* (2007). A North American *Yersinia pestis* draft genome sequence: SNPs and phylogenetic analysis. PLoS ONE 2, e220.

Vainio, A., Karden-Lilja, M., Ibrahem, S., Kerttula, A.M., Salmenlinna, S., Virolainen, A., and Vuopio-Varkila, J. (2008). Clonality of epidemic methicillin-resistant *Staphylococcus aureus* strains in Finland as defined by several molecular methods. Eur J. Clin. Microbiol. Infect. Dis. 7, 545–555.

van Belkum, A., Struelens, M., de Visser, A., Verbrugh, H., and Tibayrenc, M. (2001). Role of genomic typing in taxonomy, evolutionary genetics, and microbial epidemiology. Clin. Microbiol. Rev. *14*, 547–560.

Vergnaud, G., Li, Y., Gorge, O., Cui, Y., Song, Y., Zhou, D., Grissa, I., Dentovskaya, S.V., Platonov, M.E., Rakin, A., Balakhonov, S.V., Neubauer, H., Pourcel, C., Anisimov, A.P., and Yang, R. (2007). Analysis of the three *Yersinia pestis* CRISPR loci provides new tools for phylogenetic studies and possibly for the investigation of ancient DNA. Adv. Exp. Med. Biol. *603*, 327–338.

Vogler, A.J., Driebe, E.M., Lee, J., Auerbach, R.K., Allender, C.J., Stanley, M., Kubota, K., Andersen, G.L., Radnedge, L., Worsham, P.L., Keim, P., and Wagner, D.M. (2008). Assays for the rapid and specific identification of North American *Yersinia pestis* and the common laboratory strain CO92. Biotechniques 44, 201, 203–4, 207.

Vos, P., Hogers, R., Bleeker, M., Reijans, M., van de Lee, T., Hornes, M., Frijters, A., Pot, J., Peleman, J., Kuiper, M., *et al.* (1995). AFLP: a new technique for DNA fingerprinting. Nucleic Acids Res. 23, 4407–4414.

Wang, X., Han, Y., Li, Y., Guo, Z., Song, Y., Tan, Y., Du, Z., Rakin, A., Zhou, D., and Yang, R. (2007). Yersinia genome diversity disclosed by *Yersinia pestis* genome-wide DNA microarray. Can. J. Microbiol. 53, 1211–1221.

Welch, T.J., Fricke, W.F., McDermott, P.F., White, D.G., Rosso, M.L., Rasko, D.A., Mammel, M.K., Eppinger, M., Rosovitz, M.J., Wagner, D., Rahalison, L., Leclerc, J.E., Hinshaw, J.M., Lindler, L.E., Cebula, T.A., Carniel, E., and Ravel, J. (2007). Multiple antimicrobial resistance in plague: an emerging public health risk. PLoS ONE 2, e309.

Williams, J.G., Kubelik, A.R., Livak, K.J., Rafalski, J.A., and Tingey, S.V. (1990). DNA polymorphisms amplified by arbitrary primers are useful as genetic markers. Nucleic Acids Res. *18*, 6531–5.

Yu, D.Z., Hai, R., Dong, X.Q., Li, M., Xia, L.X., Shi, X.M., Wei, J.C., Cui, B.Z., Wang, P., Sun, L.Z., Zhang, Z.K., Hu, Y., and Zhang, E.M. (2003). Genetic analysis of *Yersinia pestis* strains isolated in China. Zhonghua Liu Xing Bing Xue Za Zhi *24*, 1005–9.

Zhou, D., Han, Y., Song, Y., Tong, Z., Wang, J., Guo, Z., Pei, D., Pang, X., Zhai, J., Li, M., *et al.* (2004a). DNA microarray analysis of genome dynamics in *Yersinia pestis:* insights into bacterial genome microevolution and niche adaptations. J. Bacteriol .186, 5138–5146.

Zhou, D., Tong, Z., Song, Y., Han, Y., Pei, D., Pang, X., Zhai, J., Li, M., Cui, B., Qi, Z., *et al.* (2004b). Genetics of metabolic variations between *Yersinia pestis* biovars and the proposal of a new biovar, microtus. J. Bacteriol. *186*, 5147–5152.

Ziebell, K., Konczy, P., Yong, I., Frost, S., Mascarenhas, M., Kropinski, A.M., Whittam, T.S., Read, S.C., and Karmali, M.A. (2008). Applicability of phylogenetic methods for characterizing the public health significance of verocytotoxin-producing *Escherichia coli* strains. Appl. Environ. Microbiol. *74*, 1671–1675.

Population Genetics of the Symbiotic Nitrogen-fixing Bacteria Rhizobia

5

Bertrand D. Eardly and Jianping Xu

Abstract

Symbiotic nitrogen-fixing Rhizobia are of global significance, both in terms of their ecological relationships and their importance as an environmentally benign source of nitrogen for crop plants. These bacteria are capable of forming mutualistic relationships with a variety of legume hosts, where they convert atmospheric nitrogen (N_2) to ammonia that is used to help meet the nitrogen needs of the host plant. In this chapter, we review our current understanding of the population genetics of this diverse group of bacteria. First we briefly describe the various types of genomic architectures that are found within rhizobial species. Next we outline the phylogenetic relationships among the recognized taxa. We then summarize the results of studies that have examined patterns of molecular genetic variation in rhizobial populations at local, regional, and global levels. The results of these studies have revealed several important insights, including the existence of extensive diversity within, as well as significant genetic differentiation between local and regional populations. The results also provide evidence for long-distance gene flow between continental populations. Several studies have also indicated the existence of low-to-intermediate levels of recombination within rhizobial populations. Fine-scale studies of specific genomic components (e.g. symbiotic plasmids) have also shown that certain genomic elements appear to be more prone to recombination than others. The results also showed that the different loci responsible for the development of the symbiosis appear to be under different forms of selection. Because most the population studies to date have focused on strains from root nodules, surprisingly little is known on the population genetics of the more numerous non-nodulating soil rhizobia. Future efforts to characterize these populations should significantly enhance our ability to manipulate rhizobial populations in agricultural ecosystems.

Introduction

Symbiotic nitrogen fixation is a mutualistic process in which bacteria reside inside plants and reduce atmospheric nitrogen to ammonia. This ammonia can then be used by the plant for the synthesis of proteins and other nitrogen-containing compounds such as nucleic acids. The Gram-negative soil bacteria that carry out this process are collectively referred to as rhizobia (from the Greek words *riza* = root and *bios* = life). The process of symbiotic nitrogen fixation is of agricultural and ecological significance because plants capable of nitrogen fixation do not need to compete for limited quantities of soil nitrogen, nor do they require expensive nitrogenous fertilizers that can be harmful to the environment.

Rhizobia are members of the phylum Proteobacteria. On the basis of 16S rRNA (rrs) sequence similarity, rhizobia have been placed in two classes, the alpha-proteobacteria and the beta-proteobacteria (Fig. 5.1). Both the alpha- and the beta- proteobacteria contain ecologically and functionally diverse members. For example, alpha-proteobacteria includes the plant pathogen agrobacteria and intracellular pathogens of humans and animals, the rickettsias (Woese, 1987). Members of certain

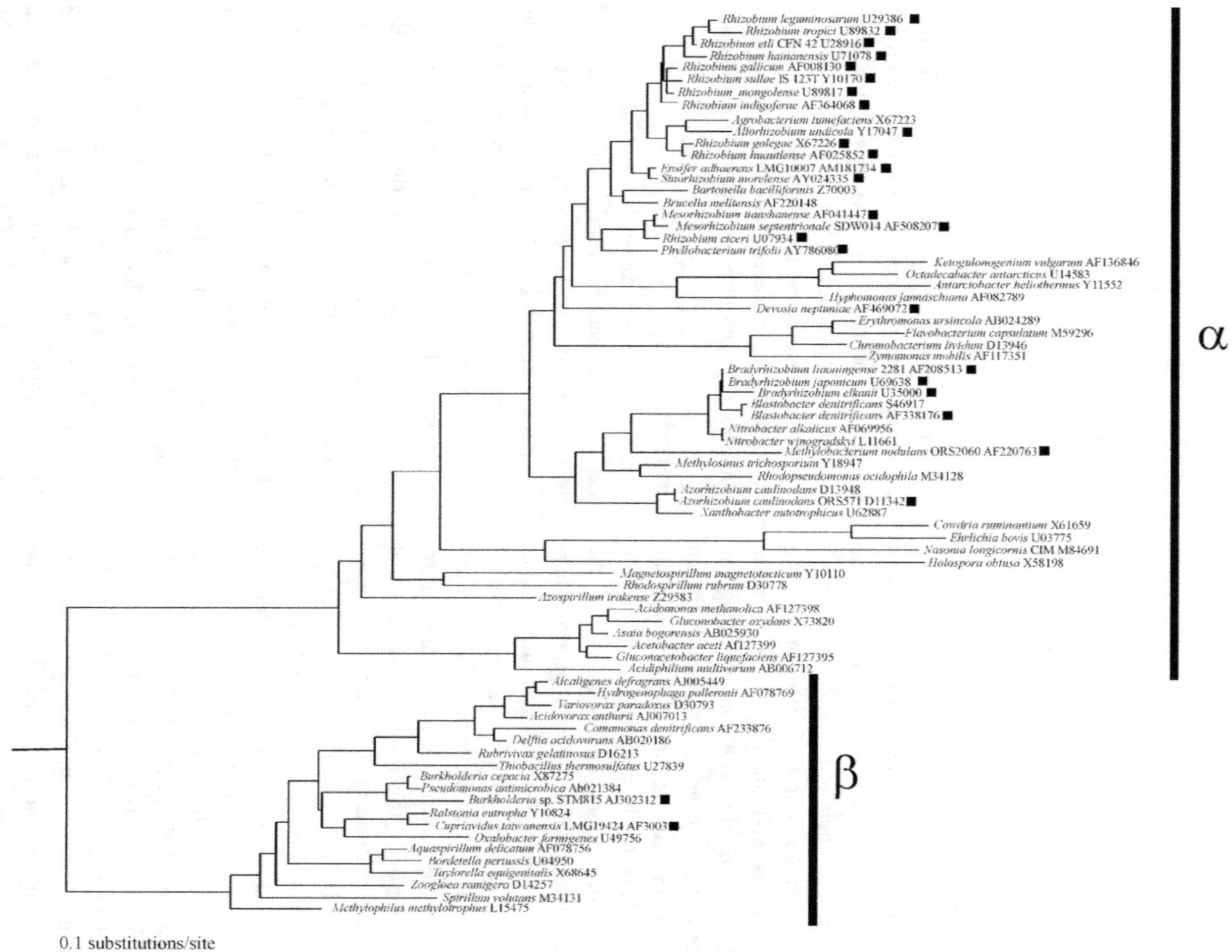

Figure 5.1 A maximum likelihood tree based on the rrs gene from 75 taxa from α- and β-subdivisions of the proteobacteria (Dressler-Nurmi *et al.*, 2007). Representatives of species capable of forming nodules are marked with a black box. (Reprinted with permission from the publishers.)

genera within the alpha-proteobacteria share the tendency to form intimate, if not intracellular associations with eukaryotic cells. Most named rhizobial species belong to alpha-proteobacteria and are distributed among genera *Allorhizobium*, *Azorhizobium*, *Bradyrhizobium*, *Mesorhizobium*, *Methylobacterium*, *Rhizobium* and *Sinorhizobium* (Andam and Parker, 2007). Symbiotic nitrogen-fixing bacteria of the class *Beta-proteobacteria* were discovered only recently, and are currently represented by the genera *Burkholderia* (Moulin *et al.*, 2001) and *Cupriavidus* (Chen *et al.*, 2001; Vandamme and Coenye, 2004; Barrett and Parker, 2006).

Rhizobial species differ in their degree of host plant specificity, which is determined largely by the complement of nodulation (or *nod*) genes expressed by a particular organism. The term biovar is used to circumscribe a subspecific cluster of strains that share a common symbiotic genotype and host range (Villegas *et al.*, 2006). Some rhizobial species are capable of forming symbiotic relationships with a very broad range of legumes (Perret *et al.*, 2000) while others establish effective relationships with only a small number of closely related species (e.g. Villegas *et al.*, 2006). Rhizobial species were originally defined on the basis of their host range and a variety of other phenotypic traits such as physiological, ecological and morphological characteristics. Reliance on host range as a primary defining factor was largely abandoned after it was determined in the 1980s that *Rhizobium* nodulation genes were carried on self-transmissible symbiotic plasmids (Johnston *et al.*, 1982; Jordan, 1984). Subsequently, rhizobial species were defined primarily on the basis of overall genomic similarity, as assessed by DNA/DNA hybridization, similar to the criteria used for defining other bacterial species (Wayne *et al.*, 1987). Recently multilocus sequence typing (or MLST) has gained popularity as a method for defining species limits among the rhizobia because MLST data can be stored and accessed easily, and because the data can be used to study the evolution of genes, gene clusters, and entire replicons to identify divergent lineages similar to reproductively isolated units or cryptic species in eukaryotes (Maiden *et al.*, 1998). Consequently many new rhizobial species are defined primarily on the basis of their multilocus chromosomal genotype.

Most rhizobial population genetics studies have focused on the agriculturally important species of the genera *Bradyrhizobium*, *Mesorhizobium*, *Rhizobium*, and *Sinorhizobium*. It has been estimated that members of these genera diverged among each other from their ancient alpha-proteobacterial ancestors somewhere between 200 and 500 million years ago (MYA) (Turner and Young, 2001; Moulin *et al.*, 2001). This divergence is thought to have occurred well before the major diversification of their legume hosts, which has been dated to approximately 59 MYA (Lavin *et al.*, 2005). It is generally assumed that the ancestors of extant rhizobia probably acquired the ability to nodulate legumes much more recently (Moulin *et al.*, 2001), and this capacity has spread horizontally among different genera (Hirsch *et al.*, 2001). This hypothesis is supported by the results of recent studies, including one in which signatures of positive selection have been detected for the nodulation receptor kinase (NORK) gene in *Medicago truncatula*, a legume host of *Sinorhizobium meliloti* (DeMita *et al.*, 2006). Additional supportive evidence was published in a recent study of *nodA* gene diversity, where the *nodA* phylogeny for a global collection of lupine isolates was found to be more strongly correlated with host phylogeography than it was with the core gene phylogeny of the isolates (Stepkowski *et al.*, 2007). Similar results were obtained previously by Kaijalainen and Lindstrom (1989).

Genomic compartmentalization in rhizobia

Regardless of the species or genus of rhizobia, it appears that most of their 'essential' (or core) genes are chromosomally encoded. Although the symbiotic genes in certain species are chromosomally encoded as well, in other species they are carried on symbiotic plasmids. For example, most of the symbiosis-specific genes in the sister genera *Rhizobium* (Young *et al.*, 2006; Gonzalez *et al.*, 2006) and *Sinorhizobium* (Galibert *et al.*, 2001) are plasmid-encoded, while in the more distantly related genera *Bradyrhizobium* (Kaneko *et al.*, 2002) and *Mesorhizobium* (Kaneko *et al.*, 2000) they are encoded on the cellular chromosome. Regardless of their genomic location, the complement of nodulation genes expressed in a particular bacterium determines its symbiotic host range.

Closely related rhizobial species tend to share certain genomic architectural features. For example, the published genomes of two strains from the sister species *Rhizobium etli* (Gonzalez *et al.*, 2006) and *Rhizobium leguminosarum* (Young *et al.*, 2006) contain circular chromosomes of 4.38 and 5.06 Mb respectively. The chromosomes contain core genes that are essential for cell growth. In addition, both sequenced strains also contain six smaller circular plasmids each and these plasmids range in size between 184 kb and 642 kb for *R. etli* and between 147 kb and 488 kb for *R. leguminosarum.* Unlike many genes on chromosomes, those on plasmids typically do not encode proteins for essential functions but are for accessory functions. However, there are overlaps between genes on chromosomes and plasmids. For example, some of the functional gene classes on the chromosomes were also found on plasmids as well in the sequenced strains. In *Rhizobium* species the plasmid genes typically do not encode entire metabolic pathways. Instead they have a pool of genes that confer adaptations to specific substrates and ecological niches. Perhaps not surprisingly these genes may be present in only some, but not all members of a soil population, even among strains within the same species. Although these accessory genes tend to have long-term associations with certain species or groups of species, they nevertheless appear to be adapted to a nomadic way of life. The genes coding for symbiotic nitrogen fixation in rhizobia possess all of the properties of accessory genes that are described above (Young *et al.*, 2006).

In contrast to the genomic architecture of the *Rhizobium* species, species of *Sinorhizobium* maintain a tripartite genomic architecture that consists of a chromosome and two megaplasmids (pSymA and pSymB). Examples include *S. meliloti* (Galibert *et al.*, 2001) and *S. medicae* (Roumiantseva *et al.*, 1999). These replicons each display certain functional features, with core (or housekeeping) genes located on the chromosome (with the exception of an essential tRNA gene in *S. meliloti* which is located on the megaplasmid pSymB), and with genes involved in secondary metabolic pathways and symbiotic interactions located primarily on megaplasmids. Although the symbiotic megaplasmids in *Sinorhizobium* species do not appear to be self-transmissible, they clearly are related to other self-transmissible plasmids – as indicated by the fact that they harbour molecular determinants involved in the conjugation process. Some examples include several *tra* loci, type IV pilus loci (Galibert *et al.*, 2001), and conjugative transfer origins (*oriT* and *mob*) (Herrara-Cervera *et al.*, 1998).

Some nitrogen-fixing bacteria can readily assimilate laterally transferred DNA into their genomes, generating recombined chromosomes and/or (mega)plasmids (Zhaxybayeva, *et al.*, 2004). Mosaicism of this type is a prominent structural feature in the symbiotic replicons of *Rhizobium* and *Sinorhizobium* species. For example, a comparison of the symbiotic plasmids in two different strains of *R. leguminosarum* biovar *viceae* revealed very similar symbiotic regions in the otherwise different plasmid backgrounds (Young *et al.*, 2006). This was interpreted as evidence of a recent transfer of the symbiosis genes between the two plasmid types. In a study of the symbiotic plasmid in another species (*R. etli*), Gonzalez *et al.* (2003) reported that the symbiotic region in the plasmid possessed features usually associated with genomic rearrangements, such as mobile genetic elements flanking blocks of functionally related sequences. In a study of single-nucleotide polymorphisms (SNPs) in the symbiotic regions of *R. etli* for strains from different geographic origins, Flores *et al.* (2005) observed that regions with unique base substitution patterns appear to have been spread by horizontal transfer and recombination. Guo *et al.* (2007) also used a SNP approach to compare patterns of sequence diversity within the pSymA replicons of *S. meliloti* and *S. medicae.* Their results showed that recombination within the pSymA replicon is common and that it has resulted in a mosaic structure similar to that observed for symbiotic regions in *R. etli* (Flores *et al.*, 2005). In a sequence-based nearest-neighbour analysis of the pSymB replicon in the *S. meliloti* genome strain 1021, Wong and Golding (2003) reported that this replicon appears to have a complex evolutionary history, with closest sequence matches to many diverse groups of bacteria. Guo *et al.* (2005) also observed this type of complexity in many other strains of *S. meliloti.*

In contrast to the plasmid-encoded symbiotic regions of *Rhizobium* and *Sinorhizobium*, the symbiotic genes in *Bradyrhizobium japonicum* (Kaneko *et al.*, 2002) and *Mesorhizobium loti* (Sullivan and Ronson, 1998) are organized on chromosomal segments called symbiotic islands. Although one might assume that the chromosomal location of these modules could limit their lateral mobility, field studies in New Zealand (Sullivan and Ronson, 1995) and Australia (Nandasena *et al.*, 2006) have shown that the symbiotic islands of *M. loti* seemed to be readily transferable under both laboratory and field conditions. The symbiosis islands of the *Mesorhizobia* are large, site-specific conjugative transposons (Sullivan *et al.*, 2002) that are similar to the pathogenicity islands of many Gram-negative bacterial pathogens (Hacker and Kaper, 1999). Pathogenicity islands and symbiotic islands have several common characteristics, including that (i) they are integrated into tRNA genes, (ii) they contain phage P4-like integrase genes, (iii) they have the potential to convert free-living environmental strains into strains capable of forming intimate associations with eukaryotic hosts, and (iv) they can be found in a broad range of bacterial species (Sullivan *et al.*, 2002). Taken together, these results indicate that the symbiotic plasmids and the symbiotic islands of rhizobia appear to have been shaped by a variety of processes, including horizontal gene transfer, transposition, recombination, and adaptation.

Initial population genetics studies of rhizobia

In the initial genetic studies of natural populations of rhizobia, considerable emphasis was placed on determining the range and extent of lateral gene transfer and recombination in natural populations. Similarly, researchers also examined the influence of other environmental and ecological variables on the genetic structure of these natural populations as well. Young (1992) reviewed the results of several of these studies that focused on soil populations of *R. leguminosarum* in England (Harrison *et al.*, 1987; Harrison *et al.*, 1988; Harrison *et al.*, 1989; Young, 1985; Young *et al.*, 1987; and Young and Wexler, 1988) and concluded: (1) that *R. leguminosarum* populations are polymorphic, (2) that *R. leguminosarum* populations are highly clonal, (3) that each plant in a cultivated crop forms nodules with a wide range of rhizobia genotypes, (4) that the three biovars of *R. leguminosarum* all share a common pool of chromosomal genotypes, (5) that the biovar *phaseoli* plasmids are found in a much less diverse range of chromosomal genotypes than the plasmids of the other two biovars, (6) that *Rhizobium (Sinorhizobium) meliloti* behaves as a separate species from *R. leguminosarum*, (7) that the levels of genetic diversity in *R. leguminosarum* populations are similar at different sites in Britain, (8) that in very acidic soils there is a lower diversity of *Rhizobium* strains than soils with higher pH values, (9) that samples of isolates obtained from the field are similar to samples obtained from soil in the greenhouse, (10) that different plant host cultivars may select a common range of electrophoretic types (ETs) of common housekeeping enzymes as determined by multilocus enzyme electrophoresis (MLEE), (11) that sampled genotypes do not seem to be influenced by gross differences in competitiveness of strains for nodulation in the laboratory, (12) that not all rhizobia strains in the soil are capable of forming nodules or have genes for nodulation, (13) that symbiotic plasmids are very polymorphic, (14) that there is little evidence for recombination between distantly related symbiotic plasmids, (15) that the relationship between plasmid variation and chromosomal variation is not completely random, (16) that there is some plasmid transfer between chromosomal classes, and (17) that the same plasmid/chromosome associations can be found at geographically distant field sites.

These population surveys and pioneering papers were followed by studies that focused more on testing specific hypotheses, including the genetic characterization of indigenous soil populations of rhizobia at the centres of domestication for certain legume species. For example, prior to its domestication, the common bean (*Phaseolus vulgaris*) and its wild relatives had already diverged into major gene pools in both Mesoamerica and in the Andes (Chacón *et al.*, 2005; Gepts, 1998). Pinero *et al.* (1988) used MLEE to examine the genetic diversity in a collection of 51 *R. leguminosarum* biovar *phaseoli* strains obtained from bean plants grown in a variety of geographic and ecological locations, largely in Mexico. Nearly every isolate they examined represented a distinct electrophoretic type (ET) and the mean genetic diversity (i.e. expected heterozygosity, He) per enzyme locus was 0.691 – one of the highest levels recorded at the time. The occurrence of strong non-random allelic associations among loci suggested a predominantly clonal population structure, reflecting infrequent recombination among the multilocus genotypes. However the very strong divergence among the ETs in the collection also suggested that the linkage disequilibrium (LD) might be due, at least in part, to the fact that the collection was a polyphyletic assemblage of strains. This explanation was supported by subsequent studies which identified three new species in the collection, including *R. tropici* (Martinez-Romero *et al.*, 1991), *R. etli* (Eardly *et al.*, 1992; Segovia *et al.*, 1993) and *R. gallicum* (Amarger *et al.*, 1997). It is known that the inclusion of phylogenetically divergent lineages in the analyses of population structure can result in non-random allelic association and high estimates of linkage disequilibrium.

Recombination and genetic structure in rhizobial populations

Obtaining evidence for recombination

It has been suggested that lateral gene transfers and host selective pressures are the key mechanisms involved in determining the genetic structure of natural populations of symbiotic microbes (Tibayrenc, 1996). A common statistical test that can provide an indication of the extent of clonality (or lack of recombination) in a multilocus dataset containing a limited number of alleles is the index of association statistic (I_A), a measure of linkage disequilibrium (LD) across multiple loci (Brown *et al.*, 1980). An I_A value that is significantly greater than zero indicates that the population has a significant clonal population structure. In contrast, an I_A value that does not differ significantly from zero suggests that a population is panmictic, with random association of alleles among loci in the genome. However, the I_A statistic has certain limitations. For example, the value of I_A can be influenced by the number of loci in the analysis. Therefore, it might be difficult to compare results from studies using different numbers of loci. One way to address this limitation is to standardize the I_A statistic based on the number of loci in an analysis (Sun *et al.*, 2006). Other complementary methods that can also be used to describe the degree of clonality and recombination in a population include the phylogenetic incompatibility test (Agapow and Burt, 2001), the partition homogeneity test (Farris *et al.*, 1994), D statistics (Lewontin, 1964; Hedrick, 1987), and split decomposition analyses (Bandelt and Dress, 1992; Hudson and Bryant, 2006).

After assessing the presence or absence of LD in a population, it is usually of interest to determine the genetic and ecological factors responsible for the observed genetic structure. For example, certain biological barriers, such as sequence divergence (Rayssiguier *et al.*, 1989; Roberts and Cohan, 1993) may result in strong LD in microbial populations such as rhizobia. This could help explain the LD values commonly observed when multiple species are inadvertently included in a population sample (see the *R. leguminosarum* example above). Geographic structuring may also be a contributing factor to the presence of LD, and this may be influenced by the sampling strategy used. Strong LD in a population may also result from a significant increase in the frequency of a particular genotype (or genotypes) as a result of selection (Maynard Smith *et al.*, 1993). This has been described as an 'epidemic' population structure where one or a few highly adaptive genotypes may be spread across many ecological niches and geographic regions in a relatively short time. One way to identify an epidemic population structure is to compare the results of I_A analyses where the genotypes of every individual in the population sample are included, with a corresponding analysis where only a single representative of each multilocus genotype is included. The appearance of strong LD in the former, but not in the latter, may be taken as evidence that the population has an epidemic population structure. Sample size is also an important factor, because a small sample size may not be representative enough for the whole population and may only reveal the predominant lineages (e.g. epidemic clones) and thus may lead to the erroneous conclusion that the entire population is highly clonal, when in fact the background non-epidemic lineages may be in equilibrium (Maynard Smith *et al.*, 1993; Silva *et al.*, 1999). While most population studies have focused on assessing the relationships among alleles at different loci, recombination can also occur within a locus, creating mosaic genetic elements. The methods described above are for estimating intergenic (or assortative) recombination. There are two comparable approaches for estimating intragenic recombination: Hudson's R parameter method (Hudson, 1987) and the C parameter method described by Hey and Wakeley (1997).

Recombination in *Rhizobium* species

The population genetic structure in *Rhizobium* populations can be influenced by a combination of ecological, geographic, and genetic factors. For example, in a study of *R. etli* field isolates associated with both wild and cultivated beans (*Phaseolus spp.*) in Mexico, Souza *et al.* (1994) observed a more limited range of chromosomal genotypes among isolates from wild *Phaseolus* species than from the common cultivated bean *Phaseolus vulgaris.* In this same study they also observed only modest LD in local populations, but stronger LD among multilocus ETs from geographically distant locations. This suggested that even though recombination may be common on a local scale, geographical barriers can limit gene flow and contribute to LD among samples from broad geographic areas. When Silva *et al.* (1999) conducted a subsequent study of isolates from beans at another site in Mexico, they discovered that two genetically distinct bean-nodulating *Rhizobium* species (*R. etli* and *R. gallicum*) were occupying the same soil (Silva *et al.*, 2003). Their analysis of the genotypic data revealed two contributing sources of LD in the sample. In the first, they identified a significant divergence and lack of recombination between the two species. Secondly, each of the two species contained signatures of the 'epidemic population structure'. Specifically, there were a few dominant (or epidemic) clones within each species, coexisting with multiple other genotypes. Analysing one representative strain from each genotype identified that alleles at different loci were largely in linkage equilibrium. Similar evidence of epidemic population structures had been obtained previously for *R. leguminosarum* populations in Oregon (Leung *et al.*, 1994; Strain *et al.*, 1994) and Australia (Demezas *et al.*, 1995). The results of these studies suggest that even though recombination may exist in *Rhizobium* soil populations, the frequency of lateral gene transfer and recombination among certain genotypes may be insufficient to prevent certain genotypes from reaching a high frequency. The predominant clones could be a result of their superior fitness and adaptation to their native hosts and/or the commercially grown agricultural crops.

Lateral gene transfer and recombination of Rhizobium *plasmids*

The evidence for the mobility of symbiotic elements in *Rhizobium* field populations was first reported by Schofield *et al* (1987) and was subsequently confirmed by several studies. In these studies it was shown that a single chromosomal genotype could be associated with a variety of symbiotic plasmid genotypes (Haukka *et al.*, 1998; Laguerre *et al.*, 2001; Mutch and Young, 2004; Parker *et al*, 2002; Quian and Parker, 2002, 2003; Schofield *et al.*, 1987; Young and Wexler, 1988). In addition, field studies of mixed populations of *R. etli* and *R. gallicum* have provided evidence that barriers to interspecific plasmid transfer in *Rhizobium* soil populations appear to be much more substantial than barriers to intraspecific symbiotic plasmid transfer (Silva *et al.*, 2003). In a study of a global collection of rhizobial isolates from 13 legume genera, Wernegreen and Riley (1999) obtained evidence that symbiotic *nod* gene transfer could occur between species. However, they found no evidence for lateral transfer of these elements across genera. It appears that rhizobial symbiotic plasmid transfer depends on a variety of genetic and ecological factors. For example, Wernegreen and Riley (1997) found only limited evidence of symbiotic plasmid transfer within populations of *R. leguminosarum* biovar. *trifolii* in non-cultivated California soils. Similar results were obtained in *R. leguminosarum* biovar. *trifolii* isolates from a pasture soil in Australia (Demezas, *et al.*, 1995). In a commentary on the study by Wernegreen and Riley (1997), Souza *et al.* (1997) suggested that the horizontal transfer of plasmid-encoded genes may be more common in intensively managed agricultural soils than in non-cultivated soils.

More recently, in a study of the diversification of DNA sequences in the symbiotic genome of *R. etli*, Flores *et al.* (2005) concluded that the majority of nucleotide substitutions that they observed were spread by recombination, and that the contribution of new mutations was relatively low.

Recombination in *Bradyrhizobium* and *Mesorhizobium* species

Evidence for the lateral transfer and recombination of symbiosis genes has also been reported for species with chromosomally encoded symbiosis genes as well. For example, the lateral transfer of a symbiosis island between two *Mesorhizobium* species was documented under both laboratory (Sullivan and Ronson, 1998) and field conditions (Sullivan *et al.*, 1995; Donate-Correa *et al.*, 2007). Several studies have also provided indirect evidence of this phenomenon by demonstrating incongruence between phylogenies based on symbiotic nitrogen fixation genes and phylogenies of a set of 'core essential genes' in isolates from *Bradyrhizobium* field populations (Parker *et al.*, 2002; Qian *et al.*, 2003; Jarabo-Lorenzo *et al.*, 2003; Moulin *et al.*, 2004; Parker and Kennedy, 2006; Vinuesa *et al.*, 2005b). Recent evidence suggests that *Bradyrhizobium* symbiotic genes may also be laterally transferred to different genera (Barcellos *et al.*, 2007) and also between different bradyrhizobial species (Vinuesa *et al.*, 1995a), in some cases to species not previously thought to be capable of symbiotic nitrogen fixation (Andam and Parker, 2007).

Core (or essential) genes in *Bradyrhizobia* also appear to be subject to lateral transfer and recombination in soil populations as well. In a MLEE study examining *Bradyrhizobium* isolates from four different host species grown in pasture soils, Bottomley *et al.* (1994) found no strong evidence for LD among the multilocus enzyme electrophoretic types that they observed (Table 5.1), despite the fact that the overall genetic diversity among the ETs was quite high (H = 0.69). In a MLEE study of sympatrically distributed *B. canariense* and *B. japonicum* populations in the Canary Islands, Vinuesa *et al.* (2005a) reported significant evidence for recombination within each of the species, but also evidence of sexual isolation between these two species.

In a more recent study, Vinuesa *et al.* (2008) developed recombination estimates based on segregating sites in four concatenated core genes in each of four species of *Bradyrhizobium*. They observed evidence of intermediate levels of recombination in each of the species, with the exception of a single clonal lineage of *B. japonicum* (*B. japonicum* Ia).

These studies provided quantitative evidence for intermediate-to-high levels of intragenic recombination within each of the species. However, they also reported and cautioned that the observed sequence polymorphisms in these isolates could be due to either mutation or recombination. Maynard Smith *et al.* (1993) estimated that to eliminate LD entirely, the frequency of localized recombination must be greater than four times the frequency of point mutation.

Recombination in *Sinorhizobium* species

Since 1990, several studies have examined the range and extent of recombination in natural populations of *Sinorhizobium*. The first such study was conducted by Eardly *et al.* (1990). They used multilocus enzyme electrophoresis (MLEE) to characterize genetic variation at 14 loci (13 of which were chromosomally encoded) in 232 strains isolated from numerous locations worldwide. This sampling approach, which is sometimes called a global population snapshot, revealed the existence of 50 distinct ETs that were clustered (by UPGMA) into two major clades and one divergent singleton. The smaller group was subsequently named *S. medicae* (Rome *et al.*, 1996) and the larger, represented by the type strain ATCC 9930, retained the *S. meliloti* epithet. An analysis of two-locus allelic associations among the loci revealed strong evidence for linkage disequilibrium between the ETs of the two species, and weaker evidence for allelic association among the isolates, but not among the ETs when a single representative strain is selected from each ET for analyses (Maynard Smith *et al.*, 1993). Bailly *et al.* (2006) concluded that these results are consistent with an epidemic population structure.

In a follow-up study examining the same strains, ten chromosomally encoded genes (including three from the MLEE analysis of Eardly

Table 5.1 Multilocus linkage disequilibrium in 15 rhizobial populations as estimated by the index of association (I_A) method

Genus and species	Strains (no.)	Sampling scale[1]	Loci (no.)	Ets (no.)	I_A[2]	Reference
Bradyrhizobium spp.	95	L	13	17	NS	Bottomley (1994)
Rhizobium etli	71	R	6	34	0.32[a]	Souza (1994)
Rhizobium etli	276	R	9	100	0.23[a]	Souza (1992)
Rhizobium etli	108	L	6	31	0.06[NS]	Silva (2003)
R. gallicum	18	L	6	12	–0.04[NS]	Silva (2003)
R. leguminosarum	200	L	13	53	0.73[c]	Leung (1994)
R. leguminosarum	176	L	13	78	0.63[c]	Strain (1994)
R. leguminosarum	70	L	13	28	0.74[c]	Strain (1994)
R. leguminosarum	112	L	13	42	1.03[c]	Strain (1994)
R. leguminosarum	200	L	13	54	0.73[c]	Strain (1994)
Sinorhizobium medicae	8	G	14	8	–0.16[NS]	Gordon (1995)
Sinorhizobium medicae	23	G	14	15	0.01[NS]	Maynard Smith (1993)
Sinorhizobium meliloti	22	G	14	22	–0.17[NS]	Gordon (1995)
Sinorhizobium meliloti	208	G	14	34	–0.24[NS]	Maynard Smith (1993)
Sinorhizobium meliloti	171	R	9	30	0.16[NS]	Silva (2007)
Sinorhizobium meliloti	147	L	9	30	0.13[NS]	Silva (2007

1 Sampling scale coded as follows: G, global; R, regional; L, local.

2 Index of Association (I_A) level of statistical significance: [a]$P<0.001$; [b]$P<0.01$; [c]$P<0.05$; NS, non-significant.

et al., 1990) were characterized using MLST (Maiden *et al.*, 1998; van Berkum *et al.*, 2006). In their analysis they identified 91 distinct multilocus allelic profiles among the strains. These are called sequence types or STs. In spite of the larger number of multilocus genotypes that were identified in this study, both this study and the previous MLEE study of Eardly *et al.* (1990) consistently identified the same two clonal complexes corresponding to *S. meliloti* and *S. medicae*, and both studies provided evidence of strong LD between the chromosomes of the two species, consistent with reproductive isolation and speciation.

Guo *et al.* (2005) also employed a multilocus approach in an analysis of 59 of the *S. meliloti* strains examined by Eardly *et al.* (1990). Rather than focusing on core chromosomal genes, Guo *et al.* (2005) examined the distribution of novel genomic sequences that were identified through representational difference analysis (RDA) (Lisitsyn *et al.*, 1993). The RDA method was used to identify 85 novel, presumably non-essential sequences that were present in the *S. meliloti* type strain (ATCC 9930), but that were absent in the model laboratory strain Rm1021, for which the entire genome sequence is available (Galibert *et al.*, 2001). Using PCR, they assayed the distribution of 12 of these novel sequences across 59 of the *S. meliloti* biovar. *meliloti* strains from the global collection. In most of the strains several of the novel marker sequences were absent and their distributions were not clustered on the phylogeny based on MLEE data. Their result suggested that gene loss and gain might be a dominant force in the evolution of the *S. meliloti* genome. An analysis of allelic associations among the novel gene markers indicated a predominantly clonal population structure. However because some of the individual markers produced incompatible single-gene phylogenies, it was concluded that the populations also have a history of low-level recombination.

In a recent study, Sun *et al.* (2006) analysed the nucleotide sequence polymorphism in 16 genes from 49 of the *S. meliloti* biovar. *meliloti* strains analysed by Guo *et al.* (2005) and originally reported by Eardly *et al.* (1990). Based on the published map for the *S. meliloti* genome (Galibert *et al.*, 2001) they designed primers to amplify gene sequences for six loci encoded by the chromosome, six on the pSymB replicon, and four on the pSymA replicon. Three different tests were used to assess the extent of genetic recombination among polymorphic nucleotide positions in the strains. The first was a standardized index of association (I_A) method, the second was a test of phylogenetic incompatibility among the loci, and the third was a partition homogeneity test, which was used to evaluate the congruency between genealogies for different loci and for different replicons. Their results indicated that the different replicons in *S. meliloti* have differing degrees of recombination. For example, they observed evidence of significant linkage disequilibrium and localized recombination among the gene pairs on the chromosome and the pSymA megaplasmid. In contrast, they observed that most of the gene pairs on the pSymB replicon were in overall linkage equilibrium. Their results, which indicate a basically clonal structure with localized recombination for the chromosomes and the pSymA replicons, is consistent with the 'clonal frames' model originally proposed by Milkman and Bridges (1990). Genealogical comparisons among trees for each of the three *Sinorhizobium* replicons revealed significant incongruence, indicating a history of assortative recombination among these elements in natural populations.

In a recent MLST study which examined possible recombination between *S. meliloti* and *S. medicae* in a small soil sample, Bailly *et al.* (2006) obtained 116 isolates from several *Medicago* species grown in a single litre of soil from one location in France. Nucleotide sequences were obtained for seven intergenic spacer regions from all of the isolates. According to the genome map of *S. meliloti* Rm1021 (Galibert *et al.*, 2001), they inferred that three of the spacers were located on the chromosome, and two each were on pSymA and pSymB. Their results indicated that despite the sympatric distribution of the two species, the chromosomes of *S. medicae* and *S. meliloti* had remained significantly diverged and sexually isolated. In addition, there was strong linkage disequilibrium between alleles from the chromosomal gene loci in both species. However, the results also revealed evidence of recombination between loci on megaplasmids in both species.

In two separate studies of localized *Sinorhizobium* populations in Tunisia, Bailly *et al.* (2007) and van Berkum *et al.* (2007) both used a MLST approach to address the question of how recombination has influenced the ability of certain *Sinorhizobium* biovars to interact with different *Medicago* host species. In both studies, isolates were obtained from both the highly selective trap host *M. laciniata*, which is nodulated by only *S. meliloti* biovar *medicaginis*, and also from the relatively non-specific trap host *M. truncatula* that is nodulated by *S. meliloti* biovar. *meliloti*. The results of both studies suggest that during the evolution of *S. meliloti*, *nod* genes specific for certain *Medicago* species have spread to different genetic backgrounds via horizontal gene transfer and recombination. This hypothesis was supported by the observation that the *nod* gene regions in the pSymA plasmid of *Sinorhizobium* appear to be 'hotspots' of recombination (Bailly *et al.*, 2007; Giuntini *et al.*, 2005). Collectively these observations support the hypothesis that the symbiotic megaplasmids and chromosomes of *S. meliloti* have evolved more or less independently, and that recombination has played a particularly important role in the evolution of the megaplasmid-encoded genes in this species. Since selection may also be an important source of LD and can cause an epidemic population structure, studies examining the effects of selection on the evolution of the rhizobial populations will be considered next.

Selection in rhizobial populations

The results of population genetics studies of *Sinorhizobium* and other species suggest that the core genes in rhizobia may be subject to selection pressures different from the accessory genes in the same strains. For example, in an MLST study of sympatric populations of *S. meliloti* and *S. medicae* in a single litre of soil, Bailly *et al.* (2006) compared patterns of sequence polymorphism among intergenic spacer (IGS) regions on the chromosome and symbiotic plasmids among 116 isolates obtained from several annual *Medicago* species. The mutation rates they observed for five (of the total of seven) loci examined did not depart significantly from the neutral model. Similar results were obtained in a study of three core loci in *Bradyrhizobium* spp. (Vinuesa *et al.*, 2005a). Collectively these results indicate that the core loci examined do not appear to be under strong selective pressures. In contrast, Bailly *et al.* (2006) observed significant evidence for balancing selection acting on an intergenic spacer (IGS) region associated with the pSymB *exo* locus in the strains mentioned above. In support of this observation, they cited previous evidence indicating that balancing selection is a common feature of bacterial genes like *exo*, that encode products that may be recognized by eukaryotic hosts (Hughes, 2005). In addition to the evidence of balancing selection acting on the *exo* locus, Bailly *et al.* (2006) also reported evidence for purifying selection acting on regions associated with *nod* loci on pSymA.

In a follow-up study, Bailly *et al.* (2007) used a similar MLST approach to examine 126 Tunisian isolates from two *Medicago* species, *M. truncatula* and *M. laciniata*, grown in the same soil. *Medicago laciniata* is unusual among *Medicago* species in that it is effectively nodulated by only *S. meliloti* biovar *medicaginis*, a biovar that is characterized by its narrow host range and its unique complement of *nod* genes (Villegas *et al.*, 2006; Bailly *et al.*, 2007). In contrast, *M. truncatula* is more typical of other *Medicago* species, in that it has a somewhat broader range of symbionts and it can be nodulated by either *S. meliloti* biovar *meliloti* or *S. medicae*. In their analysis of chromosomal and *nod* gene polymorphisms in their isolates, Bailly *et al.* (2007) observed that similar *nod* alleles could be found in isolates having substantially different chromosomal backgrounds (i.e. those of *S. meliloti* and *S. medicae*), and that *S. meliloti* is capable of carrying divergent *nod* alleles (i.e. those of either of the two biovars of *S. meliloti*). They also suggested that since multiple *Medicago* host species could be effectively nodulated by both *S. meliloti* biovar *meliloti* and *S. medicae*, their shared microsymbiont pool might have prevented the divergence between the *nod* genes of these two species. They tested this hypothesis by comparing *Ka/Ks* ratios for four *nod* gene (accessory) loci and seven chromosomal (core) gene loci in representative strains of both species. Because their results did not reveal a significant difference in the *Ka/Ks* ratios for the core genes versus the *nod* genes, they concluded that purifying selection driven by the *Medicago* hosts was probably not a major factor responsible for the low level of sequence divergence that they observed between the *nod* genes of the two species. Instead, they suggested that interspecific recombination between the *nod* genes of the two species was a more likely explanation. Evidence supporting a history of lateral gene transfer and recombination involving *nodC* genes in Tunisian *Sinorhizobium* populations was recently reported by van Berkum *et al.* (2007).

In their recent study, Bailly *et al.* (2007) also examined the possibility that disruptive selection for specific *nod* genes, particularly those associated with the ability to effectively nodulate either *M. laciniata* (Barran *et al.*, 2002) or *M. truncatula*, could explain the substantial sequence divergence they observed between the *nod* gene clusters of their respective microsymbionts, *S. meliloti* biovar *medicaginis* and *S. meliloti* biovar *meliloti*. Their results, based on *nod* cluster IGS divergence ratios for the two biovars, indicated that host-driven disruptive selection apparently has influenced polymorphism patterns in the *nod* genes in the populations they examined, and that this selection, coupled with the effects of hitchhiking, could explain the *nod* region sequence divergence they observed between the *S. meliloti* biovars.

Biogeography and the potential for gene flow in rhizobial populations

Virtually all of the current information available on gene flow in rhizobia originates from studies examining the biogeography of only a few rhizobial species. If a species can be found at many locations around the globe, then this species is said to have a cosmopolitan distribution. In contrast, if the natural distribution of a species is limited to a particular geographic area or region, then it is said to be endemic to that region (Staley and Gosink, 1999). The hypothesis that most bacterial species have a cosmopolitan distribution is attributed to the Dutch microbiologist Baas-Becking (1934), who suggested that since bacteria can be dispersed so easily by mechanisms such as wind, water, and animal vectors, that bacterial species are likely to be very widely distributed in nature.

The results of several biogeographical studies of rhizobial populations and their associated legume symbionts do not appear to support the hypothesis that rhizobial species are globally distributed, although some clearly have attained a wider distribution than others (e.g. Silva *et al.*, 2005). The results of studies examining rhizobial populations at local (Souza *et al.*, 1992), regional (Lie, 1978; Silva *et al.*, 2007; Villegas *et al.*, 2006), and global scales (Barrett and Parker, 2006; Parker and Kennedy, 2006; Eardly *et al.*, 1990; Sun *et al.*, 2006) indicate that (1) that relatively large geographic distances are required to function as effective barriers to gene flow, (2) genetic differentiation among populations generally increases with increasing geographic distance, presumably as a result of physical isolation (i.e. the commonly called 'genetic isolation by geographic distance' or simply 'isolation by distance'), and (3) the diversities of both rhizobial species and their legume hosts are correlated on a regional basis. For example, in a global study of *Sinorhizobium* strains it was reported that the most diverse populations *S. meliloti* and *S. medicae* (formerly called *S. meliloti* Division B) mapped to the regional centres of diversity for their respective host species (Eardly *et al.*, 1990). Similar results were also obtained for *Rhizobium* species that nodulate the common bean (*Phaseolus vulgaris*) (Pinero, *et al.*, 1988; Laguerre *et al.*, 1993; Eardly *et al.*, 1995). Indeed it has been hypothesized that the limited geographic distribution of a rhizobial biovar may actually limit the geographic distribution of its legume hosts (e.g. Villegas *et al.*, 2006). The correlation between the geographic distributions of rhizobia and their legume hosts has also been substantiated by the common observation that symbiotically effective rhizobia are often absent in regions where their host species are not indigenous (e.g. Lie, 1978; Sullivan *et al.*, 1995).

With one notable exception, there have been few studies that have critically examined possible mechanisms by which rhizobia can be dispersed over large geographic distances. Pérez-Ramírez *et al.* (1998) reported that five-of-the-thirteen bean cultivars that they tested harboured viable rhizobia on their seed coats. Because seeds from unopened pods did not carry any viable rhizobial cells, they concluded that the source of the rhizobia on the seeds was probably soil that had contaminated the seeds during the harvesting process.

The use of multilocus fingerprinting methods in large-scale population studies has made it possible to identify other likely cases of anthropogenic transfer of rhizobia across large geographic distances. For example, Eardly *et al.* (1990) identified a complex of *S. meliloti* isolates (subsequently named *S. medicae*; Rome *et al.*, 1996) that consist of multilocus genotypes that are very similar to the genotype of a laboratory-inoculant strain (CC169) that was originally isolated in Australia. Similar *S. medicae* genotypes were also isolated at three locations in Mexico (Silva *et al.*, 2007). Because diverse indigenous populations of *S. medicae* and their annual *Medicago* hosts only occur naturally in Southwest Asia, it is likely that ancestors of both the Australian and Mexican isolates were transferred to these countries from southwest Asia either directly or indirectly as either soil and/or seed contaminants, including perhaps livestock feeds. Similarly, Stępkowski *et al.* (2005) reported that European *Bradyrhizobium* genotypes could be isolated from the soils of both Australia and South Africa, further supporting the involvement of anthropogenic transfers of rhizobia across geographic regions.

Recent studies suggest that the dispersal of rhizobia can occur by other mechanisms as well. In a study of continental and insular populations of *Bradyrhizobium canariense*, Vinuesa *et al.* (2005a) provided evidence for significant gene flow between continental and insular populations of this species. Their conclusion was based on the observations of a high number of migrants and low levels of genetic differentiations (i.e. low F_{ST} or fixation index values) among the geographic populations. They hypothesized that the most likely vector for the transfer of the cells were dust storms originating in north-west Africa that regularly reached the Canarian Archipelago.

Significance and future directions

In many parts of the world where rhizobial inoculants are not freely available, farmers must rely on indigenous populations of rhizobia to help meet the nitrogen needs of their legume crops. Sometimes these native bacteria are highly efficient nitrogen-fixers, however sometimes they are not (e.g. Eardly *et al.*, 1985). Agriculturalists have only met with limited success in introducing highly efficient rhizobial strains into natural populations. Results have shown that even if introduced inoculant strains can persist in a soil population, there remains the possibility that the introduced strain may recombine with native strains whose nitrogen fixation efficiency may be poor (e.g. Sullivan, 1995). Thus from an agronomic point of view, a thorough understanding of the population genetics of rhizobia will continue to be an important prerequisite for the effective utilization of elite rhizobial strains for agricultural applications.

Rhizobial population genetics studies will also continue to have an important influence on the systematics and taxonomy of this important group of bacteria. During the past two decades population genetics studies have provided a taxonomic framework that is based on the most fundamental evolutionary relationships among the various taxa of rhizobia (Dressler-Nurmi *et al.*, 2007). Specifically, population genetic studies have helped reveal reproductively isolated groups (i.e. cryptic species), varieties, and geographic or host specific genotypes. These studies have also identified many fundamental questions. For example, why do certain species have more clonal population structures than other species? What are the mechanisms for the widespread but differential horizontal gene transfer among rhizobial species?

One of the most intriguing issues from rhizobial population surveys and genetic studies is the significance of non-symbiotic rhizobial populations in natural soils. Previous evidence indicates that in soil the non-symbiotic population of rhizobia may outnumber the symbiotically competent population by 40 to 1 (Segovia *et al.*, 1991). How are these populations derived? What are the roles of these non-symbiotic strains in natural population? And, how do they influence the structure of the symbiotic rhizobial populations? The answers to these and other questions will greatly enhance our understanding of this ecologically and economically important group of bacteria.

Acknowledgements

The research projects on nitrogen-fixing bacteria in our labs are supported by the Pennsylvania State University (B.E.) and Genome Canada (J.P.X.).

References

Agapow, P.M., and Burt, A. (2001). Indices of multilocus linkage disequilibrium. Mol. Ecol. Notes. *1*, 101–102.

Amarger, N., Macheret, V., and Laguerre, G. (1997). *Rhizobium gallicum* sp. nov. and *Rhizobium giardinii* sp. nov., from Phaseolus vulgaris nodules. Int. J. Syst. Bacteriol. *47*, 996–1006.

Andam, C.P., and Parker, M.A. (2007). Novel alphaproteobacterial root nodule symbiont associated with *Lupinus texensis*. Appl. Environ. Microbiol. 73, 5687–5691.

Bailly, X., Olivieri, I., De Mita, S., Cleyet-Marel, J.-C., and Bena, G. (2006). Recombination and selection shape the molecular diversity pattern of nitrogen-fixing *Sinorhizobium* sp. associated to *Medicago*. Mol. Ecol. *15*, 2719–2734.

Bailly, X., Olivieri, I., Brunel, B., Cleyet-Marel, J.-C., and Béna, J. (2007). Horizontal gene transfer and homologous recombination drive the evolution of the nitrogen-fixing symbionts of *Medicago*-species. J. Bacteriol. *189*, 5223–5236.

Bandelt, H.-J., and Dress, A.W.M. (1992). Split decomposition: a new and useful approach to phylogenetic analysis of distance data. Mol. Phylogenet. Evol. *1*, 242–252.

Barcellos, F.G., Menna, P., da Silva Batista, J.S., and Hungria, M. (2007). Evidence of horizontal transfer of symbiotic genes from a *Bradyrhizobium japonicum* inoculants strain to indigenous diazotrophs *Sinorhizobium (Ensifer) fredii* and *Bradyrhizobium elkanii* in a Brazilian savannah soil. Appl. Environ. Microbiol. 73, 2635–2643.

Barran, L.R., Bromfield, E.S.P., and Brown, D.W.C. (2002). Identification and cloning of the bacterial nodulation specificity gene in the *Sinorhizobium meliloti-Medicago laciniata* symbiosis. Can. J. Microbiol. *48*, 765–771.

Barrett, C.F., and Parker, M.A. (2006). Coexistence of *Burkholderia, Cupriavidus,* and *Rhizobium* sp. nodule bacteria on two *Mimosa* spp. in Costa Rica. Appl. Environ. Microbiol. *72*, 1198–1206.

Bass-Becking, L.G.M. (1934). Geobiologie of Inleiding Tot de Milieukunde (The Hague: Van Stockum and Zoon).

Bottomley, P.J., Cheng, H.-H., and Strain, S.R. (1994). Genetic structure and symbiotic characteristics of a *Bradyrhizobium* population recovered from a pasture soil. Appl. Environ. Microbiol. *60*, 1754–1761.

Brown, A.H.D., Feldman, M.W., and Nevo, E. (1980). Multilocus structure of natural populations of *Hordeum spontaneum.* Genetics. *96*, 523–536.

Chacón, M.I., Pickersgill, S.B., and Debouck, D.G. (2005). Domestication patterns in common bean (*Phaseolus vulgaris L.*) and the origin of the Mesoamerican and Andean cultivated races. Theor. Appl. Genet. *110*, 432–444.

Chen, W.M., Laevens, T.M., Lee, T., Coenye, T., DeVos, P., Mergeay, M., and Vandamme, P. (2001). *Ralstonia taiwanensis* sp. nov., isolated from root nodules of *Mimosa* species and sputum of a cystic fibrosis patient. Int. J. Syst. Evol. Microbiol. *51*, 1729–1735.

Demezas, D.H., Reardon, T.B., Strain, T.B., Watson, J.M., and Gibson, A.H. (1995). Diversity and genetic structure of a natural population of *Rhizobium leguminosarum* bv. *trifolii* isolated from *Trifolium subterraneum* L. Mol. Ecol. *4*, 209–220.

De Mita, S., Santoni, S., Hochu, I., Ronfort, J., and Bataillon, T. (2006). Molecular evolution and positive selection of the symbiotic gene *NORK* in *Medicago truncatula.* J. Mol. Evol. *62*, 234–244.

Donate-Correa, J., León-Barrios, M., Hernández, M., Pérez-Galdona, R., del Arco-Aguilar, M. (2007). Different *Mesorhizobium* species sharing the same symbiotic genes nodulate the shrub legume *Anagyris latifolia.* Syst. Appl. Microbiol. *30*, 615–623.

Dresler-Nurmi, A., Fewer, D.P., Räsänen, L.A., and Lindström, K. (2007). The diversity and evolution of rhizobia. Microb. Monogr. (online) DOI 10.1007/7171_2007_099.

Eardly, B.D., Hannaway, D.B., and Bottomley, P.J. (1985). Characterization of rhizobia from ineffective alfalfa nodules: ability to nodulate bean plants [*Phaseolus vulgaris* (L.) Savi.]. Appl. Environ. Microbiol. *50*, 1422–1427.

Eardly, B.D., Materon, L.A., Smith, N.H., Johnson, D.A., Rumbaugh, M.D., and Selander, R.K. (1990). Genetic structure of natural populations of the nitrogen-fixing bacterium *Rhizobium meliloti.* Appl. Environ. Microbiol. *56*, 187–194.

Eardly, B.D., Young, J.P.W., and Selander, R.K. (1992). Phylogenetic position of *Rhizobium* sp. strain Or 191, a symbiont of both *Medicago sativa* and *Phaseolus vulgaris,* based on partial sequences of *16S rRNA* and *nifH* genes. Appl. Environ. Microbiol. *58*, 1809–1815.

Eardly, B.D., Wang, F.-S., Whittam, T.S., and Selander, R.K. (1995). Species limits in *Rhizobium* populations that nodulate the common bean (*Phaseolus vulgaris*). Appl. Environ. Microbiol. *61*, 507–512.

Farris, J.S., Källersjö, M., Kluge, A.G., and Bult, C., (1994). Testing significance of incongruence. Cladistics. *10*, 315–319.

Flores, M., Morales, L., Avila, A., González, V., Bustos, P., García, D., Mora, Y., Guo, X., Collado-Vidas, J., Piñero, D., Dávila, G., Mora, J., and Palacios, R. (2005). Diversification of DNA sequences in the symbiotic genome of *Rhizobium etli.* J. Bacteriol. *187*, 7185–7192.

Galibert, F. and 55 other authors. (2001). The composite genome of the legume symbiont *Sinorhizobium meliloti.* Science *293*, 668–672.

Gepts, P., (1998). Origin and evolution of common bean: past events and recent trends. HortSci. 33, 1124–1130.

Giuntini, E., Mengoni, A., De Filippo, C., Cavalieri, D., Aubin-Horth, N., Landry, C.R., Becker, A., and Bazzicalupo, M. (2005). Large-scale variation of the symbiosis-required megaplasmid pSymA revealed by comparative genomic analysis of *Sinorhizobium meliloti* natural strains. BMC Genomics. *6*, 158.

González, V., Bustos, P., Ramírez-Romero, M.A., Medrano-Soto, A., Salgado, H., Hernández-González, I., Hernández-Celis, J.C., Quintero, V., Moreno-Hagelseib, G., Girard, L., Rodríguez, O., Flores, M., Cevallos, M.A., Collardo-Vidas, J., Romero, D. and Dávila, G. (2003). The mosaic structure of the symbiotic plasmid of *Rhizobium etli* CFN42 and its relation to other symbiotic genome compartments. Genome Biol. *4*, R36.

González, V., Santamaría, R.I., Bustos, P., Hernández-González, I., Medrano-Soto, A., Moreno-Hagelseib, G., Janga, S.C., Ramírez, M.A., Jiménez-Hacinto, Collado-Vides, V., J., and Dávila, G. (2006). The partitioned *Rhizobium etli* genome: genetic and metabolic redundancy in seven interacting replicons. Proc. Natl. Acad. Sci. U.S.A. *103*, 3834–3839.

Guo, H., Sun, S., Finan, T.M., and Xu, J.P. (2005). Novel DNA sequences from natural strains of the nitrogen-fixing symbiotic bacterium *Sinorhizobium meliloti.* Appl. Environ. Microbiol. *71*, 7130–7138.

Guo, X., Flores, M., Morales, L., Garcia, D., Bustos, P., Gonzales, V., Palacios, R., and Davila, G. (2007). DNA diversification in two *Sinorhizobium* species. J. Bacteriol. *189*, 6474–6476.

Hacker, J., and Kaper, J.B. (1999). The concept of pathogenicity islands. In Pathogenicity Islands and other Mobile Virulence Elements. Hacker, J., ed. (Washington D.C.: American Society for Microbiology), pp. 1–11.

Harrison, S.P., Young, J.P.W., and Jones, D.J. (1987). *Rhizobium* population genetics: effect of clover variety and inoculum dilution on the genetic diversity sampled from natural populations. Plant Soil *103*, 147–150.

Harrison, S.P., Jones, D.G., Schünmann, P.H.D., Forster, J.W., and Young, J.P.W. (1988). Variation in *Rhizobium leguminosarum* biovar *trifolii* Sym plasmids and the association with the effectiveness of nitrogen fixation. J. Gen. Microbiol. *134*, 2721–2730.

Harrison, S.P., Young, J.P.W., and Jones, D.J. (1989). *Rhizobium* population genetics: genetic variation within and between populations from diverse locations. J. Gen. Microbiol. *135*, 1061–1069.

Haukka, K., Lindstron, K., and Young, J.P.W. (1998). Three phylogenetic groups of *nodA* and *nifH* genes in *Sinorhizobium* and *Mesorhizobium* isolates from leguminous trees growing in Africa and Latin America. Appl. Environ. Microbiol. *64*, 419–426.

Hedrick, P.W. (1987). Gametic disequilibrium measures: proceed with caution. Genetics *117*, 331–341.

Herrera-Cervera, J.A., Sanjuan-Pinilla, J.M., Olivares, Sanjuan, J., J. (1998). Cloning and identification of conjugative transfer origins in the *Rhizobium meliloti* genome. J. Bacteriol. *180*, 4583–4590.

Hey, J., and Wakely, J. (1997). A coalescent estimator of the population recombination rate. Genetics. *145*, 833–846.

Hirsch, A.M., Lum, M.R., and Downie, A.J. (2001). What makes the rhizobia-legume symbiosis so special? Plant Physiol. *127*, 1484–1492.

Hudson, R.R. (1987). Estimating the recombination parameter of a finite population model without selection. Genet. Res. *50*, 245–250.

Hughes, A.L. (2005). Evidence for abundant slightly deleterious polymorphisms in bacterial populations. Genetics. *169*, 533–538.

Huson, D.H., and Bryant, D. (2006). Application of phylogenetic networks in evolutionary studies. Mol. Biol. Evol. *23*, 254–267.

Jarabo-Lorenzo, A., Perez-Galdona, R., Donate-Correa, J., Rivas, R., Velazquez, E., Hernandez, M., Temprano, F., Martinez-Molina, E., Ruiz-Argueso, T., and Leon-Barrios, M. (2003). Genetic diversity of bradyrhizobial populations from diverse geographic origins that nodulate *Lupinus* spp. and *Ornithopus* spp. Syst. Appl. Microbiol. *26*, 611–623.

Johnston, A.W.B., Hombrecher, G., Brewin, N.J., and Cooper, N.C. (1982). Two transmissible plasmids in *Rhizobium leguminosarum* strain 300. J. Gen. Microbiol. *128*, 85–93.

Jordan, D.C. (1984). Family III. *Rhizobiaceae* Conn 1938, 321[AL], In Bergey's Manual of Systematic Bacteriology, vol. 1, In Krieg, N.R. and Holt, J.G., eds. (Baltimore, US: Williams and Wilkins). pp. 234–254.

Kaijalainen, S., and Lindstrom, K. (1989). Restriction fragment length polymorphism of *Rhizobium galegae* strains. J. Bacteriol. *171*, 5561–5566.

Kaneko, T., Nakamura, Y., Sato, S., Minamisawa, K., Uchiumi, T., Sasamoto, S., Watanabe, A., Idesawa, K., Iriguchi, M., Kawashima, K., Kohara, M., Matsumoto, M., Shimpo, S., Tsuruoka, H., Wada, T., Yamada, M., and Tabata, S. (2002). Complete genomic sequence of nitrogen-fixing symbiotic bacterium *Bradyrhizobium japonicum* USDA110. DNA Res. *9*, 189–197.

Kaneko, T., Nakamura, Y., Sato Asamizu, S., E., Kato, T., Sasamoto, S., Watanabe, A., Idesawa, Ishikawa, K., A., Kawashima, K., Kimura, T., Kishida, Y., Kiyokawa, C., Kohara, M., Matsumoto, M., Matsuno, A., Machizuki, Y., Nakayama, S., Nakazaki, N., Shimpo, S., Sugimoto, M., Takeuchi, C., Yamada, M., and Tabata, S. (2000). Complete genome structure of the nitrogen-fixing symbiotic bacterium *Mesorhizobium loti*. DNA Res. *7*, 331–338.

Laguerre, G., Fernandez, M.P., Edel, V., Normand, P., and Amarger, N. (1993). Genomic heterogeneity among French *Rhizobium* strains isolated from *Phaseolus vulgaris* L. Int. J. Syst. Bacteriol. *43*, 761–767.

Laguerre, G., Nour, S.M., Macharet, V., Sanjuan, J., Drouin, P., and Amarger, N. (2001). Classification of rhizobia based on *nodC* and *nifH* gene analysis reveals a close phylogenetic relationship among *Phaseolus vulgaris* symbionts. Microbiology *147*, 981–993.

Lavin, M., Herendeen, P.S., and Wojciechowski, M.F. (2005). Evolutionary rates analysis of *Leguminosae* implicates a rapid diversification of lineages during the tertiary. Syst. Biol. *54*, 575–594.

Leung, K., Strain, S.R., de Bruijn, F.J., and Bottomley, P.J. (1994). Genotypic and phenotypic comparisons of chromosomal types within an indigenous soil population of *Rhizobium leguminosarum* bv. *trifolii*. Appl. Environ. Microbiol. *60*, 416–426.

Lie, T.A. (1978). Symbiotic specialization in pea plants: the requirement of specific *Rhizobium* strains for peas from Afghanistan. Ann. Appl. Biol. *88*, 462–465.

Lisitsyn, N., Lisitsyn, N. and Wigler, M. (1993). Cloning the differences between two complex genomes. Science *259*, 946–951.

Lewontin, R.C. (1964). The interaction of selection and linkage. I. General considerations, heterotic models. Genetics *49*, 49–67.

Maiden, M.C., Bygraves, J.A., Feil, E., Morelli, G., Russell, J.E., Urwin, R., Zhang, Q., Zhou, J., Zurth, K., Caugant, D.A., Feavers, I.M., Achtman, M., and Spratt, B.G. (1998). Multilocus sequence typing: a portable approach to the identification of clones within populations of pathogenic microorganisms. Proc. Natl. Acad. Sci. U.S.A. *95*, 3140–3145.

Maynard Smith, J., Smith, N.H., O'Rourke, M., and Spratt, B.G. (1993). How clonal are bacteria? Proc. Natl. Acad. Sci. U.S.A. *90*, 4384–4388.

Martínez-Romero, E., Segovia, L., Mercante, F.M., Franco, A.A., Graham, P., and Pardo, M.A. (1991). *Rhizobium tropici*, a novel species nodulating *Phaseolus vulgaris* L. beans and *Leucaena* sp. trees. Int. J. Syst. Bacteriol. *41*, 417–426.

Milkman, R., and Bridges, M.M. (1990). Molecular evolution of the *Escherichia coli* chromosome. III. Clonal Frames. Genetics *126*, 505–517.

Moulin, L., Munive, A., Dreyfus, B., and Boivin-Masson, C. (2001). Nodulation of legumes by members of the β-subclass of Proteobacteria. Nature. *411*, 948–951.

Moulin, L., Bena, G., Boivin-Masson, C., and Stepkowski, T. (2004). Phylogenetic analysis of symbiotic nodulation genes support vertical and lateral gene co-transfer within the *Bradyrhizobium* genus. Mol. Phylogenet. Evol. *30*, 720–732.

Mutch, L.A. and Young, J.P.W. (2004). Divesity and specificity of *Rhizobium leguminosarum* biovar *viceae* on wild and cultivated legumes. Mol. Ecol. *13*, 2435–2444.

Nandesena, K.G., O'Hara, G.W., Tiwari, R.P., and Howieson, J.G. (2006). Rapid in situ evolution of nodulating strains for *Biserrula pelecinus* L. through lateral transfer of a symbiosis island from the original mesorhizobial inoculants. Appl. Environ. Microbiol. *72*, 7365–7367.

Parker, M.A., Lafay, B., Burdon, J., van Berkum, P. (2002). Conflicting phylogeographic patterns in rRNA and *nifD* indicate regionally restricted gene transfer in *Bradyrhizobium*. Microbiology. *148*, 2557–2565.

Parker, M.A., and Kennedy, D.A. (2006). Diversity and relationships of bradyrhizobia from legumes native to eastern North America. Can. J. Microbiol. *52*, 1148–1157.

Pérez-Ramírez, N.O., Rogel, M.A., Wang, E., Castellanos, J.Z., and Martínez-Romero, E. (1998). Seeds of *Phaseolus vulgaris* carry *Rhizobium etli*. FEMS Microbiol. Ecol. *26*, 289–296.

Perret, X., Staehelin, C., and Broughton, W.J. (2000). Molecular basis of symbiotic promiscuity. Mol. Biol. Rev. *64*, 180–201.

Pinero, D., Martinez, E. and Selander, R.K. (1988). Genetic diversity and relationships among isolates of *Rhizobium leguminosarum* boivar *phaseoli*. Appl. Environ. Microbiol. *54*, 2825–2832.

Qian, J., and Parker, M.A. (2002). Contrasting *nifD* and ribosomal gene relationships among *Mesorhizobium* from *Lotus oroboides* in northern Mexico. Syst. Appl. Microbiol. *25*, 68–73.

Qian, J., Kwon, S.W., and Parker, M.A. (2003). rRNA and nifD phylogeny of Bradyrhizobium from sites across the Pacific Basin. FEMS Microbiol. Lett. *219*, 159–165.

Rayssiguier, C., Thaler, D.S., and Radman, M. (1989). The barrier to recombination between *Escherichia coli* and *Salmonella typhimurium* is disrupted in mismatch repair mutants. Nature. *342*, 396–401.

Roberts, M.S., and Cohan, F.M. (1993). The effect of DNA sequence divergence on sexual isolation in Bacillus. Genetics *134*, 401–408.

Rome, S., Brunel, B., Cleyet-Marel, J.-C., Rome, Fernandez, M.P., and Normand, P. (1996). *Sinorhizobium medicae* sp. nov., isolated from annual *Medicago* spp. Int. J. Syst. Bacteriol. *46*, 972–980.

Roumiantseva, M.L., Yakutkina, V.V., Damman-Kalinowski, T. *et al.* (1999). Comparative analysis of structural organization of the genome in alfalfa nodule bacteria *Sinorhizobium medicae* and *Sinorhizobium meliloti*. Russian J. Genet. *35*, 128–135.

Schofield, P.R., Gibson, A.H., Dudman, W.F., and Watson, J.M. (1987). Evidence for genetic exchange and recombination of *Rhizobium* symbiotic plasmids in a soil population. Appl. Environ. Microbiol. *53*, 2942–2947.

Segovia, L., Piñero, D., Palacios, R., and Martínez-Romero, E. (1991). Genetic structure of a soil population of nonsymbiotic *Rhizobium leguminosarum*. Appl. Environ. Microbiol. *57*, 426–433.

Segovia, L., Young, J.P.W., and Martinez-Romero, E. (1993). Reclassification of American *Rhizoium leguminosarum* biovar *phaseoli* type I strains as *Rhizobium etli* sp. nov. Int. J. Syst. Bacteriol. *43*, 374–377.

Silva, C., Eguiarte, L.E., and Souza, V. (1999). Reticulated and epidemic population genetic structure of *Rhizobium etli* biovar *phaseoli* in a traditionally managed locality in Mexico. Mol. Ecol. *8*, 277–287.

Silva, C., Vinuesa, P., Eguiarte, L.E., Martínez-Romero, E., and Souza, V. (2003). *Rhizobium etli* and *Rhizobium gallicum* nodulate common bean (*Phaseolus vulgaris*) in a traditionally managed milpa plot in Mexico: population genetics and biogeographic implications. Appl. Environ. Microbiol. *69*, 884–893.

Silva, C., Vinuesa, P., Eguiarte, L.E., Souza, V., and Martínez-Romero, E. (2005). Evolutionary genetics and biogeographic structure of *Rhizobium gallicum sensu lato*, a widely distributed bacterial symbiont of diverse legumes. Mol. Ecol. *14*, 4033–4050.

Silva, C., Kan, F.-L., and Martinez-Romero, E. (2007). Population genetic structure of *Sinorhizobium meliloti* and *S. medicae* isolated from nodules of *Medicago* spp. in Mexico. FEMS Microbiol. *60*, 477–489.

Souza, V., Nguyen, T., Hudson, R.R., Piñero, D., and Lenski, R.E. (1992). Hierarchical analysis of linkage disequilibrium in *Rhizobium* populations: evidence for localized sex? Proc. Natl. Acad. Sci. U.S.A. *89*, 8389–8393.

Souza, V., Eguiarte, L., Avila, G., Capello, R., Gallardo, C., Montoya, J., and Piñero, D. (1994). Genetic structure of *Rhizobium etli* biovar phaseoli associated with wild and cultivated bean plants (*Phaseolus vulgaris* and *Phaseolus coccineus*) in Morelos, Mexico. Appl. Environ. Microbiol. *60*, 1260–1268.

Souza, V. and Eguiarte, L.E. (1997). Bacteria gone native vs. bacteria gone awry?: plasmidic transfer and bacterial evolution. Proc. Natl. Acad. Sci. U.S.A. *94*, 5501–5503.

Staley, J.T., and Gosink, J.J. (1999). Poles apart: biodiversity and biogeography of sea ice bacteria. Annu. Rev. Microbiol. *53*, 189–215.

Stępkowski, T., Moulin, L., Krzyzanska, A., McInnes, A., Law, I.J., and Howieson, J. (2005). European origin of *Bradyrhizobium* populations infecting lupins and serradella in soils of western Australia and South Africa. Appl. Environ. Microbiol. *71*, 7041–7052.

Stępkowski, T., Hughes, C.E., Law, I.J., Markiewicz, Ł., Gurda, D., Chlebicka, A., and Moulin, L. (2007). Diversification of lupine Bradyrhizobium strains: evidence from nodulation gene trees. Appl. Environ. Microbiol. *73*, 3254–3264.

Strain, S.R., Leung, K., Whittam, T.S., de Bruijn, F.J., and Bottomley, P.J. (1994). Genetic structure of *Rhizobium leguminosarum* biovar *trifolii* and *viceae* populations found in two Oregon soils under different plant communities. Appl. Environ. Microbiol. *60*, 2772–2778.

Sullivan, J.T., Patrick, H.N., Lowther, W.L., Scott, D.B., and Ronson, C.W. (1995). Nodulating strains of *Rhizobium loti* arise through chromosomal symbiotic gene transfer in the environment. Proc. Natl. Acad. Sci. U.S.A. *92*, 8985–8989.

Sullivan, J.T. and Ronson, C.W. (1998). Evolution of rhizobia by acquisition of a 500-kb symbiosis island that integrates into a phe-tRNA. Proc. Natl. Acad. Sci. U.S.A. *95*, 5145–5149.

Sullivan, J.T., and 13 other authors. (2002). Comparative sequence analysis of the symbiosis island of *Mesorhizobium loti* strain R7A. J. Bacteriol. *184*, 3086–3095.

Sun, S., Guo, H., and Xu, J.P. (2006). Multiple gene genealogical analyses reveal both common and distinct population genetic patterns among replicons in the nitrogen-fixing bacterium *Sinorhizobium meliloti*. Microbiology. *152*, 3245–3259.

Tibayrenc, M. (1996). Towards a unified evolutionary genetics of micoroorganisms. Ann. Rev. Microbiol. *50*, 401–429.

Turner, S.L., and Young, J.P.W. (2001). The glutamine synthetases of rhizobia: phylogenetics and evolutionary implications. Mol Biol. Evol. *17*, 309–319.

van Berkum, P., Elia, P. and Eardly, B.D. (2006). Multilocus sequence typing as an approach for population analysis of *Medicago*-nodulating rhizobia. J. Bacteriol. *188*, 5570–5577.

van Berkum, P., Badri, Y., Elia, P., Aouani, M.E., and Eardly, B.D. (2007). Chromsomal and symbiotic relationships of rhizobia nodulating *Medicago truncatula* and *M. laciniata*. Appl. Environ. Microbiol. 73, 7597–7604.

Vandamme, P., and Coenye, T. (2004). Taxonomy of the genus *Cupriavidus*: a tale of lost and found. Int. J. Syst. Microbiol. *54*, 2285–2289.

Villegas, M.D.C., Rome, S., Mauré, L., Domergue, O., Gardan, L., Bailly, X., Cleyet-Marel, J.-C., and Brunel, B. (2006). Nitrogen-fixing sinorhizobia with *Medicago laciniata* constitute a novel biovar (bv. medicaginis of *S. meliloti*. Syst. Appl. Microbiol. *29*, 526–538.

Vinuesa, Leon-Barriós, P., M., Silva, C., Willems, A., Jarabo-Lorenzo, A., Pérez-Galdon, R., Werner, D., and Martínez-Romero, E. (2005a). *Bradyrhizobium canariense* sp. nov., an acid-tolerant endosymbiont that nodulates endemic genistoid legumes (Papilionoideae: Genisteae) from the Canary Islands, along with *Bradyrhizobium japonicum* bv. *genistearum*, *Bradyrhizobium* genospecies alpha, and *Bradyrhizobium* genospecies beta. Int. J. Syst. Bacteriol. *55*, 569–575.

Vinuesa, P., Silva, C., Werner, D., and Martinez-Romero, E. (2005b). Population genetics and phylogenetic inference in bacterial molecular systematic: the roles of migration and recombination in *Bradyrhizobium* species cohesion and delineation. Mol. Phylogenet. Evol. *34*, 29–54.

Vinuesa, P., Rojas-Jimenez, K., Contreras-Moreira, B., Mahna, S.K., Prasad, B.N., Moe, H., Selvaraju, S.B., Thierfelder, H., and Werner, D. (2008). Appl. Env. Microbiol. *74*, 6987–6996.

Wayne, L.G., Brenner, D.J., Colwell, R.R., Grimont, P.A.D., Kandler, O., Krichevsky, L.H., Moore, L.H., Moore, W.E.C., Murray, E., Stackebrandt, M.P., Starr, M.P., and Trüper, H.G. (1987). Report of the ad hoc committee on reconciliation of approaches to bacterial systematic. Int. J. Syst. Bacteriol. *37*, 463–464.

Wernegreen, J.J., Harding, E.E., and Riley, M.A. (1997). *Rhizobium* gone native: unexpected plasmid stability of indigenous *Rhizobium leguminosarum*. Proc. Natl. Acad. Sci. U.S.A. *94*, 5483–5488.

Wernegreen, J.J., and Riley, M.A. (1999). Comparison of the evolutionary dynamics of symbiotic and housekeeping loci: a case for genetic coherence of rhizobial lineages. Mol. Biol. Evol. *16*, 98–113.

Woese, C.R. (1987). Bacterial evolution. Microbiol. Rev. *51*, 221–271.

Wong, K., and Golding, G.B. (2003). A phylogenetic analysis of the pSymB replicon from the *Sinorhizobium meliloti* genome reveals a complex evolutionary history. Can. J. Microbiol. *49*, 269–280.

Young, J.P.W. (1985). *Rhizobium* population genetics: enzyme polymorphism in isolates from peas, clover, beans and lucerne grown at the same site. J. Gen. Microbiol. *131*, 2399–2408.

Young, J.P.W., Demetriou, L., and Apte, R.G. (1987). *Rhizobium* population genetics: enzyme polymorphism in *Rhizobium leguminosarum* from plants and soil in a pea crop. Appl. Environ. Microbiol. *53*, 397–402.

Young, J.P.W., and Wexler, M. (1988). Sym plasmid and chromosomal genotypes are correlated in field populations of *Rhizobium leguminosarum*. J. Gen. Microbiol. *134*, 2731–2739.

Young, J.P.W. (1992). Molecular population Genetics and Evolution of Rhizobia. In The Nitrogen Fixation and its Research in China. Hong, G.-F., ed. (Berlin: Spring-Verlag Publishers), pp. 366–380.

Young, J.P.W., and 31 coauthors. (2006). The genome of *Rhizobium leguminosarum* has recognizable core and accessory components. Genome Biol. *7*, R34.

Zhaxybayeva, O., Lapierre, P., and Gogarten, J.P. (2004). Genome mosaicism and organismal lineages. Trends. Genet. *20*, 254–260.

Population Genetics of Cyanobacteria

6

Scott R. Miller

Abstract

Cyanobacteria are a group of ecologically diverse photosynthetic bacteria. Because niche differentiation is ultimately the product of differences among individuals within populations, understanding the evolutionary origins of this diversity ultimately requires a population genetics perspective. This chapter considers the current state of our understanding of the mechanisms that generate variation in cyanobacteria, the distribution of this diversity and its potential functional importance, with an emphasis on recent work from our laboratory regarding diversity within and among populations of the cosmopolitan, multicellular cyanobacterium, *Mastigocladus laminosus*. It concludes with an appraisal of the potential to take a population genomics approach to address fundamental questions regarding the nature of adaptive variation and niche differentiation in these microorganisms.

Why cyanobacterial population genetics?

Cyanobacteria comprise an ancient clade of photosynthetic microorganisms belonging to the domain *Bacteria*, with a microfossil record that goes back approximately 2 Ga (Hofmann, 1976; Schopf, 1992), and possible biomarker-based evidence of their existence 2.7 Ga (Brocks *et al.*, 1999). During the course of its long evolutionary history, this group of oxygenic phototrophs has diversified to exploit a broad range of freshwater, marine and terrestrial habitats (reviewed by Whitton and Potts, 1999), from the oligotrophic open ocean to hypereutrophic lakes, from hot springs to seasonally frozen polar ponds, and from deserts hot and cold to hypersaline lagoons. Commensurate with this scope of habitat heterogeneity, cyanobacteria exhibit niche (i.e. ecotypic) differentiation at the macroevolutionary (i.e. 16S rRNA-defined) scale along environmental gradients including salinity (Garcia-Pichel *et al.*, 1998), light (Moore *et al.*, 1998), temperature (Miller and Castenholz, 2000) and sulphide (Miller and Bebout, 2004).

Making sense of this breadth of ecological diversity ultimately necessitates an understanding of the genetic mechanisms by which ecological niches evolve. Attaining this understanding requires the investigation of variation within populations, for the simple reason that genetic differences between individuals within populations provide the raw material for all evolutionary diversification, whether that variation is generated by new mutations or by the acquisition of novel DNA from a donor. Consequently, niche differentiation is fundamentally a population-level process.

This point is emphasized by insights provided by experimental evolution studies of microbial populations in the laboratory [reviewed by Elena and Lenski (2003) and Kassen and Rainey (2004)], during which clonal founding populations experiencing different or spatially heterogeneous environments frequently diversify within hundreds of generations (or fewer) in response to selection. This is in accordance with the prediction of the ecological theory of adaptive radiation that phenotypic divergence of populations is the outcome of divergent natural selection

(Schluter, 2000). Populations of microorganisms in nature should likewise be expected to adaptively differentiate along an ecological selection gradient on time scales too rapid for detection by the slowly-evolving molecular markers typically used in surveys of microbial diversity (such as the SSU ribosomal RNA gene and the downstream internal transcribed spacer).

Furthermore, accounting for the population-level component of biotic diversity is essential for developing a full understanding of how ecosystems function and for predicting ecosystem response to environmental change (Hughes *et al.*, 1997; Luck *et al.*, 2003), particularly if different population members specialize on different regions along an environmental gradient. An explicit focus on population diversity is therefore timely and necessary, and will transform our understanding of both the evolutionary origins and the maintenance of diversity in microbial systems.

Variation in cyanobacterial populations: mechanism, distribution and significance

The discipline of population genetics provides a theoretical framework to address the microevolutionary processes that, over time, have shaped the macroevolutionary patterns formed over greater than two billion years of cyanobacterial diversification. Although at present our understanding of how the microevolutionary forces of mutation (and horizontal gene transfer), natural selection, migration (gene flow) and genetic drift combine to shape cyanobacterial diversity is still limited, in the following section I focus on three fundamental areas for which recent progress has been made: (1) the genetic mechanisms that underlie evolutionary novelty in populations of cyanobacteria; (2) the geographic distribution of this variation; and (3) the possible significance of this variation with respect to niche differentiation.

The genetic mechanisms of evolutionary innovation

Mutation and recombination (i.e. horizontal gene transfer) are the potential sources of evolutionary novelty. That a number of cyanobacteria are amenable to genetic analysis in the laboratory (Flores *et al.*, 2008), either by transformation (e.g. *Synechococcus* PCC 7942) or by conjugation (e.g. *Anabaena* PCC 7120), is a testament to the potential importance of the latter as a mechanism of cyanobacterial evolution in nature. Indeed, over the past ten years, evidence for both intergenic and intragenic recombination has been reported for diverse cyanobacterial taxa (Table 6.1) using a variety of criteria, including the topological incongruence of gene genealogies (Rudi *et al.*, 1998; Barker *et al.*, 1999; Lodders *et al.*, 2005), nucleotide substitution distribution methods (Kurmayer and Gumpenberger, 2006), and a coalescent-derived maximum likelihood estimator (McVean *et al.*, 2002) of the population-scaled recombination rate (Miller *et al.*, 2007). In one case, a MLST analysis of natural populations of the marine intertidal cyanobacterium *Microcoleus chthonoplastes* (Lodders *et al.*, 2005), the extent of linkage disequilibrium among loci was among the lowest that has been reported for a bacterium and suggest that these populations are nearly freely recombining.

Table 6.1 Taxa for which recombination has been reported for natural populations of Cyanobacteria

Genus	Category of recombination event	Reference(s)
Mastigocladus	Intragenic, intergenic	Miller *et al.* (2007)
Microcoleus	Intergenic	Lodders *et al.* (2005)
Microcystis	Intragenic	Tanabe *et al.* (2004)
Nodularia	Intergenic	Barker *et al.* (1999), Barker *et al.* (2000), Batley and Hayes (2003)
Nostoc	Intragenic, intergenic	Rudi *et al.* (1998)
Planktothrix	Intragenic	Kurmayer and Gumpenberger (2006)

Recent work is beginning to provide perspective on the relative contributions of recombination and mutation to cyanobacterial diversification. We have investigated this issue for the cosmopolitan, multicellular cyanobacterium, *Mastigocladus* (*Fischerella*) *laminosus* (Fig. 6.1) over a range of geographic scales. Globally, it appears that recombination has historically made an important contribution to the origins of novel genetic variation in this group (Miller *et al.*, 2007). Estimated ratios of the intragenic recombination to mutation rate (r/μ) for a sample of thirty-seven laboratory strains of *M. laminosus* isolated from throughout its range varied between 0.14 and 0.47 (in units of events per nucleotide site). These values are comparable to estimates for human pathogens renowned for their capacity to recombine, such as *Neisseria meningitidis* (Jolley *et al.*, 2005). Further, if one makes simplifying assumptions about recombinational fragment size, then, based on the above estimates of r/μ, intergenic recombination would be about 2–5 times more important than mutation for generating novel genome combinations in *M. laminosus* (Miller *et al.*, 2007).

A

B

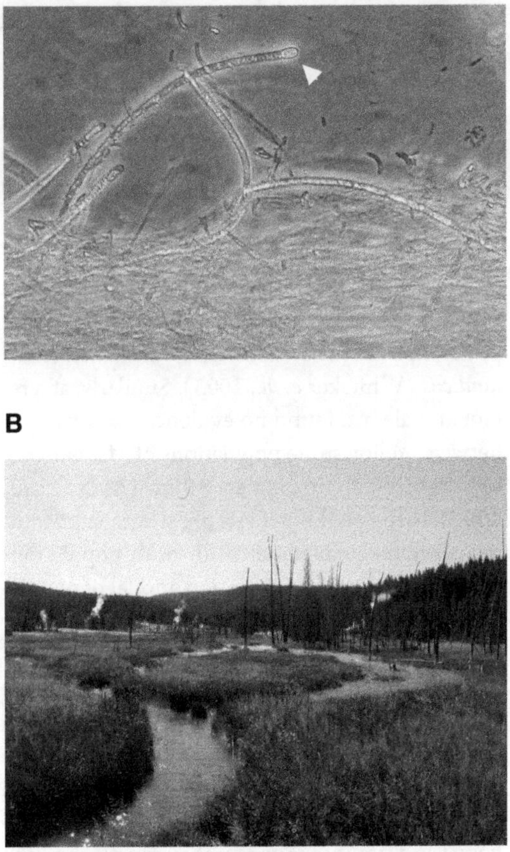

Figure 6.1 (A) Phase-contrast image of a microbial mat sample collected from White Creek in the Lower Geyser Basin of Yellowstone National Park. *Mastigocladus laminosus* (foreground) is the predominant cyanobacterium at White Creek where mean annual temperature is between 39 and 54°C. Under conditions of nitrogen limitation such as observed at White Creek, this multicellular bacterium develops heterocysts (arrow) for spatially separating the incompatible processes of nitrogen fixation and oxygenic photosynthesis. Trichome width is approximately 3 µm. (B) White Creek, looking upstream. Mean annual temperature is 43°C at this position along the channel.

Recombination appears to be comparable to mutation for generating diversity within populations of *M. laminosus* as well. Recently, we have investigated *M. laminosus* variation in White Creek, a nitrogen-limited, thermal-discharge fed stream in the Lower Geyser Basin of Yellowstone National Park (Fig. 6.1; Miller *et al.*, 2009). Cyanobacterial biomass is dominated by this bacterium where mean annual temperature is between 39 and 54°C, a region that stretches greater than a kilometre. We genetically characterized 153 strains directly isolated from five locations spanning the range of *M. laminosus* habitat in White Creek. Although strains were identical in sequence at the SSU rRNA gene and the downstream internal transcribed spacer, they exhibited low levels of polymorphism at more rapidly evolving loci (*narB, devH, nifH* and *nirA*): approximately 0.6% of all nucleotide sites were variable, for a total of 15 biallelic single nucleotide polymorphisms (SNPs).

Each of the strains belonged to one of 15 multi-locus haplotypes whose relationships could be described by a genealogical network (Fig. 6.2). A traditional bifurcating phylogeny frequently fails to accurately describe the genealogical relationships of closely related individuals (alleles) for a variety of reasons (Posada and Crandall, 2001), including that the relationship between ancestor and descendant may not be hierarchical (because the ancestral allele is not extinct) and that some nodes may be multifurcating rather than bifurcating (because older alleles are expected to leave more descendants). Networks relax these assumptions, as well as explicitly

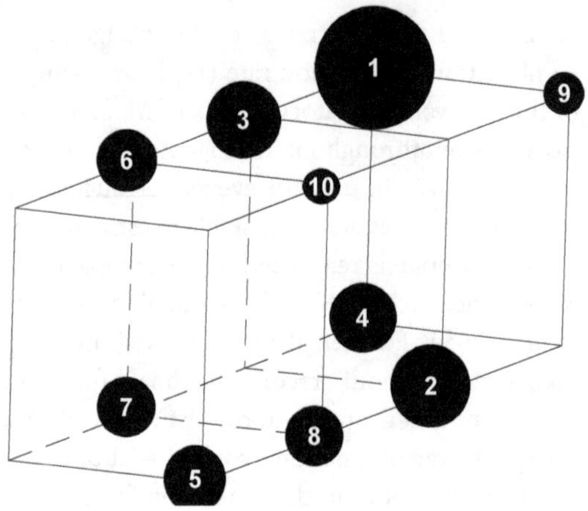

Figure 6.2 The genealogical relationships among the ten most abundant four-locus haplotypes ($N=146$ strains) observed for the *M. laminosus* population at White Creek, depicted as a three-dimensional network. Each edge connecting a haplotype pair represents either a mutation event creating a new allele at a locus, or a recombination event creating a new combination of alleles. Area of a node is proportional to the relative abundance of the haplotype, and unlabeled nodes represent allele combinations that were not observed in the sample.

accommodate recombination, with alleles related by recombination connected by loops in the network. The lattice of loops in the White Creek *M. laminosus* network is indicative of extensive recombination (Fig. 6.2). In addition, the recombination algorithm of Myers and Griffith (2003) provided evidence for recombination between all pairs of loci and estimated that a minimum of seven intergenic recombination events is required to explain the nucleotide site patterns observed among haplotypes. Together, the SNP data and recombination analysis support the conclusion that both mutation and the shuffling of allele combinations by genetic exchange have made important contributions to the standing variation of the *M. laminosus* population at White Creek.

How is population genetic diversity distributed?

Because genetic drift and local selective sweeps remove variation within, but not between, populations in the absence of significant migration, genetic differentiation at neutral loci is expected to increase between populations with physical distance when gene flow is restricted (Wright, 1943). Whether gene flow among microbial populations is in fact restricted by barriers to dispersal has been a topic of considerable recent interest (Finlay, 2002; Whitaker *et al.*, 2003; Martiny *et al.*, 2006). Although there is ample evidence for the restricted geographic distributions of certain thermophilic cyanobacterial taxa, based on their presence in or absence from various geothermal areas around the world (Castenholz, 1996; Papke *et al.*, 2003), little attention has been paid to the subject of barriers to gene flow among cyanobacteria from a population perspective.

Recently, we have addressed this question of microbial biogeography for *M. laminosus* at a variety of geographic scales. At a global scale, we have observed a strong positive association between the amount of genetic differentiation between populations (measured as nucleotide diversity) and their physical distance (Miller *et al.*, 2007), consistent with the expectation that there are dispersal barriers. This pattern of genetic isolation by geographic distance has also been observed for populations of the acidothermophile *Sulfolobus islandicus* (Whitaker *et al.*, 2003). Similarly, at a regional scale, we found no evidence for migration between Yellowstone populations of *M. laminosus* separated by approximately 50 km (Miller *et al.*, 2006), based on the analysis of concatenated four-locus sequence alignments for both populations by the 'isolation with migration' model of population divergence (Nielsen and Wakeley, 2001; Hey and Nielsen, 2004). Substantial geographic structure was also observed for southern Michigan lake populations of *Microcystis aeruginosa*, with approximately half of the genetic variation in the sample partitioned among populations (Wilson *et al.*, 2005). However, the pattern of limited dispersal and significant geographic structuring can't be generalized to all cyanobacteria. For example, a recent investigation of the population genetic diversity of *Microcoleus chthonoplastes* in the Baltic and North Sea regions did not find compelling evidence of geographic structure (Lodders *et al.*, 2005).

Things appear to be more complicated locally. At White Creek, where *M. laminosus* occupies a kilometre-long thermal gradient, an analysis of molecular variance (AMOVA) indicated significant genetic differentiation (i.e. F_{ST}) between upstream and downstream regions of the gradient, amounting to 13% of the total variation in the sample (Miller *et al.*, 2009). The observed geographic structure cannot be explained by spatially limited migration, however, as analyses of the relationship between genetic differentiation and geographic distance instead suggest that the subpopulations along the gradient have in fact been historically highly genetically connected. With one exception, there was no conclusive evidence of genetic isolation by distance for the analysed loci,. The exception was the *nifH* locus (which encodes the iron protein of nitrogenase) that exhibited strong differentiation ($r = 0.36$, $P < 0.001$), with one allele predominant at lower temperatures (average frequency of 82% at two downstream sites), and a second allele predominant at higher temperature sites (average frequency of 80% at three upstream sites). These results suggest that the *nifH* marker (or a locus in linkage disequilibrium with it) is not selectively neutral. We conclude that gene flow occurs along the gradient but is prevented in certain regions of the *M. laminosus* genome by selection. The possible reasons for this non-random spatial distribution will be discussed in the next section.

What is the functional importance of population genetic variation?

Though still limited in number, surveys of the diversity of natural populations of cyanobacteria typically reveal evidence of genetic variation (Barker *et al.*, 1999, 2000; Lodders *et al.*, 2005; Miller *et al.*, 2006). Is this variation selectively neutral, or, rather, the product of niche differences among genetically divergent individuals? Addressing this issue of what maintains variation in natural populations is not only a central question in evolutionary biology (Kimura, 1983; Gillespie, 1991), but also a matter with implications for how microbial populations function and how they respond to environmental changes.

At White Creek, we believe that selection on ecological variation within the *M. laminosus* population has contributed to the standing variation along its temperature gradient. Three haplotypes (labelled 1, 3 and 6 in Fig. 6.2), which account for nearly half of the strains in the White Creek strain collection, are ecological specialists that have diverged in thermal niche and occupy different positions along the gradient (Miller *et al.*, 2009). Haplotype 3, which exhibits peak relative abundance near the upstream population boundary, is a high-temperature specialist with a fitness maximum in laboratory culture at 55 °C, but is absent from downstream sites and exhibits reduced performance at lower temperatures. Conversely, haplotype 6 is found only at downstream sites, and strains maximize performance at lower temperature (37°C) at the cost of reduced high temperature performance. Haplotype 1 outperforms the other haplotypes at intermediate temperature (50°C), where it is most abundant in White Creek. These differences in fitness among haplotypes, which include trade-offs at population boundary extremes, bear directly on our understanding the evolution of niche breadth in microbial populations, as most evolutionary theory incorporates the idea that adaptation to one environment entails a fitness trade-off elsewhere (e.g. Lynch and Gabriel, 1987; Huey and Kingsolver, 1993).

A final example further illustrates the links to be made between the mechanisms that generate population genetic variation, its distribution and its functional importance. Comparative genomics of lineages of the marine cyanobacterium *Prochlorococcus* which exhibit less than 1% sequence divergence at the SSU rRNA gene has revealed that most differences in gene content reside on five genomic islands (Coleman *et al.*, 2006). The presence in these islands of cyanophage genes such as integrases, together with evidence that these phage can carry host genes (Lindell *et al.*, 2004; Sullivan *et al.*, 2005), lend support to the idea that horizontal gene transfer via transduction may be an important mechanism of evolution within these populations. Many of these island genes are homologous to known stress response genes and are differentially expressed under nutrient and light stress (Coleman *et al.*, 2006). The idiosyncratic repertoire of island genes harboured by particular individuals may therefore contribute to niche-defining traits that distinguish lineages of these bacteria. In a similar vein, comparative genomic and metagenomic data collected from Yellowstone hot spring microbial mat communities suggest that differences in gene content may contribute to the ecological differences among related lineages of the thermophilic cyanobacterium *Synechococcus* (Bhaya *et al.*, 2007). Drawing a comparison between the genomic islands of *Prochlorococcus* and the pathogenicity islands of many human pathogens, Coleman *et al.* (2006) suggest that the acquisition of novel genes by transduction may be a general mechanism for niche differentiation in bacteria, a provocative guiding hypothesis for future investigations of how niches evolve in microbial populations.

Future directions: population genomics of cyanobacteria

In light of the above, a fundamental challenge that remains for investigators of the evolutionary origins, maintenance and distribution of microbial diversity is to develop a better understanding of the nature of adaptive variation and niche differentiation. For example, are niche differences among microorganisms primarily determined by differences in gene content or by divergence at homologous loci? Which makes a greater contribution to niche evolution: mutation or horizontal gene transfer? How many genes are involved? With respect to the latter, no general conclusion has been reached regarding whether niche differentiation is determined by many genes of individually small effect or by one or few genes of major effect (Orr, 1998). Some fitness-related phenotypes are known to have a simple genetic basis (e.g. antibiotic resistance), but it is an open question whether many evolutionary changes are required for significant modification of complex ecological traits with a polygenic basis (e.g. thermal performance). Can we even distil general answers to these questions, or are they taxon specific?

A family of approaches which has come to be referred to as population genomics has great potential to address these issues. Population genomics – a term coined by Gulcher and Stefansson (1998) that encompasses a variety of approaches which use genome-wide sampling of population genetic variation to identify the loci responsible for phenotypic differences among related organisms – has transformed the study of adaptation and genetic disease in both model and non-model eukaryotic systems (Hoh and Ott, 2003; Luikart *et al.*, 2003; Wall and Pritchard, 2003; Storz, 2005).

Population genomics approaches share three key principles: (1) the pattern of variation for neutral loci will reflect a population's demographic history; (2) in contrast, loci under selection will exhibit outlier patterns of variation against this neutral background – for example, at a locus that underlies an ecological trait, selection for different alleles between environments is expected to increase the amount of genetic differentiation at this locus relative to that observed for unlinked neutral loci (Barton, 1999); and (3) provided that a genetic map is available, the chromosomal positions of outlier marker loci statistically associated with phenotypic variation can be identified, which both enables inferences regarding the genetic architecture of the trait (e.g. the number of loci involved and their relative contributions) and localizes selected loci to specific chromosomal regions for finer-scale analysis of its molecular basis.

The expected difference in the pattern of variation for neutral and selected loci presupposes that recombination breaks up non-random associations between particular alleles at nearby loci due to physical linkage. The utility of the approach therefore depends upon sufficient recombination, and would not be appropriate for a bacterial taxon which does not readily exchange DNA. Though originally applied to systems for which experimental crossing designs are tractable or sampled individuals are of known pedigree, population genomics principles have more recently been used to investigate natural populations of non-model organisms, in this case taking advantage of naturally occurring variation resulting from historic recombination events (Schlötterer, 2002; Luikart *et al.*, 2003; Morin *et al.*, 2004; Storz, 2005). Examples of the diverse natural systems for which the genetic basis of fitness-related traits have been inferred include: drug resistance in *Plasmodium falciparum* (Wootton *et al.*, 2002); warfarin resistance in rats (Kohn *et al.*, 2000); ecological diversification along a salinity gradient in the intertidal snail *Littorina saxatilis* (Wilding *et al.*, 2001); and ecological specialization along an altitudinal gradient in deer mice (Storz and Dubach, 2004).

A schematic population genomics approach is illustrated in Fig. 6.3, and its potential application illustrated below for the investigation of the genetic basis of evolved differences in thermal

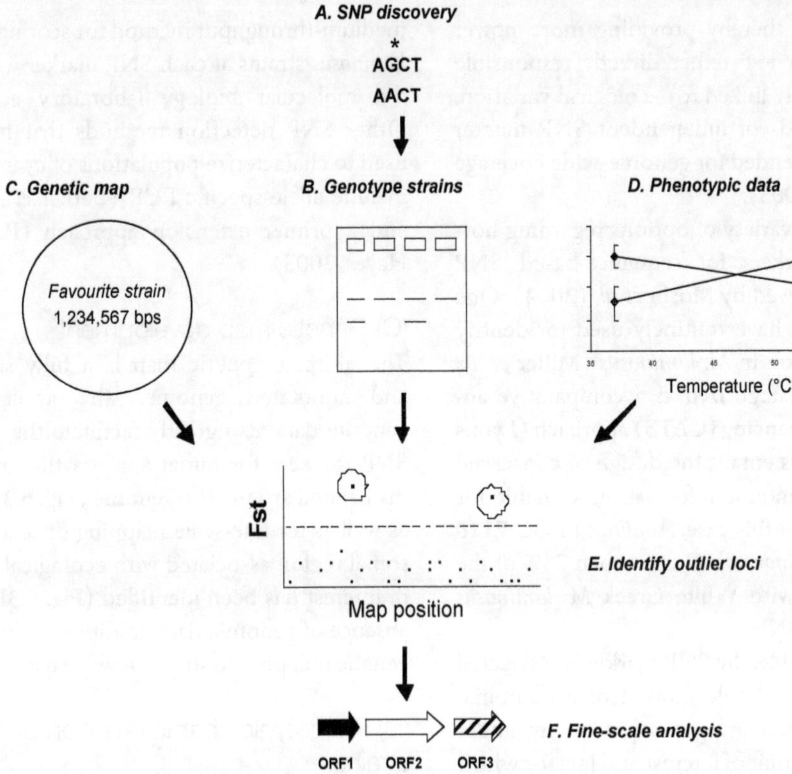

Figure 6.3 Flow chart depicting a population genomics approach for identifying the genetic basis of niche differentiation in *M. laminosus*. See text for details.

performance among specialist lineages of *M. laminosus* at White Creek (see above). The approach is feasible, given that recombination in *M. laminosus* reduces the extent of linkage between loci, thereby increasing the power to resolve specific chromosomal regions under selection. What is specifically described is an example of an 'outlier analysis' (Luikart *et al.*, 2003). The interested reader should consult several recent reviews for related alternative population genomics approaches for identifying genetic variation of ecological importance (Schlötterer, 2002; Storz, 2005; Vasemägi and Primmer, 2005; Falush and Bowden, 2006; Whitaker and Banfield, 2006).

The main steps of the approach (Fig. 6.3) are (A) the discovery of variable molecular markers segregating in the population(s) under investigation for (B) genome-wide typing of individual strains; (C) the development of a high-density genetic map for identifying the position of markers along the chromosome; (D) phenotypic characterization of strains with respect to an ecological trait of interest; (E) the identification of outlier markers that exhibit exceptional genetic divergence between ecologically differentiated lineages; and (F) fine-scale mapping of chromosomal regions associated with ecological variation to identify the likely targets of selection. I briefly consider each in turn.

(A) SNP discovery

Because they can provide broader genome coverage and higher quality data than microsatellites or AFLPs, SNPs have rapidly become the molecular markers of choice for the genome-wide sampling required by population genomics approaches (Brumfield *et al.*, 2003; Morin *et al.*, 2004). The purpose of genotyping strains at multiple markers is two-fold. First, a greater number of loci can better resolve chromosomal regions associated with phenotypic variation to a manageable size for subsequent fine-scale analyses (see F.). Second, multiple markers provide a more accurate estimate of the baseline patterns of variation for unselected loci, thereby providing more power to detect outlier loci either directly responsible for (or genetically linked to) ecological variation. Tens to hundreds of independent SNP marker loci are recommended for genome-wide coverage (Luikart *et al.*, 2003).

There are a variety of options regarding how to develop markers for sequence-based SNP discovery (reviewed by Morin *et al.* (2004). One strategy that we have routinely used to identify sequence variation in *M. laminosus* (Miller *et al.*, 2006) may be categorized as a comparative anchor tagged sequencing (CATS) approach (Lyons *et al.*, 1997). This entails the design of conserved primers from genomic information available for close relatives (in this case, *Anabaena* PCC 7120 and *Nostoc punctiforme*). Greater than 75% of the markers tested with White Creek *M. laminosus* have contained SNPs.

Where possible, the SNP marker loci selected for development should consist of a combination of both candidate genes potentially under selection for the trait of interest and loci for which there is no expectation of a relationship with fitness (to provide information on baseline levels of genetic differentiation for the discrimination of outlier locus behaviour). Although the success of a population genomics approach is *not* contingent upon identification of potential candidates in advance, including candidate gene loci based on one's understanding of ecological variation in the population may facilitate identification of genes under selection. Choice of candidates, of course, depends on the system under investigation. In the case of *M. laminosus*, these might include genes known to contribute to thermal physiology in cyanobacteria, such as *hik33*, a histidine kinase involved in regulation of the cold-temperature response (Murata and Los, 2006), and *hik34*, involved in acquired thermotolerance in cyanobacteria (Suzuki *et al.*, 2005).

(B) SNP detection

There is a dizzying and ever-growing array of options regarding SNP detection methods [see Kim and Misra (2007) for a recent review], the choice of which should be guided by factors such as the scale of genotyping required, cost-effectiveness and access to technology. We have found PCR-RFLP ('snip-SNP') to be a reasonable low-tech, medium-throughput method for scoring ~150 *M. laminosus* strains at each SNP marker with standard molecular biology laboratory equipment. Other SNP detection methods that have been used to characterize populations of cyanobacteria include allele-specific PCR (Barker *et al.*, 2000) and a primer extension approach (Batley and Hayes, 2003).

(C) Genetic map development

The ultimate genetic map is a fully sequenced and annotated genome. The availability of genome data also greatly facilitate the design of SNP markers for initial analysis that are evenly distributed around the genome (Fig. 6.3A and E), as well as for fine-scale mapping once a chromosomal region associated with ecological variation of interest has been identified (Fig. 6.3F). In the absence of genomic data, traditional physical and genetic mapping strategies may be used.

(D) Phenotypic characterization of strains

Comparing patterns of genetic differentiation among samples of individuals that are grouped according to their phenotypic differences can be a particularly powerful way to identify genomic regions underlying ecological variation (Storz, 2005). In the case of *M. laminosus*, for example, this would entail the comparison of ecologically divergent strain groups assigned based on rank performance at different temperatures. It is recommended that one assay cryopreserved strain stocks that have not evolved under (and perhaps adapted to) routine maintenance conditions in the laboratory.

(E) Genome scans to identify and map outlier loci associated with ecological variation

Divergent natural selection between different environments is expected to produce a specific distribution pattern of contemporary sequence variation. Whereas the demographic history of a population (e.g. population growth) will similarly affect all regions of the genome, selection will result in increased genetic differentiation at selected (and linked) loci between groups compared with the neutral background (Charlesworth *et al.*, 1997; Barton, 1999). Genome scans (Luikart *et al.*, 2003; Storz, 2005) of SNP marker data can be used to identify the chromosomal position of outlier loci that exhibit extraordinary genetic differentiation (i.e. high F_{ST} values) between groups of samples from different geographic locations or which differ in phenotype. The statistical identification of outlier loci may be accomplished by either a model-based approach which simulates null distributions according to assumptions about population structure (e.g. Beaumont and Nichols, 1996) or a model-free approach based on an empirical null distribution generated for the test statistic e.g. (Akey *et al.*, 2002; Storz *et al.*, 2004). Use of the latter avoids any (possibly violated) assumptions about population history, but does require the use of many (tens to hundreds) markers to generate a robust empirical distribution for identifying outlier loci (Luikart *et al.*, 2003). The end product of a genome scan is an estimate of the minimum number of loci that putatively contribute to ecological variation (i.e. the 'How many genes?' question). In Fig. 6.3E, two loci (circled) exhibit outlier behaviour and would be selected for fine-scale analysis (Fig. 6.3F).

(F) Fine-scale mapping of genomic regions associated with ecological variation

M. laminosus genome size has been estimated to be approximately 7 Mbp based on DNA reassociation kinetics (Herdman and Carr, 1974). 100 SNP markers would thereby achieve an average resolution of approximately 70 kbp. Following initial mapping, a more precise identification of the target(s) of selection (i.e. positional cloning) within chromosomal regions containing an outlier marker can be accomplished through the development of additional gene-targeted SNP markers to localize the centre of maximal genetic differentiation within the region. Sequencing through the localized region for ecologically divergent groups enables inference of the nature of the variation that underlies phenotypic divergence, thus getting at the heart of fundamental questions introduced above: whether, for example, this variation consists of allelic differences at homologous loci, or, alternatively, differences in gene content.

Tight association between genetic and phenotypic variation, in combination with the molecular signatures of selection, constitute strong evidence for adaptation, yet formal genetic proof remains the ultimate arbiter of the functional significance of variation. Though at present we do not have a system for the genetic manipulation of *M. laminosus*, we are optimistic based on the evidence for extensive genetic exchange in this organism that one will be realized in the future, as has been accomplished for its close relatives *Anabaena* PCC 7120 and *N. punctiforme*.

References

Akey, J.M., Zhang, G., Zhang, K., Jin, L., and Shriver, M.D. (2002). Interrogating a high-density SNP map for signatures of natural selection. Genome Res. *12*, 1805–1814.

Barker, G.L.A., Hayes, P.K., O'Mahony, S.L., Vacharapiyasophon, P., and Walsby, A.E. (1999). A molecular and phenotypic analysis of *Nodularia* (Cyanobacteria) from the Baltic Sea. J. Phycol. *35*, 931–937.

Barker, G.L.A., Handley, B.A., Vacharapiyasophon, P., Stevens, J.R., and Hayes, P.K. (2000). Allele-specific PCR shows that genetic exchange occurs among genetically diverse *Nodularia* (Cyanobacteria) filaments in the Baltic Sea. Microbiol. *146*, 2865–2875.

Barton, N. (1999). Clines in polygenic traits. Genet. Res. *74*, 223–236.

Batley, J., and Hayes, P.K. (2003). Development of high throughput single nucleotide polymorphism genotyping for the analysis of *Nodularia* (Cyanobacteria) population genetics. J. Phycol. *39*, 248–252.

Beaumont, M.A., and Nichols, R.A. (1996). Evaluating loci for use in the genetic analysis of population structure. Proc. Roy. Soc. Lond. B 263, 1619–1626.

Bhaya, D., Grossman, A.R., Steunou, A.-S., Khuri, N., Cohan, F.M., Hamamura, N., Melendrez, M.C., Bateson, M.M., Ward, D.M., and Heidelberg, J.F. (2007). Population level functional diversity in a microbial community revealed by comparative genomic and metagenomic analyses. ISME J. *1*, 703–713.

Brocks, J.J., Logan, G.A., Buick, R., and Summons, R.E. (1999). Archean molecular fossils and the early rise of eukaryotes. Science *285*, 1033–1036.

Brumfield, R.T., Beerli, P., Nickerson, D.A., and Edwards, S.V. (2003). The utility of single nucleotide polymorphisms in inferences of population history. Trends Ecol. Evol. *18*, 249–256.

Castenholz, R.W. (1996). Endemism and biodiversity of thermophilic cyanobacteria. Nova Hedwigia Beih. *112*, 33–47.

Charlesworth, B., Nordborg, M., and Charlesworth, D. (1997). The effects of local selection, balanced polymorphism and background selection on equilibrium patterns of genetic diversity in subdivided populations. Genet. Res. *70*, 155–174.

Coleman, M.L., Sullivan, M.B., Martiny, A.C., Steglich, C., Barry, K., DeLong, E.F., and Chisholm, S.W. (2006).

Genomic islands and the ecology and evolution of *Prochlorococcus.* Science *311*, 1768–1770.

Elena, S.F., and Lenski, R.E. (2003). Evolution experiments with microorganisms: The dynamics and genetic bases of adaptation. Nature Rev. Genet. *4*, 457–469.

Falush, D., and Bowden, R. (2006). Genome-wide association mapping in bacteria? Trends Microbiol. *14*, 353–355.

Finlay, B.J. (2002). Global dispersal of free-living microbial eukaryote species. Science *296*, 1061–1063.

Flores, E., Muro-Pastor, A.M., and Meeks, J.C. (2008). Gene transfer to cyanobacteria in the laboratory and in nature. In The Cyanobacteria: Molecular Biology, Genetics and Evolution, Herrero, A., and Flores, E., eds. (Norfolk, UK: Caister Academic Press) pp. 45–57.

Garcia-Pichel, F., Nubel, U., and Muyzer, G. (1998). The phylogeny of unicellular, extremely halotolerant cyanobacteria. Arch. Microbiol. *153*, 344–351.

Gillespie, J.H. (1991). The Causes of Molecular Evolution (New York: Oxford Univ. Press).

Gulcher, J., and Stefansson, K. (1998). Population genomics: Laying the groundwork for genetic disease modeling and targeting. Clin. Chem. Lab. Med. *36*, 523–527.

Herdman, M., and Carr, N. (1974). Estimation of the genome size of blue-green algae from DNA renaturation rates. Arch. Microbiol. *99*, 251–254.

Hey, J., and Nielsen, R. (2004). Multilocus methods for estimating population sizes, migration rates and divergence time, with applications to the divergence of *Drosophila pseudoobscura* and *D. persimilis.* Genetics *167*, 747–760.

Hofmann, H.J. (1976). Precambrian microflora, Belcher Islands, Canada: Significance and systematics. J. Paleontol. *50*, 1040–1073.

Hoh, J., and Ott, J. (2003). Mathematical multi-locus approaches to localizing complex human trait genes. Nature Rev. Genet. *4*, 701–709.

Huey, R.H., and Kingsolver, J.G. (1993). Evolution of resistance to high temperature in ectotherms. Am. Nat. *142*, S21–S46.

Hughes, J.B., Daily, G.C., and Ehrlich, P.R. (1997). Population Diversity: Its Extent and Extinction. Science *278*, 689–692.

Jolley, K.A., Wilson, D.J., Kriz, P., McVean, G., and Maiden, M.C.J. (2005). The influence of mutation, recombination, population history, and selection on patterns of genetic diversity in *Neisseria meningitidis.* Mol. Biol. Evol. *22*, 562–569.

Kassen, R., and Rainey, P.B. (2004). The ecology and genetics of microbial diversity. Annu. Rev. Microbiol. *58*, 207–231.

Kim, S., and Misra, A. (2007). SNP genotyping: Technologies and biomedical applications. Annu. Rev. Biomed. Eng. *9*, 289–320.

Kimura, M. (1983). The Neutral Theory of Molecular Evolution (Cambridge: Cambridge University Press).

Kohn, M.H., Pelz, H.-J., and Wayne, R.K. (2000). Natural selection mapping of the warfarin-resistance gene. Proc. Natl. Acad. Sci. U.S.A. *97*, 7911–7915.

Kurmayer, R., and Gumpenberger, M. (2006). Diversity of microcystin genotypes among populations of the filamentous cyanobacteria *Planktothrix rubescens* and *Planktothrix agardhii.* Mol. Ecol. *15*, 3849–3861.

Lindell, D., Sullivan, M.B., Johnson, Z.I., Tolonen, A.C., Rohwer, F., and Chisholm, S.W. (2004). Transfer of photosynthesis genes to and from *Prochlorococcus* viruses. Proc. Natl. Acad. Sci. U.S.A. *101*, 11013–11018.

Lodders, N., Stackebrandt, E., and Nübel, U. (2005). Frequent genetic recombination in natural populations of the marine cyanobacterium *Microcoleus chthonoplastes.* Environ. Microbiol. *7*, 434–442.

Luck, G.W., Daily, G.C., and Ehrlich, P.R. (2003). Population diversity and ecosystem services. Trends Ecol. Evol. *18*, 331–336.

Luikart, G., England, P.R., Tallmon, D., Jordan, S., and Taberlet, P. (2003). The power and promise of population genomics: From genotyping to genome typing. Nat. Rev. Genet. *4*, 981–994.

Lynch, M., and Gabriel, W. (1987). Environmental tolerance. Am. Nat. *129*, 283–303.

Lyons, L.A., Laughlin, T.F., Copeland, N.G., Jenkins, N.A., Womack, J.E., and O'Brien, S.J. (1997). Comparative anchor tagged sequences (CATS) for integrative mapping of mammalian genomes. Nat. Genet. *15*, 47–56.

Martiny, J.B.H., Bohannan, B.J.M., Brown, J.H., Colwell, R.K., Fuhrman, J.A., Green, J.L., Horner-Devine, M.C., Kane, M., Krumins, J.A., Kuske, C.R., *et al.* (2006). Microbial biogeography: putting microorganisms on the map. Nature Rev. Microbiol. *4*, 102–112.

McVean, G., Awadalla, P., and Fearnhead, P. (2002). A coalescent-based method for detecting and estimating recombination from gene sequences. Genetics *160*, 1231–1241.

Miller, S.R., and Castenholz, R.W. (2000). Evolution of thermotolerance in hot spring cyanobacteria of the genus *Synechococcus.* Appl. Environ. Microbiol. *66*, 4222–4229.

Miller, S.R., and Bebout, B.M. (2004). Variation in sulfide tolerance of photosystem II in phylogenetically diverse cyanobacteria from sulfidic habitats. Appl. Environ. Microbiol. *70*, 736–744.

Miller, S.R., Purugganan, M.D., and Curtis, S.E. (2006). Molecular population genetics and phenotypic diversification of two populations of the thermophilic cyanobacterium *Mastigocladus laminosus.* Appl. Environ. Microbiol. *72*, 2793–2800.

Miller, S.R., Castenholz, R.W., and Pedersen, D. (2007). Phylogeography of the thermophilic cyanobacterium *Mastigocladus laminosus.* Appl. Environ. Microbiol. *73*, 4751–4759.

Miller, S.R., Williams, C., Strong, A.L. and Carvey, D. (2009). Ecological specialization in a spatially structured population of the thermophilic cyanobacterum *Mastigocladus laminosus.* Appl. Environ. Microbiol. *75*, 729–734.

Moore, L.R., Rocap, G., and Chisholm, S.W. (1998). Physiology and molecular phylogeny of coexisting *Prochlorococcus* ecotypes. Nature *393*, 464–467.

Morin, P.A., Luikart, G., Wayne, R.K., and the S.N.P workshop group. (2004). SNPs in ecology, evolution and conservation. Trends Ecol. Evol. *19*, 208–216.

Murata, N., and Los, D. (2006). Histidine kinase Hik33 is an important participant in cold-signal transduction in cyanobacteria. Physiol. Plant. *126*, 17–27.

Myers, S.R., and Griffiths, R.C. (2003). Bounds on the minimum number of recombination events in a sample history. Genetics *163*, 375–394.

Nielsen, R., and Wakeley, J. (2001). Distinguishing migration from isolation: A Markov chain Monte Carlo approach. Genetics *158*, 885–896.

Orr, H.A. (1998). The population genetics of adaptation: the distribution of factors fixed during adaptive evolution. Evolution *52*, 935–949.

Papke, R.T., Ramsing, N.B., Bateson, M.M., and Ward, D.M. (2003). Geographic isolation in thermophilic cyanobacteria. Environ. Microbiol. *5*, 650–659.

Posada, D., and Crandall, K.A. (2001). Intraspecific gene genealogies: Trees grafting into networks. Trends Ecol. Evol. *16*, 37–45.

Rudi, K., Skulberg, O.M., and Jakobsen, K.S. (1998). Evolution of cyanobacteria by exchange of genetic material among phyletically related strains. J. Bacteriol. *180*, 3453–3461.

Schlötterer, C. (2002). Towards a molecular characterization of adaptation in local populations. Curr. Opin. Genet. Dev. *12*, 683–687.

Schluter, D. (2000). The Ecology of Adaptive Radiation (Oxford: Oxford University Press).

Schopf, J.W. (1992). Proterozoic prokaryotes: Affinities, geologic distribution and evolutionary trends. In The Proterozoic Biosphere: A Multidisciplinary Study, Schopf, J.W., and Klein, C., eds. (Cambridge: Cambridge University Press), pp. 195–232.

Storz, J.F. (2005). Using genome scans of DNA polymorphism to infer adaptive population divergence. Mol. Ecol. *14*, 671–688.

Storz, J.F., and Dubach, J.M. (2004). Natural selection drives altitudinal divergence at the albumin locus in deer mice, *Peromyscus maniculatus*. Evolution *58*, 1342–1352.

Storz, J.F., Payseur, B.A., and Nachman, M.W. (2004). Genome scans of DNA variability in humans reveal evidence for selective sweeps outside of Africa. Mol. Biol. Evol. *21*, 1800–1811.

Sullivan, M.B., Coleman, M.L., Weigele, P., Rohwer, F., and Chisholm, S.W. (2005). Three *Prochlorococcus* cyanophage genomes: Signature features and ecological interpretations. PLoS Biol. *3*, e144.

Suzuki, I., Kanesaki, Y., Hayashi, H., Hall, J., Simon, W., Slabas, A., and Murata, N. (2005). The histidine kinase hik34 is involved in thermotolerance by regulating the expression of heat shock genes in *Synechocystis*. Plant Physiol. *138*, 1409–1421.

Vasemägi, A., and Primmer, C.R. (2005). Challenges for identifying functionally important genetic variation: The promise of combining complementary research strategies. Mol. Ecol. *14*, 3623–3642.

Wall, J.D., and Pritchard, J.K. (2003). Haplotype blocks and linkage disequilibrium in the human genome. Nature Rev. Genet. *4*, 587–597.

Whitaker, R.J., and Banfield, J.F. (2006). Population genomics in natural microbial communities. Trends Ecol. Evol. *21*, 508–516.

Whitaker, R.J., Grogan, D.W., and Taylor, J.W. (2003). Geographic barriers isolate endemic populations of hyperthermophilic *Archaea*. Science *301*, 976–978.

Whitton, B.A., and Potts, M., eds. (1999). Ecology of Cyanobacteria: Their Diversity in Time and Space (Dordrecht, The Netherlands: Kluwer Acad. Publ.).

Wilding, C.S., Butlin, R.K., and Grahame, J. (2001). Differential gene exchange between parapatric morphs of *Littorina saxatilis* detected using AFLP markers. J. Evol. Biol. *14*, 611–619.

Wilson, A.E., Sarnelle, O., Neilan, B.A., Salmon, T.P., Gehringer, M.M., and Hay, M.E. (2005). Genetic variation of the bloom-forming cyanobacterium *Microcystis aeruginosa* within and among lakes: Implications for harmful algal blooms. Appl. Environ. Microbiol. *71*, 6126–6133.

Wootton, J.C., Feng, X., Ferdig, M.T., Cooper, R.A., Mu, J., Baruch, D.I., Magill, A.J., and Su, X.-z. (2002). Genetic diversity and chloroquine selective sweeps in *Plasmodium falciparum*. Nature *418*, 320–323.

Wright, S. (1943). Isolation by distance. Genetics *28*, 114–138.

Population Genetics of Microalgae

7

Jim Provan

Abstract

Algae are a highly diverse group of protists, ranging from simple, unicellular organisms to complex, multicellular entities with a range of differentiated tissues and distinct organs. They are found among diverse aquatic ecosystems and play important roles by supplying carbon and energy as well as providing habitat to other members of the biological communities. Some microalgae cause significant environmental and health problems. However, despite their importance, relatively little is known about this group of organisms. The first section of this chapter briefly summarizes our current understanding of the diversity and evolution of the algae. The second section reviews the molecular markers used to address microalgal population genetic issues. The third section summarizes the population genetic analyses of three microalgal groups: the dinoflagellates, the diatoms and the haptophytes. With the application of genomics information and technology, the field of microalgal population genetics is approaching an exciting period of development and expansion.

Introduction: Taxonomic overview of algae

Algae are a highly diverse group of protists, ranging from simple, unicellular organisms to complex, multicellular entities with a range of differentiated tissues and distinct organs. The main feature uniting all groups of algae, their primarily photoauotrophic mode of nutrition, led them to be grouped with plants in early systems of classification, but it is now accepted that they are polyphyletic (Fig. 7.1). Cyanobacteria were originally also classified with algae, being described variously as 'cyanophytes' or 'blue-green algae', but are prokaryotes and now properly classified as one of the major groups of bacteria. At the other end of the evolutionary scale, although the chlorophytes, or green algae, contained the original ancestors of 'higher' plants, they differ from plants in several respects including a lack of specialization of body plan and apical meristems, absence of gametangia, and lack of a dependent embryonic stage during development.

The phylogeny of eukaryotes is constantly being revised (e.g. Cavalier-Smith, 2002; Stechman and Cavalier-Smith, 2003; Keeling *et al.*, 2005; Burki *et al.*, 2007, 2008; Hackett *et al.*, 2007; Lane and Archibald, 2008), and the accumulating evidence suggests that many of the major groups of algae are monophyletic or at least paraphyletic. Four of these, the rhodophytes (red algae), chlorophytes (green algae), phaeophytes (brown algae) and charophytes are primarily multicellular and include the various seaweeds. The rhodophytes, chlorophytes and charophytes are classed in eukaryotic supergroup Archaeplastida (Lane and Archibald, 2008), whilst the phaeophytes belong to supergroup Chromalveolata. The eight algal groups that are primarily or exclusively comprised of microalgae are distributed across supergroups Archaeplastida (the glaucocystophytes), Excavata (the euglenids), Rhizaria (the chlorarachniophytes) and Chromalveolata (the chrysophytes, bacillariophytes, haptophytes, cryptophytes and dinoflagellates). A summary of the characteristics of the main microalgal groups

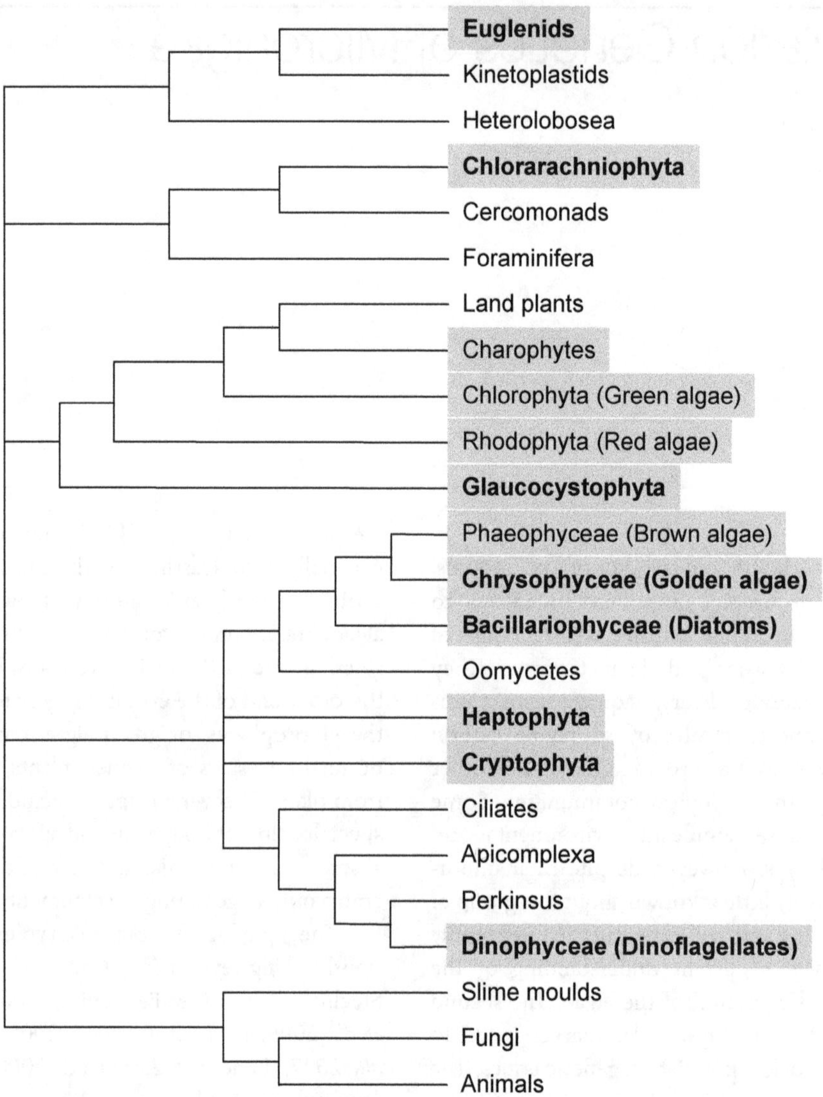

Figure 7.1 Phylogenetic tree showing the relationships between major taxonomic groups of eukaryotes. Algal groups are boxed, and groups comprising primarily or exclusively of microalgae are represented in bold (data from Palmer *et al.*, 2004).

is given in Table 7.1, and the bacillariophytes, haptophytes and dinoflagellates are discussed in further detail in the section on case studies.

Plastid evolution and diversity

It is now accepted that the plastids of algae and higher plants evolved originally via an endosymbiotic event involving a cyanobacterium *ca.* 1.5 GYA (Hedges *et al.*, 2004). In the late nineteenth century, the botanist Andreas Schimper highlighted the resemblance between plastids and cyanobacteria (Schimper, 1883), but it was not until the early 20th century that the theory of an endosymbiotic origin of plastids from cyanobacteria was first proposed (Mereschkowsky, 1905, 1910). Mereschkowsky originally proposed seven independent endosymbiotic events that gave rise to the plastids of various algal groups, but molecular evidence now suggests that a single endosymbiotic event gave rise to the three 'primary' plastids of chlorophytes, rhodophytes and glaucocystophytes, which have a double membrane corresponding to the inner and outer membranes of the symbiont. Plastids with three

Table 7.1 Characteristics of the major groups of microalgae

Supergroup	Algal taxon	Main characteristic(s)	Typical plastid pigment(s)[1]	Example genera
Excavata	Euglenids	Pellicle (submembrane proteinoid structure)	Chl. *a*/*b*, diadinoxanthin	*Euglena*
Rhizaria	Chlorarachniophyta	Typically amoeboid Plastids contain nucleomorph	Chl. *a*/*b* and violaxanthin	*Chlorarachnion*/*Bigelowiella*
Archaeplastida	Glaucocystophyta	Plastid has peptidoglycan layer	Chl. *a* and phycobiliproteins	*Glaucocystis*/*Cyanophora*
Chromalveolata	Chrysophyceae	Primarily freshwater flagellates	Chl. $a/c_1/c_2$ and fucoxanthin	*Ochromonas*/*Chrysamoeba*
	Bacillariophyceae	Frustule (hydrated silica shell wall)	Chl. $a/c_1/c_2$ and fucoxanthin	*Thalassiosira*/*Pseudo-nitzschia*
	Haptophyta	Haptonema (unique flagellum)	Chl. $a/c_1/c_2$ and fucoxanthin	*Emiliania*/*Prymnesium*
	Cryptophyta	Plastids contain nucleomorph	Chl. a/c_2 and phycobiliproteins	*Cryptomonas*
	Dinophyceae	Two distinct flagella	Chl. a/c_2 and peridinin	*Alexandrium*/*Symbiodinium*

1 Plastid pigments taken from Larkum and Vesk (2003).

or more membranes are believed to have evolved via one or more secondary endosymbioses, where a prokaryote engulfed a eukaryotic cell that already possessed a plastid.

Although there are notable differences between the three types of primary plastid, including pigmentation and the presence of a peptidoglycan cell wall in the plastids of glaucocystophytes, a further relic of their cyanobacterial ancestry, there is a large body of evidence which suggests that they subsequently diverged following a single endosymbiotic event. Most compelling of these lines of evidence is the recovery of a monophyletic clade of plastids in a wide range of phylogenetic studies (reviewed in Douglas, 1998; Palmer, 2000). Studies using plastid genes have provided strong support for monophyly (e.g. Bhattacharya and Medlin, 1995; Turner *et al.*, 1999), although this was not as pronounced when nuclear or mitochondrial genes were studied. In general, non-recovery of a monophyletic clade of plastids was attributed to lack of sampling, either of plastids or of their cyanobacterial ancestors (Douglas, 1998).

The plastids of euglenids, chlorarachniophytes, phaeophytes, chrysophytes, bacillariophytes, haptophytes, cryptophytes and dinophytes contain more than two membranes and have been gained via secondary endosymbiosis (Archibald and Keeling, 2002; Gould *et al.*, 2008). Euglenids and chlorarachniophytes have plastids that were derived via two independent symbiotic events with green algal precursors. The remaining secondary plastids appear to have diversified following a single endosymbiosis involving a red alga (the 'Chromalveolate' hypothesis; Cavalier-Smith, 2003; Harper *et al.*, 2005; Patron *et al.*, 2007). Interestingly, the non-photosynthetic apicomplexans harbour a vestigial plastid thought to have been gained during the same process, and there is also evidence that some ciliates may also have possessed plastids but subsequently lost them (Cavalier-Smith, 2003). The chlorarachniophytes and cryptophytes also contain the remnants of the endosymbiont nucleus, called the nucleomorph, in the region of the plastid which would previously have been the endosymbiont's cytosol (Greenwood *et al.*, 1977; Hibberd and Norris, 1984). Many of the nucleomorph genes have been transferred over time to the nucleus of the host, but the nucleomorph itself still retains sufficient features to betray its endosymbiotic ancestry (Gilson and McFadden, 1996; Douglas *et al.*, 2001).

Molecular markers in microalgal population genetics

The development of allozyme electrophoresis in the mid-1960s provided the first methodological approach to studying the levels and patterns of molecular genetic variation within and between natural populations of organisms (Lewontin and Hubby, 1966). Although not without their drawbacks, primarily low levels of observed variation, possible non-neutrality of loci and the need for relatively large amounts of fresh tissue, allozymes remained the marker of choice for population-level studies until the advent of polymerase chain reaction- (PCR)-based marker techniques (reviewed in Lewontin, 1991; May, 1998). The first allozyme study in microalgae was published in 1976 and described low levels of genetic variation in clones of the freshwater diatoms *Thalassiosira pseudonana* and *T. fluviatilis* (Murphy and Guillard, 1976). The observed limited variation was attributed at least in part to the limited number of loci analysed (seven), but subsequent allozyme studies that utilized similar numbers of loci revealed higher levels of variation (e.g. Murphy and Guillard, 1976; Gallagher, 1980, 1982; Skov *et al.*, 1997).

Since these original studies, a wide range of molecular marker methodologies have been utilised to examine genetic variation within and between populations of microalgae. Fig. 7.2 shows an overview of the molecular approaches used over the last two decades. The data are split into 6-year periods since 1991 (bibliographic data from before this time are incomplete, frequently lacking abstracts) and reveal several trends worthy of further comment.

As mentioned previously, allozyme electrophoresis has largely been superseded by PCR-based approaches. Similarly, the use of restriction fragment length polymorphism (RFLP), including PCR-RFLP, has decreased, probably as a result of increased affordability of DNA sequencing. 'Anonymous' marker techniques, including randomly amplified polymorphic DNA (RAPD; Williams *et al.*, 1993), inter-simple sequence

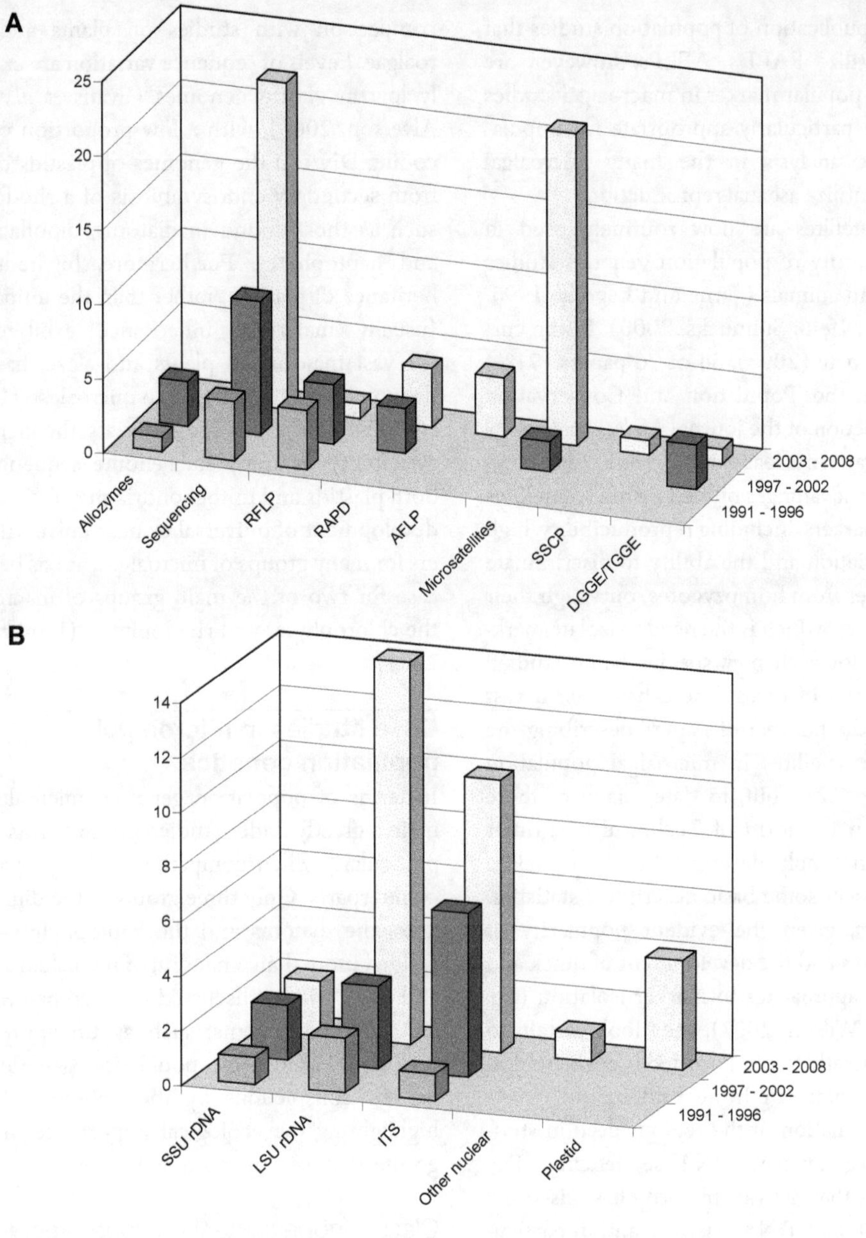

Figure 7.2 Use of molecular markers in published microalgal population genetic studies from 1991 onwards. Data were obtained by searching in either the 'Title' OR 'Topic' fields in ISI Web of Science with the following expression: (microalga* OR diatom* OR bacillariophy* OR dinoflage* OR dinophy* OR coccolith* OR prymnesi* OR haptophy* OR chlorarachnio* OR cryptophy* OR chrysophy*) AND ((population* AND genetic*) OR (allozyme* OR isozyme* OR ((enzyme* OR protein) AND electrophore*) OR microsat* OR AFLP* OR RAPD* OR RFLP* OR sequenc* OR SSCP* OR DGGE* OR TGGE*)) AND population*. (A) shows the molecular approach used, whilst (B) shows regions used in sequencing studies.

repeat polymorphism (ISSR; Zietkiewicz *et al.*, 1994) and amplified fragment length polymorphism (AFLP; Vos *et al.*, 1995) have not seen the same levels of popularity as in studies on macroalgae and higher plants. The drawbacks of these techniques include their dominant nature (i.e. the inability to discriminate heterozygotes), limited transferability across studies and, particularly in the case of RAPDs, a lack of robustness and reproducibility. As a result, several population genetics journals, including *Molecular Ecology* and *Conservation Genetics*, have implemented policies against the publication of population studies that primarily utilize RAPDs. AFLPs, however, are becoming a popular marker in macroalgal studies and may be particularly appropriate for population genetic analysis in the many microalgal groups that utilize asexual reproduction.

Microsatellites are now routinely used in the vast majority of population genetics studies on plants and animals (Jarne and Lagoda, 1996; Powell *et al.*, 1996; Sunnucks, 2000). In the current year to date (2008), 54 of 76 papers (71%) published in the 'Population and Conservation Genetics' section of the journal *Molecular Ecology* that investigated nuclear variation utilized microsatellites. The advantages offered by microsatellites as genetic markers, including reproducibility, high levels of variation and the ability to discriminate heterozygotes from homozygotes, outweigh their main drawback, which is the need to isolate markers *de novo* for each new species being studied in the majority of cases. There has been a vast increase in the number of papers describing the use of microsatellites in microalgal population studies (Fig. 7.2a), but, to date, many of these have been in the form of Technical or Primer Notes which simply describe the microsatellite loci and present some basic descriptive statistics. Nevertheless, given the evident popularity of microsatellites and the development of quick and inexpensive approaches to marker isolation (e.g. Provan and Wilson, 2007), they look certain to contribute greatly to our knowledge of microalgal population genetics in the near future.

An examination of the regions used in studies that have involved DNA sequencing (Fig. 7.2b) reveals that the vast majority have used the ribosomal DNA (rDNA) cistron, and in particular the large subunit (LSU or 28S) rDNA or the internal transcribed spacer (ITS). The popularity of the rDNA cistron arises from the availability of universal primers, originally designed for fungi (White *et al.*, 1990), and the fact that molecular studies on several groups of dinoflagellates and diatoms use 'standard' rDNA clade designations to delineate taxa e.g. the various 'ribotypes' of the *Alexandrum tamarense* species complex (Scholin *et al.*, 1994, 1995) and the A-H designations of *Symbiodinium* spp. (LaJeunesse, 2001; Pochon *et al.*, 2006) in dinoflagellates. Organellar regions, however, have been vastly under-utilized in comparison with studies on plants and macroalgae. Levels of sequence variation are extremely low in plastid genomes (Evans *et al.*, 2007; Alverson, 2008), with a low proportion of noncoding DNA in the genomes of plastids derived from secondary endosymbiosis of a rhodophyte such as those found in diatoms, dinoflagellates and haptophytes. Furthermore, biparental inheritance of plastids, rather than the uniparental (usually maternal) inheritance exhibited by the vast majority of plants and algae, has been documented in at least some microalgae (Ghiron *et al.*, 2008). Despite these caveats, the increasing availability of complete genome sequences for both plastids and mitochondria may facilitate the development of universal or near-universal primers for many groups of microalgae, as has been the case for two of the main groups of macroalgae, the chlorophytes and rhodophytes (Provan *et al.*, 2004).

Case studies in microalgal population genetics

In terms of population genetics, microalgae remain a greatly under-studied group of organisms, particularly when compared to their macroalgal counterparts. Only three groups – the dinoflagellates, the diatoms and the haptophytes – have yielded more than a handful of published studies, and these will be discussed in detail below. Even for model organisms, such as *Chlamydomonas* spp. and *Volvox* spp., population genetic studies are conspicuous by their absence, further highlighting the ecological importance of those groups that have been studied.

Class Dinophyceae (Dinoflagellates)

The dinoflagellates are a large group of algae characterized by two unequal flagella used for motility. They are found in both marine and freshwater habitats, can be free-living or symbiotic and range in mode of nutrition from obligate heterotrophs through mixotrophs to photosynthetic autotrophs with a wide diversity of plastids. These plastids are among the most reduced of all the photosynthetic eukaryotes and have been gained by secondary and tertiary endosymbioses from both red and green plastid-containing lineages (Hackett *et al.*, 2004). Beyond the evolution of plastid genomes, interest in dinoflagellates is mainly focused on their roles in two important ecological processes, namely the production of toxic blooms ('red tides') that can have serious impacts on many species including fish, shellfish, seabirds and humans (reviewed in Hallegraeff, 1993), and as endosymbionts in mutualistic associations with reef-forming corals (reviewed in Trench, 1993). Consequently, the vast majority of population genetic studies in dinoflagellates have concentrated on these two areas.

Harmful algal blooms (HABs) are primarily caused by aggregations of dinoflagellates that produce neurotoxins, often leading to outbreaks of paralytic, diarrhetic and neurotoxic shellfish poisoning. Filter feeders, such as mussels and scallops, can accumulate these toxins in such quantities that subsequent consumption by humans can lead to illness and, in some cases, death. Population genetic approaches have been used to study the spread of toxic dinoflagellates in an attempt to elucidate the relative roles of natural dispersal and human-mediated introductions via ballast water of ships. Studies on *Alexandrium tamarense* and related species using ribosomal DNA sequences have revealed at least five geographically distinct evolutionary lineages, or ribotypes, that comprise the '*tamarensis* species complex' (Scholin *et al.*, 1994, 1995). The importance of anthropogenic dispersal in the spread of these lineages, however, remains somewhat equivocal. Sebastian *et al.* (2005) concluded that a human-mediated introduction of *A. catenella* into South Africa was unlikely, but conceded that the ribosomal DNA region used was lacking in resolution. AFLPs, on the other hand, had previously been shown to be too variable (John *et al.*, 2004), but a subsequent study using microsatellites revealed that although geographic distances between populations in Asia were generally correlated with genetic distances, suggesting natural dispersal with restricted gene flow, there was evidence for human-mediated dispersal between populations (Nagai *et al.*, 2007). Similarly, anthropogenic transfer from Japan to France was implicated as the cause of an outbreak of paralytic shellfish poisoning in Thau Lagoon, France, based on molecular and toxin data (Lilly *et al.*, 2002), whilst a combined analysis of molecular, biochemical and palaeobiogeographical data also suggested anthropogenic introduction of *A. catanella* and *A. tamarense* from Asia to Australasia within the last 100 years (Bolch *et al.*, 2007). The same study highlighted the possible introduction of another toxic dinoflagellate species, *Gymnodinium catenatum*, from Japan to Australia, although earlier studies on the same species using RAPDs had failed to identify the source of these putative introduced populations (Adachi *et al.*, 1997; Bolch *et al.*, 1999). A study on an *Alexandrium* bloom from the Orkney Islands highlighted genetic similarities with a North American population but concluded that the level of sequence divergence suggested natural, rather than anthropogenic, dispersal (Medlin *et al.*, 1997).

Many species of dinoflagellates belonging to the genus *Symbiodinium*, collectively known as zooxanthellae, form mutualistic associations with tropical reef-building corals (Trench, 1987). The algae, which are also found in symbiotic associations with sponges, foraminiferans, anemones and ciliates among other taxa, fix carbon from the atmosphere via photosynthesis and transfer it to the host, which in turn provides nutrients, metabolites and protective habitat for the alga (Trench, 1993; Marshall, 1996). Loss of the symbionts or their pigments due to extremes of light and/or temperature gives rise to the phenomenon of 'coral bleaching', which can cause large-scale loss or degradation of reefs (Brown *et al.*, 2000). All of these symbiotic dinoflagellates were originally considered to belong to a single species, *S. microadriaticum*, but, as with toxic dinoflagellates, current taxonomic divisions within *Symbiodinium* are based on rDNA sequence variation and classify the genus into at least eight major clades, designated A-H (LaJeunesse, 2001; Pochon *et al.*, 2006), at least five of which (A, B, C, D and G – very rarely, possibly F) are found associated with corals. Due to the difficulty in identifying *Symbiodinium* species morphologically, molecular genetic approaches have provided the only means of addressing issues such as host/symbiont specificity and ecological niche specialization.

Theoretical expectations would suggest that symbiotic dinoflagellates should harbour less genetic variation than their free-living counterparts, but early work using 18S rDNA revealed variation within the genus *Symbiodinium* consistent with that found at the ordinal level in other dinoflagellates (Rowan and Powers, 1992). Likewise, although early molecular studies on the relationships between host coral species and symbiotic zooxanthellae had suggested a one host – one symbiont relationship (Rowan and Powers, 1991a,b), subsequent work has indicated the occurrence of multiple *Symbiodinium* types within a species and even within individual colonies (Rowan and Knowlton, 1995; Rowan *et al.*, 1997; van Oppen *et al.*, 2001; Ulstrup and van Oppen, 2003; LaJeunesse *et al.*, 2004; Magalon *et al.*, 2006). A series of population genetic studies now seems to indicate that the distribution of zooxanthellae within and between regions and populations is most likely due to a combination of biogeographic, adaptive and reproductive factors rather than being primarily taxon-specific.

A comparison between the two main groups of corals, the octocorals and the scleractinians, has revealed both taxonomic and biogeographic patterns in the distribution of zooxanthellae types (van Oppen *et al.*, 2005). In the Great Barrier Reef, both groups are characterized by the dominance of clade C and clade D rDNA types, with clades A and B being much rarer and clade G types only being found occasionally in octocorals (van Oppen *et al.*, 2001; LaJeunesse *et al.*, 2003). Clades C and D are also found in eastern Pacific scleractinian corals but octocorals from the same region have been found to be azooxanthellate. Finally, Caribbean scleractinians harbour a much wider range of symbionts than octocorals, with clades A, B and C all being common in scleractinian corals but only clade B symbionts being found at high frequency in octocorals (LaJeunesse, 2002; Goulet and Coffroth, 2004; Santos *et al.*, 2004). The use of higher-resolution markers, such as the ribosomal DNA cistron ITS and microsatellites, has also revealed further genetic structuring at smaller geographic scales (e.g. Loh *et al.*, 2001; Santos *et al.*, 2003, 2004; Magalon *et al.*, 2006).

One explanation for the distribution of zooxanthellae symbionts within and between reefs was that differentially photo-adapted symbionts were found in regions with differing regimes of irradiance. This was originally proposed in the first studies to demonstrate the occurrence of multiple *Symbiodinium* types in a single host, where the composition of symbiont diversity was found to be correlated with levels of irradiance (Rowan and Knowlton, 1995; Rowan *et al.*, 1997). Subsequent studies have also found light-dependent distribution of clades and, using higher-resolution markers, subtypes, but have also found that host specificity, thermal tolerance and biogeography play important roles, further confirming the role of multiple factors in determining the diversity and distribution of dinoflagellate symbionts in corals (Ullstrup and van Oppen, 2003; Frade *et al.*, 2008; Thornhill *et al.*, 2008). Interestingly, recent work on the relative benefits of separate symbiont types to their hosts has even suggested that clade A zooxanthellae may be interacting parasitically, rather than mutualistically, with their host corals, and possibly compromising host fitness via reduced carbon provision (Stat *et al.*, 2008).

It has been hypothesized that two different forms of transmission of zooxanthellae in corals might also influence the patterns of genetic structuring observed in symbiont populations. Some species transmit zooxanthellae 'vertically' to offspring, whereas others form azooxanthellate oocytes, and the larvae subsequently acquire symbionts 'horizontally' from the environment. Consequently, broadcast spawning coral species should harbour higher levels of genetic diversity due to repeated uptake of symbionts through successive generations, with low levels of population substructuring. In contrast, species exhibiting maternal inheritance of zooxanthellae should have low intracolony diversity but high differentiation between colonies. Patterns of genetic diversity consistent with these hypotheses were demonstrated in a comparison between the coral species *Seriatopora hystrix*, a vertical transmitter, and *Acropora longicyathus*, which has an open-spawning system with horizontal transmission (Loh *et al.*, 2001), but a subsequent study that compared *Acropora* spp. with another closed-system genus, *Montipora*, found similar levels of dinoflagellate diversity in the two genera (van Oppen, 2004). This lack of a direct correlation between mode of transmission and genetic structure may be due in part to a more flexible mode of zooxanthellae inheritance, where species that exhibit vertical transmission may also subsequently acquire symbionts from the environment (Magalon *et al.*, 2006). Such a strategy could offer a selective advantage in terms of flexible response to episodes of environmental change, including disturbance and bleaching events (Lewis and Coffroth, 2004).

Class Bacillariophyceae (diatoms)

Diatoms are probably the most species-rich group among all the algae, numbering at least *ca.* 100,000 species belonging to more than 200 genera (Round and Crawford, 1990; Canter-Lund and Lund, 1995). They are primarily marine organisms, but are also found in freshwater and terrestrial habitats and are important components of the global cycling of both carbon and silicon (Mann, 1999). Diatom cells are characterized by a cell wall comprising two overlapping silica plates, called frustules, and are often found in colonial assemblages. During asexual reproduction, daughter cells inherit one of these frustules each, and grow a smaller second plate. Successive rounds of asexual reproduction lead to smaller and smaller cell sizes until a minimum viable cell size is reached. At this stage, the individual escapes the 'miniaturisation trap' by sexual production of an auxospore, which expands in size and allows subsequent asexual reproduction. The chloroplasts of diatoms have been gained by secondary endosymbiosis from a rhodophyte progenitor and, like the other heterokont algae (the phaeophytes and the chrysophytes), contain chlorophylls *a* and *c* and several carotenoid pigments.

If population genetics can be defined as the study of genetic variation below the species level, the definition of what constitutes a population genetics study in diatoms is rendered somewhat difficult as a result of the immense taxonomic confusion within the class Bacillariophyceae. Although the topics of phylogenetic relationships and species concepts in diatoms are largely beyond the scope of this review [see Mann (1999) for an excellent overview of the latter], many population genetic studies have revealed the existence of cryptic or pseudo-cryptic species. Analyses of genetic variation in the ribosomal DNA cistron, particularly the internal transcribed spacer (ITS), and at microsatellite loci, have identified cryptic species in the genera *Pseudo-nitzschia* (Orsini *et al.*, 2004; Evans *et al.*, 2004; Lundholm *et al.*, 2006), *Skeletonema* (Medlin *et al.*, 1991; Sarno *et al.*, 2005; Zingone *et al.*, 2005) and *Cyclotella* (Beszteri *et al.*, 2005; Beszteri *et al.*, 2007) and in many cases, pseudo-cryptic ultrastructural variation has subsequently been found between genetically delineated groups (e.g. Amato *et al.*, 2007). Crossing studies have also shown reproductive isolation between distinct groups of genetically related individuals (Orsini *et al.*, 2004; Evans *et al.*, 2004; Amato *et al.*, 2007; Vanormelingen *et al.*, 2008), further confirming the occurrence of cryptic species defined by both the biological and phylogenetic species concepts (Mann, 1999; Alverson. 2008). Interestingly, it has been demonstrated that variation in ITS secondary structure is correlated with reproductive isolation in at least some *Pseudo-nitzschia* species (Amato *et al.*, 2007), as had been shown previously in the chlorophycean alga *Gonium pectorale* (Coleman, 2000).

The distributions of microbial organisms have been of great interest and debate since the Dutch microbiologist Lourens Baas Becking stated that 'Everything is everywhere, but the environment selects' (Baas Becking, 1934), and the implications of the 'EisE hypothesis' have recently been extended to address to concept of variation below the species level (e.g. Medlin, 2007). The discovery of cryptic taxa within previously described diatom species has forced a reappraisal of the distribution of species over a range of spatial and temporal scales, from global to within individual blooms. In several cases, species that had previously been assumed to have a cosmopolitan distribution were subsequently shown to be assemblages of biogeographically distinct cryptic species. Findlay *et al.* (2002) analysed reports of 83 diatom species and classified *Cyclotella meneghiniana*, *Gomphonema parvulum*, *Navicula cryptocephala* and *Nitzschia palea* as cosmopolitan. A reanalysis by Mann (2007), however, highlighted that in each case taxonomic misidentification in at least part of the range had rendered these designations invalid and that none of the four were truly cosmopolitan. Nevertheless, recent genetic, morphological and reproductive compatibility analyses suggest that *Pseudo-nitzschia pungens* represents a truly global diatom species (Evans *et al.*, 2005; Casteleyn *et al.*, 2008), although genetic studies on other morphologically defined diatom species that were assumed to have global distributions revealed the existence of discrete gene pools (Evans *et al.*, 2004; Rynearson and Armbrust, 2004; Lundholm *et al.*, 2006; Kooistra *et al.*, 2008).

At more local scales, molecular studies have provided insights into the population genetic processes associated with the formation of diatom blooms. Under appropriate irradiance conditions and with sufficient nutrients, diatoms undergo rapid phases of asexual reproduction and increase in numbers to the point where they can account for around 90% of the local phytoplankton (Wasmund *et al.*, 1998). In many cases, species produce domoic acid, which can lead to outbreaks of shellfish poisoning due to increased levels during bloom periods. Early genetic studies suggested that blooms were largely comprised of a single, dominant clonal genotype due to selective sweeps within a population of genotypes with differential fitness (Gallagher *et al.*, 1980; Orsini *et al.*, 2004). Other studies have suggested that this is not always the case, and that bloom periods do not necessarily lead to genetic bottlenecks (Skov *et al.*, 1997; Rynearson and Armbrust, 2000, 2005).

Phylum Haptophyta (haptophytes)

The haptophytes were originally classified with the heterokont algae (phaeophytes, chrysophytes and bacillariophytes) due to the similarities of their plastids, but have subsequently been shown to constitute a distinct monophyletic group (Bhattacharya *et al.*, 1992; Yoon *et al.*, 2002; Harper and Keeling, 2003). They are characterized by having a uniquely organized flagellum-like structure called a haptonema, which differs from normal eukaryotic flagella in its arrangement of microtubules and which is believed to function in attachment and acquisition of nutrients. Best known among the haptophytes are the coccolithophores, unicellular algae that possess intricately structured calcium carbonate plates known as coccoliths that form microfossil deposits. They constitute a substantial proportion of oceanic phytoplankton and are a key component in global carbon cycling (Westbroek *et al.*, 1989). As with many diatoms, molecular analyses have revealed population substructuring indicative of restricted gene flow (Iglesias-Rodríguez *et al.*, 2006a; Medlin, 2007) and the possible existence of cryptic or species or subspecies, as has been found in other coccolithophores (Medlin *et al.*, 1996; Sáez *et al.*, 2003). At both global levels and within populations, dominant (Barker *et al.*, 1994; Medlin *et al.*, 1996) and microsatellite markers (Iglesias-Rodríguez *et al.*, 2002, 2006a,b) have revealed high levels of diversity, suggesting that sexual reproduction plays a major role in bloom events. The existence of discrete gene pools and possibly cryptic species have also been reported in the prymnesiophyte genus *Phaeocystis*, an important component in oceanic carbon- and sulphur-cycling (Gaebler *et al.*, 2007; Medlin and Zingone, 2007).

Conclusions and prospects

The field of microalgal population genetics is approaching an exciting period of development and expansion. Microsatellites, the staple tool of population genetic studies in a range of other organisms, are now being developed for a range of microalgal species and will undoubtedly provide new levels of insight into population dynamics at a range of geographical scales, from determining the relative roles of sexual and asexual reproduction in the formation of localized blooms to revealing levels of connectivity between populations on a global scale. The ability to genotype individuals, rather than having to culture samples in order to extract enough DNA for analysis, will allow these population studies to be carried out with an even greater degree of resolution (Marín *et al.*, 2001; Sebastian and O'Ryan, 2001; Orsini *et al.*, 2004). Furthermore, many species are not amenable to growth in culture and for those that can be cultured, there is the real possibility that 'selective' clone death during the culturing process may lead to samples that are not truly representative of field material (reviewed in Medlin *et al.*, 2000).

Microsatellites undoubtedly offer many advantages as markers for population genetics studies, but it appears increasingly likely that they will soon be superseded by single nucleotide polymorphisms (SNPs). Although SNPs are not yet routinely used in population genetics, they offer many advantages over microsatellites including an almost limitless number of loci that can be studied, potentially more efficient and accurate genotyping, simple mutational models, less homoplasy and the ability to directly compare results across species (reviewed in Brumfield *et al.*, 2003; Morin *et al.*, 2004). Like microsatellites, SNPs development can be somewhat time-consuming as it involves sequencing multiple regions of the nuclear genome in an ascertainment panel of representative individuals to identify polymorphisms. Many aspects of the development process can be circumvented via the availability of genome sequence information either from the taxon under study or a related species. Genome sequencing projects are under way for a variety of microalgae, including the diatoms *Thalassiosira pseudonana, Cyclotella meneghiniana* and *Phaeodactylum tricornutum*, the dinoflagellates *Alexandrium tamarense, Amphidinium operculatum, Heterocapsa triquetra* and *Karenia brevis*, and the haptophytes *Emiliania huxleyi* and *Pavlova lutheri*. Even in the absence of complete genome sequence data, the increasing availability of expressed sequence tag (EST) information for a variety of species means that SNP development will be quicker, cheaper and simpler than microsatellite development. Given the wealth of information potentially at their disposal, perhaps microalgal population geneticists should 'seize the day' and adopt SNPs as their marker of choice from the outset.

References

Adachi, M., Sako, Y., and Ishida, Y. (1997). Analysis of *Gymnodinium catenatum* dinophyceae using sequences of the 5.8S rDNA-ITS regions and random amplified polymorphic DNA. Fisheries Sci. *63*, 701–707.

Amato, A., Kooistra, W.H.C.F., Ghiron, J.H.L., Mann, D.G., Pröschold, T., and Montresor, M. (2007). Reproductive isolation among sympatric cryptic species in marine diatoms. Protist *158*, 193–207.

Alverson, A.J. (2008). Molecular systematics and the diatom species. Protist *159*, 339–353.

Archibald, J.M., and Keeling, P.J. (2002). Recycled plastids: a 'green movement' in eukaryotic evolution. Trends Genet. *18*, 577–584.

Baas Becking, L.G.M. (1934). Geobiologie of inleiding to de milieukunde. (The Hague: W.P. van Stockurn & Zoon).

Barker, G.L.A., Green, J.C., Hayes, P.K., and Medlin, L.K. (1994). Preliminary results using the RAPD analysis to screen bloom populations of *Emiliania huxleyi* (Haptophyta). Sarsia 79, 301–306.

Beszteri, B., Acs, E., and Medlin, L.K., (2005). Ribosomal DNA sequence variation among sympatric strains of the *Cyclotella meneghiniana* species complex (Bacillariophyceae) reveals cryptic diversity. Protist *156*, 317–333.

Beszteri, B., John, U., and Medlin, L.K., (2007). An assessment of cryptic genetic diversity within the *Cyclotella meneghiniana* species complex (Bacillariophyta) based on nuclear and plastid genes, and amplified fragment length polymorphisms. Eur. J. Phycol. *42*, 47–60.

Bhattacharya, D., Medlin, L., Wainright, P.O., Ariztia, E.V., Biseau, C., Stickle, S.K., and Slogin, M.L. (1992). Algae containing chlorophylls *a+c* are paraphyletic: molecular evolutionary analysis of the Chromophyta. Evolution, 46, 1801–1817.

Bhattacharya, D., and Medlin, L. (1995). The phylogeny of plastids – a review based on comparisons of small-subunit ribosomal RNA coding regions. J. Phycol. *31*, 498–498.

Bolch, C.J.S., Blackburn, S.I., Hallegraeff, G.M., and Vaillancourt, R.E. (1999). Genetic variation among strains of the toxic dinoflagellate *Gymnodium catenatum* (Dinophyceae). J. Phycol. 35, 356–367.

Bolch, C.J.S., and de Salas, M.F. (2007). A review of the molecular evidence for ballast water introduction of the toxic dinoflagellates *Gymnodinium catenatum* and the *Alexandrium 'tamarensis* complex' to Australasia. Harmful Algae 6, 465–485.

Brown, B.E., Dunne, R.P., Goodson, M.S., and Douglas, A.E. (2000). Bleaching patterns in reef corals. Nature *404*, 142–143.

Brumfield, R.T., Beerli, P., Nickerson, D.A., and Edwards, S.V. (2003). The utility of single nucleotide polymorphisms in inferences of population history. Trends Ecol. Evol. *18*, 249–256.

Burki, F., Shalchian-Tabrizi, K., Minge, M., Skjaeveland, A., Nikolaev, S.I., Jakobsen, K.S., and Pawlowski, J. (2007). Phylogenomics reshuffles the eukaryotic supergroups. PLoS One 2, e790.

Burki, F., Shalchian-Tabrizi, K., and Pawlowski, J. (2008). Phylogenomics reveals a new 'megagroup' including most photosynthetic eukaryotes. Biol. Lett. *4*, 366–369.

Canter-Lund, H., and Lund, J.W.G. (1995). Freshwater Algae (Bristol: Biopress Limited).

Casteleyn, G., Chepurnov, V.A., Leliaert, F., Mann, D.G., Bates, S.S., Lundholm, N., Rhodes, L., Sabbe, K., and Vyverman, W. (2008). *Pseudo-nitzschia pungens* (Bacillariophyceae): A cosmopolitan diatom species? Harmful Algae 7, 241–257.

Cavalier-Smith, T. (2002). The phagotrophic origin of eukaryotes and phylogenetic classification of protozoa. Int. J. Syst. Evol. Microbiol. *52*, 297–354.

Cavalier-Smith, T. (2003). Genomic reduction and evolution of novel genetic membranes and protein-targeting machinery in eukaryote-eukaryote chimeras (meta-algae). Philos. Trans. R. Soc. London B 358, 109–133.

Coleman, A.W. (2000). The significance of a coincidence between evolutionary landmarks found in mating affinity and a DNA sequence. Protist *151*, 1–9.

Douglas, S.E. (1998). Plastid evolution: origins, diversity, trends. Curr. Opin. Genet. Dev. *8*, 655–661.

Douglas, S., Zauner, S., Fraunholz, M., Beaton, M., Penny, S., Deng, L.-T., Reith, M., Cavalier-Smith, T., and Maier, U.-G. (2001). The highly reduced genome of an enslaved algal nucleus. Nature *410*, 1091–1096.

Evans, K.M., Bates, S.S., Medlin, L.K., and Hayes, P.K. (2004). Microsatellite marker development and genetic variation in the toxic marine diatom *Pseudo-nitzschia multiseries* (Bacillariophyceae). J. Phycol. *40*, 911–920.

Evans, K.M., Kühn, S.F., and Hayes, P.K. (2005). High levels of genetic diversity and low levels of genetic differentiation in North Sea *Pseudo-nitzschia pungens* (Bacillariophyceae) populations. J. Phycol. *41*, 506–514.

Evans, K.M., Wortley, A.H., and Mann, D.G. (2007). An assessment of potential diatom 'barcode' genes (*cox*1, *rbc*L, 18S and ITS rDNA) and their effectiveness in determining relationships in *Sellaphorella* (Bacillariophyta). Protist *158*, 349–364.

Finlay, B., Monaghan, E.B., and Maberly, S.C. (2002). Hypothesis: the rate and scale of dispersal of freshwater diatom species is a function of their global abundance. Protist *153*, 261–273.

Frade, P.R., de Jongh, F., Vermuelen, F., van Bleijswijk, J., and Bak, R.P.M. (2008). Variation in symbiont distribution between closely related coral species over large depth ranges. Mol. Ecol. *17*, 691–703.

Gaebler, S., Hayes,.P.K., and Medlin, L.K. (2007). Methods used to reveal genetic diversity in the colony-forming prymnesiophytes *Phaeocystis antarctica, P. globosa* and *P. pouchetti* – preliminary results. Biogeochemistry *83*, 19–27.

Gallagher, J.C. (1980). Population genetics of *Skeletonema costatum* (Bacillariophyceae) in Narrangasett Bay. J. Phycol. *16*, 464–474.

Gallagher, J.C. (1982). Physiological variation and electrophoretic banding patterns of genetically different seasonal populations of *Skeletonema costatum* (Bacillariophyceae). J. Phycol. *18*, 148–162.

Ghiron, J.H.L., Amato, A., Montresor, M., and Kooistra, W.H.C.F. (2008). Plastid inheritance in the planktonic raphid pinnate diatom *Pseudo-nitzschia delicatissima* (Bacillariophyceae). Protist *159*, 91–98.

Gilson, P.R., and McFadden, G.I. (1996). The miniaturized nuclear genome of a eukayotic endosymbiont contains genes that overlap, genes that are cotranscribed, and the smallest known spliceosomal introns. Proc. Natl. Acad. Sci. U.S.A. *93*, 7737–7742.

Gould, S.B., Waller, R.F., and McFadden, G.I. (2008). Plastid evolution. Ann. Rev. Plant Biol. *59*, 491–517.

Goulet, T.L., and Coffroth, M.A. (2004). The genetic identity of dinoflagellate symbionts from Caribbean coral reefs. Coral Reefs *23*, 465–472.

Greenwood, A.D., Griffiths, H.B., and Santore, U.J. (1977). Chloroplasts and cell compartments in Cryptophyceae. Br. Phycol. J. *12*, 119.

Hackett, J.D., Anderson, D.M., Erdner, D.L., and Bhattacharya, D. (2004). Dinoflagellates: a remarkable evolutionary experiment. Am. J. Bot. *91*, 1523–1534.

Hackett, J.D., Yoon, H., Li, S., Reyes-Prieto, A., Rümele, S., and Bhattacharya, D. (2007). Phylogenomic analysis supports the monophyly of cryptophytes and haptophytes and the association of rhizaria with chromalveolates. Mol. Biol. Evol. *24*, 1702–1713.

Hallegraeff, G.M. (1993). A review of harmful algal blooms and their apparent global increase. Phycologia *32*, 79–99.

Harper, J.T., and Keeling, P.J. (2003). Nucleus-encoded, plastid-targeted glyceraldehyde-3-phosphate dehydrogenase (GAPDH) indicates a single origin for chromalveolate plastids. Mol. Biol. Evol. *20*, 1730–1735.

Harper, J.T., Waanders, E., and Keeling, P.J. (2005). On the monophyly of chromalveolates using a six-protein phylogeny of eukaryotes. Int. J. Syst. Evol. Microbiol. *55*, 487–496.

Hedges, S.B., Blair, J.E., Venturi, M.L., and Shoe, J.L. (2004). A molecular timescale of eukaryotic evolution and the rise of complex multicellular life. BMC Evol. Biol. *4*, 2.

Hibberd, D.J., and Norris, R.E. (1984). Cytology and ultrastructure of *Chlorarachnion reptans* (Chlorarachniophyta *division nova* Chlorarachniophyceae *classis nova*). J. Phycol. *20*, 310–330.

Iglesias-Rodríguez, D.M., Sáez, A.G., Groben, R., Edwards, K.J., Batley, J., Medlin, L.K., and Hayes, P.K. (2002). Polymorphic microsatellite loci in global populations of the marine coccolithophorid *Emiliania huxleyi*. Mol. Ecol. Notes 2, 495–497.

Iglesias-Rodríguez, D.M., Schofield, O.M., Batley, J., Medlin, L.K., and Hayes, P.K. (2006a). Intraspecific genetic diversity in the marine coccolithophore *Emiliania huxleyi* (Prymnesiophyceae): the use of microsatellite analysis in marine phytoplankton studies. J. Phycol. *42*, 526–536.

Iglesias-Rodríguez, D.M., Probert, I., and Batley, J. (2006b). Microsatellite cross-amplification in coccolithophores: application in population diversity studies. Hereditas *143*, 99–102.

Jarne, P., and Lagoda, P.J.L. (1996). Microsatellites, from molecules to populations and back. Tr. Ecol. Evol. *11*, 424–430.

John, U., Groben, R., Beszteri, N., and Medlin, L. (2004). Utility of amplified fragment length polymorphisms (AFLP) to analyse genetic structures within the *Alexandrium tamarense* species complex. Protist *155*, 169–179.

Keeling, P.J., Burger, G., Durnford, D.G., Lang, B.F., Lee, R.W., Pearlmen, R.E., Roger, A.J., and Gray, M.W. (2005). The tree of eukaryotes. Tr. Ecol. Evol. *20*, 670–676.

Kooistra, W.H.C.F., Sarno, D., Balzano, S., Gu, H., Andersen, R.A., and Zingone, A. (2008). Global diversity anfd biogeography of *Skeletonema* species. Protist *159*, 177–193.

LaJeunesse, T.C. (2001). Investigating the biodiversity, ecology and phylogeny of endosymbiotic dinoflagellates in the genus *Symbiodinium* using the ITS region: in search of a 'species' level marker. J. Phycol. 37, 866–880.

LaJeunesse, T.C. (2002). Diversity and community structure of symbiotic dinoflagellates from Caribbean coral reefs. Mar. Biol. *141*, 387–400.

LaJeunesse, T.C., Loh, W.K.W., van Woesik, R., Hoegh-Gulberg, O., Schmidt, G.W., and Fitt, W.K. (2003). Low symbiont diversity in southern Great Barrier

Reef corals, relative to those of the Carbbean. Limnol. Oceanogr. *48*, 2046–2054.

LaJeunesse, T.C., Bhagooli, R., Hidaka, M., de Vantier, L., Done, T., Schmidt, G.W., Fitt, W.K., and Hoegh-Gulberg, O. (2004). Closely related *Symbiodinium* spp. Differ in relative dominance in coral reef host communities across environmental, latitudinal and biogeographic gradients. Mar. Ecol. Prog. Ser. *284*, 147–161.

Lane, C.E., and Archibald, J.M. (2008). The eukaryotic tree of life: endosymbiosis takes its TOL. Tr. Ecol. Evol. *23*, 268–275.

Larkum, A.W.D., and Vesk, M. (2003). Algal plastids: their fine structure and properties. In Photosynthesis In Algae, Douglas, S.E., Larkum, A.W.D., and Raven, J.A., eds. (Kluwer Academic Publishers, Dordrecht), pp. 11–28.

Lewis, C.L., and Coffroth, M.A. (2004). The acquisition of exogenous algal symbionts by an octocoral after bleaching. Science *304*, 1490–1492.

Lewontin, R.C., and Hubby, J.L. (1966). A molecular approach to the study of genetic heterozygosity in natural populations. II. Amount of variation and degree of heterozygosity in natural populations of *Drosophila pseudoobscura*. Genetics 54, 595–609.

Lewontin, R.C. (1991). 25 years ago in genetics – electrophoresis in the development of evolutionary genetics – milestone or millstone. Genetics *120*, 849–852.

Lilly, E.L., Kulis, D.M., Gentien, P., and Anderson, D.M. (2002). Paralytic shellfish poisoning toxins in France linked to a human –introduced strain of *Alexandrium catanella* from the western pacific: evidence from DNA and toxin analysis. J. Plank. Res. *24*, 443–452.

Loh, W.K.W., Loi, T., Carter, D., and Hoegh-Guldberg, O. (2001). Genetic variability of the symbiotic dinoflagellates from the wide-ranging coral species *Seriatopora hystrix* and *Acropora longicyanthus* in the Indo-West Pacific. Mar. Ecol. Prog. Ser. *222*, 97–107.

Lundholm, N., Moestrup, Ø., Kotaki, Y., Hoef-Emden, K., Scholin, C., and Miller, P. (2006). Inter- and intraspecific variation in the *Pseudo-nitzschia delcatissima* complex (Bacillariophyceae) illustrated by rRNA probes, morphological data and phylogenetic analyses. J. Phycol. *42*, 464–481.

Magalon, H., Baudry, E., Husté, A., Adjeroud, M., and Veuille, M. (2006). High genetic diversity of the symbiotic dinoflagellates in the coral *Pocillopora meandrina* from the South Pacific. Mar. Biol. *148*, 913–922.

Mann, D.G. (1999). The species concept in diatoms. Phycologia *38*, 437–495.

Mann, D.G. (2007). Naïvety, hazard, ecology or history: what determines the biogeography of freshwater microalgae (especially diatoms). Proceedings of the 4th European Phycological Congress, Oviedo, Spain. pp 65–66.

Marín, I., Aguilera, A., Reguera, B., and Abad, J.P. (2001). Preparation of DNA suitable for PCR amplification from fresh or fixed single dinoflagellate cells. Biotechniques *30*, 89–93.

Marshall, A. (1996). Calcification in hermatypic and ahermatypic corals. Science *271*, 1788–1792.

May, B. (1998). Starch gel electrophoresis of allozymes. In Molecular Genetic Analysis of Populations. Hoelzel, A.R., ed. (Oxford, UK: Oxford University Press), pp. 1–28.

Medlin, L.K., Elwood, H.J., Stickel, S., and Sogin, M.L. (1991). Morphological and genetic variation within the diatom *Skeletonema costatum* (Bacillariophyta): evidence for a new species *Skeletonema pseudocostatum*. J. Phycol. 27, 514–524.

Medlin, L.K., Barker, G.L.A., Campbell, L., Green, J.C., Hayes, P.K., Marie, D., Wrieden, S., and Vaulot, D. (1996). Genetic characterisation of *Emeliania huxleyi* (Haptophyta). J. Mar. Syst. *9*, 13–31.

Medlin, L.K., Lange, M., Wellbrock, U., Donner, G., Elbrachter, M., Hummert, C., and Luckas, B. (1997). Sequence comparisons link toxic European isolates of *Alexandrium tamarense* from the Orkney Islands to toxic North American stocks. Eur. J. Protistology *34*, 329–335.

Medlin, L.K., Lange, M., and Nöthig, E.-M. (2000). Genetic diversity in the marine phytoplankton: a review and a consideration of Antarctic phytoplankton. Antarctic Sci. *12*, 325–333.

Medlin, L.K. (2007). If everything is everywhere, do they share a common gene pool? Gene *406*, 180–183.

Medlin, L.K., and Zingone, A. (2007). A taxonomic review of the genus *Phaeocystis*. Biogeochemistry *83*, 3–18.

Mereschkowsky, C. (1905). Über Natur und Ursprung der Chromatophoren im Pflanzenreiche. Biol. Zentralbl. *25*, 593–604.

Mereschkowsky, C. (1910). Theorie der zwei Plasmaarten als Grundlage der Symbiogenesis, einer neuen Lehr even der Entstehung der Organismen. Biol. Zentralbl. *30*, 278–367.

Morin, P.A., Luikart, G., Wayne, R.K., and the S.N.P workshop group. (2004). SNPs in ecology, evolution and conservation. Trends Ecol. Evol. *19*, 208–216.

Murphy, L.S., and Guillard, R.R.L. (1976). Biochemical taxonomy of marine phytoplankton by electrophoresis of enzymes I. Centric diatoms *Thalassiosira pseudonana* and *Thalassiosira fluviatilis*. J. Phycol. *12*, 9–13.

Nagai, S., Lian, C., Yamaguchi, S., Hamaguchi, M., Matsuyama, Y., Itakura, S., Shimada, H., Kaga, S., Yamauchi, H., Sonda, Y., Nishikawa, T., Kim, C.H., and Hogetsu, T. (2007). Microsatellite markers reveal population genetic structure of the toxic dinoflagellate *Alexandrium tamarense* (Dinophyceae) in Japanese coastal waters. J. Phycol. *43*, 43–54.

van Oppen, M.J.H. (2004). Mode of zooxanthella transmission does not affect zooxanthella diversity in acroporid corals. Mar. Biol. *144*, 1–7.

van Oppen, M.J.H., Palstra, F.P., Piquet, A.M.-T., and Miller, D.J. (2001). Patterns of coral-dinoflagellate associations in *Acropora*: significance of local availability and physiology of Symbiodinium strains and host-symbiont selectivity. Proc. R. Soc. Lond. B 268, 1759–1767.

van Oppen, M.J.H., Mieog, J.C., Sánchez, C.A., and Fabricius, K.E. (2005). Diversity of algal symbionts (zooxanthellae) in octocorals: the roles of geography and host relationships. Mol. Ecol. *14*, 2403–2417.

Orsini, L., Procaccini, G., Sarno, D., and Montresor, M. (271). Multiple rDNA ITS-types within the diatom *Pseudonitzschia delicatissima* (Bacillariophyceae) and their relative abundance across a spring bloom in the gulf of Naples. Mar. Ecol. Prog. Ser. *271*, 87–98.

Palmer, J.D. (2000). A single birth of all plastids? Nature *405*, 32–33.

Palmer, J.D., Soltis, D.E., and Chase, M.W. (2004). The plant tree of life: an overview and some points of view. Am. J. Bot. *91*, 1437–1445.

Patron, N.J., Inagaki, Y., and Keeling, P.J. (2007). Multiple gene phylogenies support the monophyly of cryptomonad and haptophyte host lineages. Curr. Biol. *17*, 887–891.

Pochon, X., Montoya-Burgos, J.I., Stadelmann, B., and Pawlowski, J. (2006). Molecular phylogeny, evolutionary rates and divergence timing of the symbiotic dinoflagellate genus *Symbiodinium*. Mol. Phyl. Evol. *38*, 20–30.

Powell, W., Machray, G.C., and Provan, J. (1996). Polymorphism revealed by simple sequence repeats. Trends Plant Sci. *1*, 215–222.

Provan, J., Murphy, S., and Maggs, C.A. (2004). Universal plastid primers for Chlorophyta and Rhodophyta. Eur. J. Phycol. *39*, 43–50.

Provan, J., and Wilson, P.J. (2007). Development of microsatellites for the peat moss *Sphagnum capillifolium* using ISSR cloning. Mol. Ecol. Notes 7, 254–256.

Round, F.E., and Crawford, R.M. (1990). The Diatoms. Biology and Morphology of the Genera (Cambridge: Cambridge University Press).

Rowan, R., and Knowlton, N. (1995). Intraspecific diversity and ecological zonation in coral-algal symbiosis. Proc. Natl. Acad. Sci. U.S.A. *92*, 2850–2853.

Rowan, R., and Powers, D.A. (1991a). Molecular genetic identification of symbiotic dinoflagellates (zooxanthellae). Mar. Ecol. Prog. Ser. *71*, 65–73.

Rowan, R., and Powers, D.A. (1991b). A molecular genetic classification of zooxanthellae and the evolution of animal-algal symbioses. Science *4999*, 1348–1351.

Rowan, R., and Powers, D.A. (1992). Ribosomal RNA sequences and the diversity of symbiotic dinoflagellates (zooxanthellae). Proc. Natl. Acad. Sci. U.S.A. *89*, 3639–3643.

Rowan, R., Knowlton, N., Baker, A., and Jara, J. (1997). Landscape ecology of algal symbionts creates variation in episodes of coral bleaching. Nature *388*, 265–269.

Rynearson, T.A., and Armbrust, E.V. (2000). DNA fingerprinting reveals extensive genetic diversity in a fiels population of the centric diatom *Ditylum brightwellii.* Limnol. Oceanogr. *46*, 1329–1340.

Rynearson, T.A., and Armbrust, E.V. (2004). Genetic differentiation among populations of the planktonic marine diatom *Ditylum brightwellii* (Bacillariophyceae). J. Phycol. *40*, 34–43.

Rynearson, T.A., and Armbrust, E.V. (2005). Maintenance of clonal diversity during a spring bloom of the centric diatom *Ditylum brightwelli*. Mol. Ecol. *14*, 1631–1640.

Sáez, A.G., Probert, I., Giesen, M., Quinn, P., Young, J.R., and Medlin, L.K. (2003). Pseudo-cryptic speciation in coccolithophores. Proc. Natl. Acad. Sci. U.S.A. *100*, 7163–7168.

Santos, S.R., Gutiérrez-Rodriguez, C., Lasker, H.R., and Coffroth, M.A. (2003). *Symbiodinium* sp. Associations in the gorgonian *Pseudopterogorgia elisabethae* in the Bahamas: high levels of genetic variability and population structure in symbiotic dinoflagellates. Mar. Biol. *143*, 111–120.

Santos, S.R., Shearer, T.L., Hannes, A.R., and Coffroth, M.A. (2004). Fine-scale diversity and specificity in the most prevalent lineage of symbiotic dinoflagellates (*Symbiodinium*, Dinophyceae) of the Caribbean. Mol. Ecol. *13*, 459–469.

Sarno, D., Kooistra, W.H.C.F., Medlin, L.K., Percopo, I., and Zingone, A. (2005). Diversity in the genus *Skeletonema* (Bacillariophuceae) II. An assessment of the taxonomy of *S. costatum*-like species with the description of four new species. J. Phycol. *41*, 151–176.

Schimper, A.F.W. (1883). Über die Entwicklung der Chlorophyllkörner und Farbkörper. Bot. Zeitung. 41, 105–14, 121–131, 137–146, 153–162.

Scholin, C.A., Herzog, M., Sogin, M., and Anderson, D.M. (1994). Identification of group-specific and strain-specific genetic markers for globally distributed *Alexandrium* (Dinophyceae). 2. Sequence analysis of a fragment of the LSU ribosomal RNA gene. J. Phycol. *30*, 999–1011.

Scholin, C.A., Hallegraeff, G.M., and Anderson, D.M. (1995). Molecular evolution of the *Alexandrium tamarense* 'species complex' (Dinophyceae): dispersal in the North American and West Pacific regions. Phycologia *34*, 472–485.

Sebastian, C.R., and O'Ryan, C. (2001). Single-cell sequencing of dinoflagellate (Dinophyceae) nuclear ribosomal genes. Mol. Ecol. Notes *1*, 329–331.

Sebastian, C.R., Etheridge, S.M., Cook, P.A., O'Ryan, C., and Pitcher, G.C. (2005). Phylogenetic analysis of toxic *Alexandrium* (Dinophyceae) isolates from South Africa: implications for the global phylogeography of the *Alexandrium tamarense* species complex. Phycologia *44*, 49–60.

Skov, J., Lundholm, N., Pocklington, R., Rosendahl, S., and Moestrup, Ø. (1997). Studies on the marine planktonic diatom *Pseudo-nitzschia.* 1. Isozyme variation among isolates of *P. pseudodelicatissima* during a bloom in Danish coastal waters. Phycologia *36*, 374–380.

Stat, M., Morris, E., and Gates, R.D. (2008). Functional diversity in coral-dinoflagellate symbiosis. Proc. Natl. Acad. Sci. U.S.A. *105*, 9256–9261.

Stechmann, A., and Cavalier-Smith, T. (2003). Phylogenetic analysis of eukaryotes using heat shock protein Hsp90. J. Mol. Evol. *57*, 408–419.

Sunnucks, P. (2000). Efficient genetic markers for population biology. Trends Ecol. Evol. *15*, 199–203.

Thornhill, D.J., Kemp, D.W., Bruns, B.U., Fitt, W.K., and Schmidt, G.W. (2008). Correspondence between cold tolerance and temperate biogeography in a western Atlantic *Symbiodinium* (Dinophyta) lineage. J. Phycol. DOI 10.1111/j.1529-8817.2008.00567.x.

Trench, R.K. (1987). Dinoflagellates in non-parasitic symbioses. In The Biology of Dinoflagellates, Taylor, F.J.R., ed. (Oxford, UK: Blackwell Scientific Publishing), pp. 530–570.

Trench, R.K. (1993). Macroalgal-invertebrate symbioses: a review. Endocytobiosis Cell Res. *9*, 135–175.

Turner, S., Pryer, K.M., Miao, V.P.W., and Palmer, J.D. (1999). Investigating deep phylogenetic relationships among cyanobacteria and plastids by small subunit rRNA sequence analysis. J. Eukaryot. Microbiol. *46*, 327–338.

Ulstrup, K.E., and van Oppen, M.J.H. (2003). Geographic and habitat partitioning of genetically distinct zooxanthellae (*Symbiodinium*) in *Acropora* corals on the Great Barrier Reef. Mol. Ecol. *12*, 3477–3484.

Vanormelingen, P., Chepurnov, V.A., Mann, D.G., Sabbe, K., and Vyverman, W. (2008). Genetic divergence and reproductive barriers among morphologically heterogeneous sympatric clones of *Eunotia bilunaris* sensu lato (Bacillariophyta). Protist *159*, 73–90.

Vos, P., Hogers, R., Bleeker, M., Reijans, M., van de Lee, T., Hornes, M., Frijters, A., Pot, J., Peleman, J., Kuiper, M., and Zabeau, M. (1995). AFLP: a new technique for DNA fingerprinting. Nucleic Acids Res. *23*, 4407–4414.

Wasmund, N., Nausch, G., and Matthaus, W. (1998). Phytoplankton spring blooms in the southern Baltic Sea – spatio-temporal development and long-term trends, J. Plank. Res. *20*, 1099–1117.

Westbroek, P., Young, J.R., and Linschooten, K. (1989). Coccolith production (biomineralization) in the marine alga *Emiliania huxleyi*. J. Protozoology *36*, 368–373.

White, T.J., Bruns, T., Lee, S., and Taylor, J. (1990). Amplification and direct sequencing of fungal ribosomal RNA genes for phylogenetics. In PCR Protocols: A Guide to Methods and Applications, Innis, M., Gelfland, D., Sninsky, J., and White, T. eds. (San Diego, USA: Academic Press) pp. 315–322.

Williams, J.G.K., Hanafey, M.K., Rafalski, J.A., and Tingey, S.V. (1993). Genetic analysis using random amplified polymorphic DNA markers. Methods in Enzymology *218*, 704–740.

Yoon, H.S., Hackett, J.D., Pinto, G., and Bhattacharya, D. (2002). The single, ancient origin of chromist plastids. Proc. Natl. Acad. Sci. U.S.A. *99*, 15507–15512.

Zietkiewicz, E., Rafalski, J.A., and Labuda, D. (1994). Genome fingerprinting by simple sequence repeat (SSR)- anchored polymerase chain reaction amplification. Genomics *20*, 176–183.

Zingone, A., Percopo, I., Sims, P.A. and Sarno, D. (2005). Diversity in the genus Skeletonema (Bacillariophyceae). I. A reexamination of the type material of S. costatum with the description of *S. grevillei* sp. nov. J. Phycol. *41*, 140–150.

8

Population Genetics of Fungal Mutualists of Plants

Teresa E. Pawlowska

Abstract

Mutualisms are reciprocal exploitations that nonetheless increase the fitness of each interacting partner. Two groups of fungal mutualists of plants, epichloë endophytes of grasses and arbuscular mycorrhizal fungi, were selected as focal systems to discuss population-level processes that contribute to the establishment and maintenance of mutualistic interactions. These two classes of fungal cooperators of plants are subject to different and often conflicting selective pressures and represent distinct trajectories of mutualism evolution. Yet, in both cases population structure of symbionts is a source of information critical for understanding how these fungi interact with their hosts.

Introduction

The goal of this chapter is to explore the patterns of microevolution in two groups of fungi that engage in mutualistic interactions with plants: epichloë grass endophytes and arbuscular mycorrhizal fungi. Mutualisms are reciprocal exploitations that nonetheless increase the fitness of each partner (Herre *et al.*, 1999; Sachs *et al.*, 2004). This modern view of mutualisms emphasizes conflicting interests of the interacting partners, which may lead to accepting benefits without reciprocating, i.e. cheating. Such conflicts of interests may be disruptive to the interaction and affect its ecological and evolutionary stability. However, despite these challenges, we see an overwhelming diversity of mutualistic interactions in the biosphere. This paradox makes it difficult for the ecological and evolutionary theories to explain the establishment and maintenance of mutualisms. Consequently, there is no general theory of mutualism but rather a collection of models developed to explain specific types of interactions. In most of these models, the population structure of the interacting partners is thought to be an important determinant of the stability of the mutualistic association. With this consideration in mind, I will focus on two examples of fungal lineages, the *Epichloë/Neotyphodium* fungi that engage in grass endosymbiosis (Schardl *et al.*, 2004) and Glomeromycota that form arbuscular mycorrhizae with the majority of land plants (Smith and Read, 2008). Services and goods acquired from these fungal cooperators by their plant hosts range from synthesis of defence alkaloids to facilitation of mineral nutrient uptake from the soil. In return, the fungi are provisioned with sources of energy and a habitat for growth and development. The fungi involved in these two mutualisms represent divergent taxonomic lineages, life styles, and reproduction modes. Yet, at least in the pooid grasses, the two interactions can coexist within the same host plant (Guo *et al.*, 1992; Müller, 2003; Mack and Rudgers, 2008). I will examine data on population structures of these fungi in the context of the existing theoretical framework in an attempt to understand the evolution and maintenance of the two mutualisms. Both associations attract much attention because of their significance for ecology, agriculture, and as model systems to study mutualisms.

Evolutionary stability of mutualisms

While there is no general theory of mutualism, the *geographic mosaic theory of coevolution* provides a broad framework that aims to explain the establishment and maintenance of mutualisms by examining coevolutionary forces at different spatial scales (Thompson, 2005). In contrast to the models exploring mutualisms only within isolated local communities (Axelrod and Hamilton, 1981; Bull and Rice, 1991; Frank, 1994a; Noë and Hammerstein, 1994; Doebeli and Knowlton, 1998; Foster and Kokko, 2006; Foster and Wenseleers, 2006), the geographic mosaic theory of coevolution looks at entire species, which are viewed as collections of genetically differentiated populations interacting across their geographic ranges. Interactions between partners are subject to reciprocal selection leading to coevolution only in some local communities known as coevolutionary hotspots. The hotspots are embedded in a matrix of coevolutionary coldspots where local selection is not reciprocal (Gomulkiewicz *et al.*, 2000). Variation in the patterns of natural selection between the habitats (Gomulkiewicz *et al.*, 2003), together with gene flow, genetic drift, mutations, migration, and local extinctions, contribute to a dynamic geographic mosaic of coevolution (Nuismer *et al.*, 2000). This global process results in a pattern where traits are well matched between the interacting species in some locations but not in others.

A number of factors that generally favour local establishment and maintenance of mutualisms have been identified by theoretical and empirical studies (Bull and Rice, 1991; Noë and Hammerstein, 1994; Doebeli and Knowlton, 1998; Sachs *et al.*, 2004; Yamamura *et al.*, 2004; Foster and Wenseleers, 2006). In interactions between two species, these factors include: (i) a high ratio of benefit to costs, (ii) a high degree of within-species relatedness, and (iii) a high level of fidelity between the interacting species (Foster and Wenseleers, 2006). These factors operate through several feedback mechanisms, such as: (i) *cooperator associations* generated by spatial co-occurrence of cooperative genotypes of the interacting species, (ii) *partner-fidelity feedback* that results from coupling the fitness of the two cooperators through repeated interactions, and (iii) *partner choice*, realized when an individual has the ability to reward the cooperative partners and sanction the cheaters (Foster and Wenseleers, 2006).

When the framework of partner choice is expanded from dichotomous decisions of rewarding versus sanctioning partners in two–species interactions (Bull and Rice, 1991) to decisions about the degree of partner utilization, it can accommodate stable multi-species cooperation where an individual selectively interacts with multiple unrelated partners (Kummel and Salant, 2006). However, this expansion is only possible for mutualistic systems that can be described by *biological market models* (Noë and Hammerstein, 1994; Noë and Hammerstein, 1995; Schwartz and Hoeksema, 1998; Simms and Taylor, 2002; Hoeksema and Schwartz, 2003). The biological market framework is based on the principles of economic theory and requires that partners exchange commodities and have a choice of trading partners who offer the same kind of commodity in an outbidding competition (Noë and Hammerstein, 1994; Kummel and Salant, 2006). The ability to select and evaluate partners is expected to increase the stability of mutualisms by constraining cheating.

Most symbiotic mutualisms, i.e. interactions where partners remain in a long-term intimate association, fit neatly into the conceptual framework developed for a single pair of locally interacting species. The *Neotyphodium*–grass symbiosis discussed in the next section is an example of such symbiotic mutualism. In contrast, symbiotic interactions like mycorrhizae, where symbionts are transmitted horizontally, have proven more difficult for theoretical analyses (Egger and Hibbett, 2004; Kiers and van der Heijden, 2006). Mycorrhizal associations combine prolonged intimate association typical for symbiotic mutualisms with the capacity of the cooperators to interact with multiple individuals of several species during their lifetimes, which is an attribute of mutualisms among free-living non-symbiotic species (Thompson, 2005). One of the characteristics of non-symbiotic mutualisms, which include, for example, animal-mediated plant pollination and seed dispersal, is a seemingly diffuse structure of the association that nonetheless is a product of reciprocal selection (Stanton, 2003; Thompson,

2005). Representing these plant–animal mutualisms as *networks of interacting partners* offers a framework to better understand their coexistence and coevolution (Bascompte *et al.*, 2003; Bascompte and Jordano, 2007). For example, plant–pollinator and plant–frugivore networks are heterogeneous and highly nested, with most species involved in few interactions and a few species that are involved in many interactions (Bascompte *et al.*, 2003). Interactions within networks tend to be weak and asymmetric, with one partner specialized more than the other. Such network structure is consistent with a pattern of selection that favours genetically variable multispecies groups, in which the species convergently adapt and specialize on a suite of mutualistic traits rather than directly on other species (Waser *et al.*, 1996; Thompson, 2005). For example, many tropical plants representing diverse phylogenetic lineages rely on bats for pollination and share floral traits of a 'bat pollination syndrome' of fragrant dull-coloured flowers that open at night and contain copious amounts of nectar and protein-rich pollen (Fleming and Muchhala, 2008). These flowers attract a range of chiropteran pollinators that share cranial and dental adaptations to nectar feeding (Baker, 1961; Freeman, 1995).

Nearly all models of inter-species cooperation emphasize the population structures of the cooperators. On a local scale, a high within-species relatedness and cooperator associations generated by the spatial structure of cooperator populations promote two-species mutualisms in individual communities (Ewald, 1987; Frank, 1994a; Doebeli and Knowlton, 1998; Yamamura *et al.*, 2004; Foster and Wenseleers, 2006). On a geographic scale, the spatial structure of the coevolving species, together with mixing of the locally coevolving traits, which is mediated by gene flow, mutation, genetic drift, migration, and local population extinctions, contribute to a shifting mosaic of coevolution by altering the spatial distribution of alleles in local populations (Thompson, 2005). Modelling studies suggest that gene flow between communities of coevolving populations can facilitate maintenance of allelic polymorphism and stabilize fixed mutualisms in the face of local antagonistic selection (Nuismer *et al.*, 1999; Gomulkiewicz *et al.*, 2000; Nuismer *et al.*, 2000). However, it can also lead to local maladaptation and trait mismatching.

While the predictive power of the local models of coevolution is well established, the empirical tests of the geographic-scale models have begun only recently in a handful of systems, including the *Neotyphodium*–grass symbiosis (Sullivan and Faeth, 2004, 2008).

Epichloë endophytes of grasses

Epichloë fungi belong to the *Epichloë/Neotyphodium* complex (phylum Ascomycota, order Hypocreales, family Clavicipitaceae) where *Epichloë* species reproduce sexually and *Neotyphodium* species are their asexual derivatives; the informal term 'epichloë' refers to both sexual and asexual species (Clay and Schardl, 2002; Schardl *et al.*, 2004). The *Epichloë/Neotyphodium* fungi engage in symbiotic associations with grasses from the Poaceae family (Clay and Schardl, 2002; Schardl *et al.*, 2004). These associations range from parasitism to mutualism. Epichloë symbionts are perennial systemic endophytes that proliferate in intercellular spaces parallel to the long axes of the aboveground grass structures. Without forming haustoria, these fungi subsist on nutrients released into the plant apoplast. Most endophytes produce alkaloids that contribute to the host defence against herbivores. Some of these alkaloids can cause toxicoses in livestock grazed on endophyte-infected forage grasses.

Epichloë reproductive parasites of grasses

Epichloë species reproduce sexually and transmit horizontally in a process that interferes with the development of the sexual reproductive structures of the host plant (the choke disease). Fungal reproductive stromata are formed on leaf sheaths that surround young inflorescence primordia before the inflorescences emerge. Successful completion of the fungal sexual process relies on the transfer of conidia, which serve as spermatia, between stromata of opposite mating types and is mediated by flies from the genus *Botanophila* (order Diptera, family Anthomyiidae) (Bultman and White, 1988). *Botanophila* spp. flies visit developing stromata and consume, transport, and deposit conidia while laying eggs. Their larvae feed and develop on the fertilized fungal fruiting structures. Fungal fruiting bodies are the sole source of nourishment for the *Botanophila* larvae (until they start feeding on subtending plant tissue), which suggests a degree of mutual dependence between the partners, even though this pollination-like association is not highly specific (Bultman and Leuchtmann, 2003; Leuchtmann, 2007; Steinebrunner *et al.*, 2008a; Steinebrunner *et al.*, 2008b). Once the fungal perithecia embedded in the stroma mature, meiotic ascospores are forcefully ejected into the air and infect neighbouring plants.

Dual-transmission *Epichloë* symbionts

In many species of *Epichloë*, expression of sexual structures varies among individual infected plants and their tillers (Bucheli and Leuchtmann, 1996; Meijer and Leuchtmann, 1999). In addition to supporting fungal sexual structures, which interfere with plant sexual reproduction, the infected plants may form normal fertile inflorescences and seeds that are asymptomatically penetrated by fungal hyphae and enable vertical transmission of the fungus. These dual-transmission symbiotic systems offer important insights into the evolution of symbiosis between the grass hosts and the endophytes from the *Epichloë/Neotyphodium* species complex. Vertical parasite transmission is expected to align the reproductive interests of the host and the parasite, and favour reduction of the negative effects of the parasite on the host fitness in comparison to horizontal transmission (Frank, 1996c; Frank, 1996b).

Studies of the population structure of *E. sylvatica* in a European woodland grass *Brachypodium sylvaticum* revealed that the expression of the fungal sexual structures is controlled by the fungus genotype (Bucheli and Leuchtmann, 1996; Meijer and Leuchtmann, 1999). In these endophytes, stroma-forming strains and the asymptomatic strains did not belong to a single random mating population but were genetically differentiated into two separate subpopulations despite their occasional co-occurrence on individual host plants. The subpopulation of the stroma-forming strains was in linkage equilibrium expected among outcrossing sexual individuals (Bucheli and Leuchtmann, 1996). In contrast, the subpopulation of asymptomatic vertically transmitted strains exhibited significant linkage disequilibrium.

In addition to the fungal genotype, the expression of the sexual structures of the endosymbiont is affected by the host genotype as well as environmental factors (Meijer and Leuchtmann, 2000; Meijer and Leuchtmann, 2001; Tintjer *et al.*, 2008). For example, *E. festucae* is known to form stromata on fine fescue *Festuca rubra* (Leuchtmann *et al.*, 1994). However, two separate surveys of populations of this species associated with *F. rubra* and *F. ovina* in grasslands of the European Subarctic and the Mediterranean semidesert, did not yield an instance of fruiting body formation but instead revealed predominantly clonal population structures (Arroyo García *et al.*, 2002; Wäli *et al.*, 2007). These observations suggest that under specific conditions entire populations of *Epichloë* fungi can thrive as asymptomatic endophytes that depend on their host for dispersal.

Emergence of the *Neotyphodium*–grass mutualism

Neotyphodium (formerly *Acremonium* sect. *Albo-lanosa*) endophytes are asexual forms derived from sexual species of *Epichloë* (Clay and Schardl, 2002; Schardl *et al.*, 2004). In contrast to *Epichloë* species, *Neotyphodium* species never exhibit sexual reproductive structures. They exist asymptomatically in cool-season grasses in the subfamily Poöideae and rely on the host for seed-mediated transmission from parent to the offspring. Reports of epiphyllous conidia in *Neotyphodium* endophytes suggest that some of these endophytes may also have the ability for plant-to-plant spread and their dependence on the host-mediated transmission may not be absolute (Moy *et al.*, 2000; Tadych *et al.*, 2007).

Two evolutionary scenarios explain emergence of asexual species of *Neotyphodium* from the sexual *Epichloë* species. Some of the *Neotyphodium* endophytes are believed to have evolved from sexual ancestors as a consequence of host shifts that resulted in the loss of the ability to undergo sexual differentiation (Brem and Leuchtmann, 2003). Most asexual species are, however, interspecific hybrids between different species of *Epichloë* or *Neotyphodium* (Tsai *et al.*, 1994; Moon *et al.*, 2000; Moon *et al.*, 2002; Moon *et al.*, 2004; Gentile *et al.*, 2005; Moon *et al.*, 2007). Vegetative interactions between endophytes that co-occur within the same host are believed to be responsible for the hybridization events (Moon *et al.*, 2004). Relatively weak intra- and interspecific vegetative incompatibility barriers exhibited by *Epichloë* fungi may be one of the factors contributing to the prevalence of interpecific hybrids among the *Neotyphodium* species (Chung and Schardl, 1997). Duplicate gene copies present in the *Neotyphodium* hybrid genomes are expected to protect these asexual organisms from the advancement of the Muller's ratchet (Clay and Schardl, 2002; Schardl *et al.*, 2004; Selosse and Schardl, 2007), i.e. from the accumulation of deleterious mutations accompanied by a progressive and irreversible loss of mutation-free individuals that leads to population extinction (Muller, 1964; Kondrashov, 1984; Lynch and Gabriel, 1990; Gabriel *et al.*, 1993).

Stability of the *Neotyphodium*–grass mutualism

The direct link between the reproductive successes of the host and the endophyte creates conditions for a partner-fidelity feedback mechanism where both partners are expected to enhance each other's fitness (Sachs *et al.*, 2004). Certain host benefits are related to the fungal production of secondary compounds (particularly alkaloids) that act as chemical defences against herbivores, though beneficial effects on the host physiology are also documented (reviewed in Schardl *et al.*, 2004). The investment by the fungus into the secondary compounds feeds back into the protection of the host that provisions nutrients to the fungus and enables its dispersal. In addition, the fungal infection increases the host competitive ability through increased pathogen resistance (Lehtonen *et al.*, 2006; Wali *et al.*, 2006), drought tolerance (Malinowski and Belesky, 2000), and uptake of nutrients (Malinowski and Belesky, 1999; Malinowski *et al.*, 1999; Malinowski *et al.*, 2000).

The *Neotyphodium*–grass symbiosis is considered a classic example of a partner-fidelity feedback mechanism to maintain interspecific cooperation (Sachs *et al.*, 2004). This view holds well in the cases of the associations of *Neotyphodium* with agronomically important grasses like tall fescue (*Lolium arundinaceum*, formerly *Festuca arundinacea*) and perennial ryegrass (*L. perenne*), which are native to Eurasia and cultivated in North and South America, New Zealand, and Australia as turf and forage grasses. However, several recent papers have suggested that *Neotyphodium* endophytes of native grasses in natural populations may not always have positive effects on the plant performance or its competitive ability (Faeth and Sullivan, 2003; Faeth *et al.*, 2004; Faeth and Hamilton, 2006; Saikkonen *et al.*, 2006). This conclusion is based on the lack of strong mutualistic outcomes in populations of Arizona fescue (*F. arizonica*) in the studies conducted in semiarid grasslands of the southwestern US (Faeth and Sullivan, 2003; Faeth *et al.*, 2004; Faeth and Hamilton, 2006). The Arizona fescue plants were colonized by hybrid and non-hybrid forms of *Neotyphodium* (Sullivan and Faeth, 2004), whose alkaloid production was reported as variable and apparently not sufficient to deter herbivores (Saikkonen *et al.*, 1999; Faeth *et al.*, 2002). Consequently, the presence of the fungus was probably a draw on the plant's resources without providing the benefit of facilitating the defence against herbivores.

A possible explanation for the observed parasitic effects of *Neotyphodium* species in *F. arizonica* may be related to the relaxation of the reproductive dependence of *Neotyphodium* on the host due to the horizontal transmission mediated by epiphyllous conidia. Epiphyllous conidiation in the *Neotyphodium* fungi has not received much attention, and its contribution to the proliferation and dispersal of these organisms may be underappreciated. Horizontal transmission could be a source of multiple infections of the same host leading to an among-symbiont competition for host resources and loss of mutualistic outcomes. However, considering that no multiple infections have been detected in *F. arizonica*, this explanation is not very plausible (Sullivan and Faeth, 2004). An alternative explanation of the apparent lack of mutualistic outcomes in the *Neotyphodium*–Arizona fescue system can be derived from a model developed to explain coexistence of endophyte-infected and endophyte-free plants in a structured metapopulation (Saikkonen *et al.*, 2002). According to this model, strictly vertically transmitted endophytes are expected to persist in a metapopulation, despite their negative effect on host fitness, as long as there are environment patches favourable to the association. The predictions of this model are generally consistent with the predictions of the geographic mosaic theory of coevolution (Thompson, 2005). However, testing whether the geographic mosaic theory of coevolution is applicable to a particular system is not a trivial matter. A rigorous set of steps required to test this hypothesis has been recently outlined by Gomulkiewicz *et al.* (Gomulkiewicz *et al.*, 2007). These steps include (i) identification of traits that could be under reciprocal selection, based on information on the distribution and natural history of the interacting partners, (ii) positive evidence of reciprocal selection, (iii) identification of processes that contribute to trait remixing, including gene flow, mutation, drift, and migration and extinction events, and (iv) determination whether the trait remixing and selection mosaic occur at comparable scales. Several empirical data sets have been collected to test the applicability of the geographic mosaic theory of coevolution to the *Neotyphodium*–Arizona fescue system. These data sets provide evidence that *Neotyphodium* species have divergent effects on hosts at different geographic locations, evidence of local adaptations in marginal habitats, and evidence of limited gene flow among different coevolving populations (Sullivan and Faeth, 2004; Sullivan and Faeth, 2008). While not entirely conclusive, they together suggest that the processes expected to generate the geographic mosaic of coevolution operate in the *Neotyphodium*–Arizona fescue system.

One must remember that the mutualism with the *Neotyphodium* fungi represents only a subset of the complex set of interactions in which the grass hosts are engaged. Another, equally important and far less understood mutualism involves arbuscular mycorrhizal fungi that, unlike the aboveground endophytes, are sequestered in the roots and are a source of additional selective pressures on the host (Guo *et al.*, 1992; Müller, 2003; Mack and Rudgers, 2008).

Arbuscular mycorrhizae

Arbuscular mycorrhizal (AM) fungi (phylum Glomeromycota) are the oldest, most widespread, and one of the least understood symbiotic associates of vascular plants (Smith and Read, 2008). Fossil record indicates that structures resembling modern mycorrhizal root colonization existed in rhizomes of early Devonian land plants (Remy *et al.*, 1994). Although the exact function of these arbuscule-like structures remains unknown, their modern counterparts translocate mineral nutrients from the fungus to the host in exchange for plant-assimilated carbon (Smith and Read, 2008). Other benefits of AM root colonization may include improved drought tolerance (Augé, 2004) and resistance to root pathogens (Borowicz, 2001). Even though 60 to 90% of plant species on the planet are believed to support mycorrhizal root colonization, the significance of this association to the host fitness is not always obvious (Jones and Smith, 2004). In contrast, Glomeromycota are obligately dependent on the plant-derived carbon, and cannot complete their life cycle without engaging in the symbiosis. However, unlike the *Neotyphodium* fungi, Glomeromycota are exclusively horizontally transmitted, which together with the uncertainties concerning their genetic makeup and transmission mode, compounds the difficulties in explaining the evolution of arbuscular mycorrhizae.

Transmission mode of AM fungi

AM fungi disperse from host to host by asexual spores. New hosts can be also colonized by the expanding hyphal networks that connect roots of the same and/or different plant species (Chiariello *et al.*, 1982; Newman *et al.*, 1994). Even though the transmission mode of the interacting organisms has been identified by all models as a major determinant of mutualism stability, different models identify different modes of transmission as the most conducive to interspecific cooperation (Noë and Hammerstein, 1994; Herre *et al.*, 1999). In game-theoretical models, which rely on repeated interactions between partners, the need for reassembling the association in every successive generation of the host and symbiont is a potential source of instability that can unravel the mutualism (Axelrod and Hamilton, 1981; Doebeli and Knowlton, 1998; Herre *et al.*, 1999; Wilkinson and Sherratt, 2001). Likewise, in genetic kinship models, horizontal transmission reduces the symbiont relatedness and is expected to favour increased among-symbiont competition for host resources, which could be detrimental to the host (Frank, 1994b; Frank, 1996b; Frank, 1996a; Frank, 1997). In contrast, in biological market models, horizontal transmission is viewed as a critical prerequisite for partner choice (Noë and Hammerstein, 1994; Noë and Hammerstein, 1995; Wilkinson and Sherratt, 2001; Simms and Taylor, 2002; Kummel and Salant, 2006). It exposes an individual to a range of potential partners that can be evaluated for suitability as cooperators.

In an attempt to reconcile the apparent incongruity between different theoretical frameworks, Wilkinson (Wilkinson, 1997; Wilkinson and Sherratt, 2001) proposed that horizontal transmission under conditions of spatial structure and dispersal limitations experienced by cooperating partners may result in pseudo-vertical transmission. This localized horizontal transmission is expected to align reproductive interests of cooperators, increase their genetic homogeneity, and reduce selection for traits disruptive to the interspecific cooperation while facilitating a degree of partner choice. Based on seed dispersal limitations characteristic for many mycorrhizal host plants, Wilkinson (1997) suggested that the pseudo-vertical transmission model could explain evolutionary stability of arbuscular mycorrhizae where young plants are likely to encounter the same fungal genotypes that were associated with their parents. However, the applicability of the pseudo-vertical transmission model to arbuscular mycorrhizae has been questioned recently by Kiers and van der Heijden (2006). These authors pointed out that the empirical evidence of significantly different communities of AM fungal species associated with seedlings versus adults of the same tropical tree species (Husband *et al.*, 2002a,b) is incongruent with the predictions of the pseudo-vertical transmission model.

Equally difficult to explain under the pseudo-vertical transmission model is the co-occurrence of multiple symbiont species within the root system of the same individual, which is typical for arbuscular mycorrhizae. The coexistence of multiple unrelated symbionts in a host is expected to result in increased selection for traits that improve symbiont competitiveness for host resources, leading to the disruption of mutualism (Frank, 1994b, 1996b). Competition among AM fungi colonizing a single root system has been recently demonstrated experimentally (Jansa *et al.*, 2008). However, despite the apparent competition, the co-occurring AM fungal species exhibited functional complementarity in facilitating host phosphorus uptake from the soil. This complementarity is consistent with the recently developed model of symbiont coexistence based on the principles of biological markets (Kummel and Salant, 2006). According to this model, simultaneous colonization of a single root system by several unrelated mycorrhizal fungi is possible under specific conditions that govern the exchange of the host-derived carbon for the symbiont-derived mineral nutrients. These conditions are defined by the marginal cost of mineral nutrient acquisition (i.e. the increase in the total cost needed to acquire one additional mineral nutrient unit) and the exchange functions specific to each symbiont, which describe the number of units of carbon required by this symbiont from the host in exchange for a given number of mineral nutrient units. Multiple partners are favoured if the marginal costs of their services are equal and their exchange functions are convex. This prediction is important for population biology of Glomeromycota as it suggests that multiple genetically differentiated individuals representing a single fungal species could coexist in one root system as long as they were utilized selectively by the host. While significant phenotypic and functional inter-isolate differences have been documented (Koch *et al.*, 2004; Munkvold *et al.*, 2004), possible co-occurrence of genetically different individuals of the same fungal species in a single host has not been studied. Addressing simultaneous occurrence of non-identical symbiont genotypes in the same host may not be trivial, however, given the uncertainty about the organization of genetic variation in individuals of Glomeromycota.

Genetic makeup of AM fungal individuals

Conflicting hypotheses describe the genome organization in Glomeromycota. They have been reviewed comprehensively elsewhere (Pawlowska, 2005, 2007; Rosendahl, 2008) and I give here only a brief summary of these arguments.

Spores of Glomeromycota contain hundreds of nuclei (Cooke *et al.*, 1987; Bécard and Pfeffer, 1993; Pawlowska and Taylor, 2004). They also harbour several different sequences of rRNA genes within the same individual, which is uncommon in other organisms (Sanders *et al.*, 1995). This pattern of rRNA gene polymorphism contained within the multinucleate structure of the same spore inspired a hypothesis that the spore nuclei may be genetically different and that the Glomeromycota are unique, multigenomic organisms (Kuhn *et al.*, 2001; Hijri and Sanders, 2005). The multigenomic mycelia of Glomeromycota were proposed to be products of hyphal fusions between genetically different individuals (Bever and Morton, 1999; Bever and Wang, 2005), even though, until recently, such hyphal fusions have not been observed (Giovannetti *et al.*, 1999, 2003). A new study by Croll *et al.* (2009) presented evidence of hyphal fusions between genetically divergent isolates of *Glomus intraradices*, which lends support to the hypothesis that gene exchanges are possible in populations of Glomeromycota. However, whether the nuclear heterogeneity generated by hyphal fusions is, in fact, maintained in the cytoplasm of the coenocytic hyphae, remains unclear.

An alternative explanation for the intra-individual rRNA gene sequence polymorphism in Glomeromycota suggests a departure from the molecular pattern of concerted evolution of rRNA genes (Pawlowska and Taylor, 2004, 2005; Pawlowska, 2005; Pawlowska, 2007). Concerted evolution is a universal recombination-driven process responsible for rRNA gene sequence homogeneity within an individual and among individuals of a recombining population (Arnheim *et al.*, 1980; Dover, 1982). Coexistence of polymorphic rRNA gene sequences observed in individual nuclei of Glomeromycota suggests that the process of concerted evolution may be impaired in this lineage (Pawlowska and Taylor, 2004). The pattern of relaxed concerted evolution is apparent not only at the level of individual nuclei but also at the level of individual species and phylogenetic lineages. For example, species in the *Glomus etunicatum* lineage contain distinctly paralogous variants of rRNA genes that share less sequence similarity within a taxon than with their orthologues in sister taxa (Rodriguez *et al.*, 2005). Such pattern would not be expected if divergent rRNA gene variants were distributed among different nuclei. Even though intraindividual and intraspecifc patterns of rRNA gene polymorphisms are characteristic features of Glomeromycota, there is also evidence that the processes responsible for gene homogenization are not abolished entirely in these organisms. Recombinant rRNA gene sequences were found in multiple taxa, including *G. etunicatum*, which is the species where departures from concerted evolution were described in detail (Kuhn *et al.*, 2001; Gandolfi *et al.*, 2003; Pawlowska and Taylor, 2004; Rodriguez *et al.*, 2005).

The uncertainty concerning the distribution of gene copies among the nuclei of Glomeromycota extends beyond the rRNA genes and applies to any multicopy DNA sequence. This uncertainty is a serious impediment for population structure studies aimed at inferring the reproductive mode in the extant populations of Glomeromycota because of the possible confounding effects of cryptic intraindividual polymorphisms that may be overlooked when single-copy population genetic markers are developed. In contrast to nuclear rRNA genes, no evidence of rampant intraindividual heterogeneity is apparent in mitochondrial rRNA gene sequences of Glomeromycota (Raab *et al.*, 2005). Since the mode of mitochondrial inheritance in Glomeromycota is unknown and the patterns of the mitochondrial transmission vary among fungi from biparental to strictly uniparental (Yan and Xu, 2005), the utility of the mitochondrial rRNA gene intraindividual homogeneity for understanding the reproductive mode of AM fungi remains unclear. Nonetheless, mitochondrial DNA markers are regarded as a reliable system to investigate population differentiation and phylogeography, and utilized in a growing number of studies (Croll *et al.*, 2008).

Population differentiation in AM fungi

Some AM fungal populations sampled in a spatially explicit manner, including *G. mosseae* and *G. caledonium* from agricultural sites in Denmark as well as *G. intraradices* from fields in Switzerland, exhibited population differentiation (Stukenbrock and Rosendahl, 2005; Koch *et al.*, 2006; Croll *et al.*, 2008; Rosendahl and Matzen, 2008). While the sampled populations were different between fields subjected to different cultivation regimes, the spatial heterogeneity within the fields could not be readily explained (Croll *et al.*, 2008; Rosendahl and Matzen, 2008). The recognition that natural populations of AM fungi are genetically heterogeneous and spatially structured is the critical first step towards understanding factors that contribute to the evolutionary stability of arbuscular mycorrhizae. The fact that many species of AM fungi exhibit global distribution, together with the frequently observed lack of strong mutualistic outcomes in mycorrhizal associations, suggest that arbuscular mycorrhizae may be shaped by processes typical for the geographic mosaic of coevolution. However, huge knowledge gaps need to be filled before this hypothesis could be tested. Among them, identification of processes that contribute to trait remixing in Glomeromycota is of particular importance because of the long-standing and insufficiently tested claim that these organisms do not engage in genetic recombination.

Reproductive mode of AM fungi

With no evidence of sexual structures and processes, modern Glomeromycota are considered to be strictly asexual (Pawlowska, 2005; Pawlowska, 2007; Rosendahl, 2008). For dispersal, they rely on asexual spores whose morphology has not changed since the Ordovician (Redecker *et al.*, 2000). The conservation of morphological structures apparent in the fossil record (Remy *et al.*, 1994; Redecker *et al.*, 2000) gave rise to a speculation that Glomeromycota remained asexual throughout their evolutionary history (Normark *et al.*, 2003). While attractive, this hypothesis is inconsistent with the patterns found in other groups of fungi, including species of *Neotyphodium*, where asexual lineages are of a recent evolutionary origin (Taylor *et al.*, 1999). It also challenges the predictions of multiple evolutionary models developed to explain the adaptive significance of sexual recombination (Kondrashov, 1993). The ephemeral nature of asexual taxa is attributed to the accumulation of deleterious mutations accompanied by the advancement of the Muller's ratchet, and leading to population extinction (Muller, 1964; Kondrashov, 1984; Lynch and Gabriel, 1990; Gabriel *et al.*, 1993). Recombination is expected to facilitate purging of deleterious mutations from a population and regeneration of mutation-free individuals (Muller, 1964; Kondrashov, 1984).

Even though vegetative non-self hyphal fusions are possible in Glomeromycota (Croll *et al.*, 2009) and several studies aimed at detecting recombination in populations of various species of Glomeromycota have been conducted so far, unequivocal evidence of a history of gene exchanges in these fungi remains elusive. The majority of natural and experimental populations of *Glomus* species examined in early studies turned out to be clonal, based on AFLP (amplified fragment length polymorphism) and ISSR (intersimple sequence repeat) fingerprinting followed by comparisons of indices of association between the observed and artificially recombined data sets (Rosendahl and Taylor, 1997; Vandenkoornhuyse *et al.*, 2001). These populations included experimental populations of *G. mosseae* and *G. caledonium* as well as two field populations of *G. claroideum* and *Glomus* sp. from agricultural sites in Europe. The hypothesis of recombination could not be rejected only in populations *G. claroideum* and *Glomus* sp. sampled from a site exposed to long-term sewage-sludge inputs (Vandenkoornhuyse *et al.*, 2001). It is possible that the sewage-sludge to which the fungi were exposed could have created conditions of genotoxic or environmental stress. Environmental factors are known to affect rates of recombination in sexually reproducing fungi with a trend towards increased expression of sexuality with increasing stress (Towe and Stadler, 1964; Lu, 1974; Grishkan *et al.*, 2003). The uncertainties concerning identities of positive as well as negative alleles intrinsic to fingerprinting methods, which made the conclusions of these early studies open to critique (Rosendahl, 2008), were circumvented by multilocus sequence typing of nuclear genes used in several recent studies. Again, the majority of investigated populations, including three populations of *G. mosseae* and three populations *G. caledonium* from agricultural sites in Denmark turned out to be clonal (Stukenbrock and Rosendahl, 2005; Rosendahl and Matzen, 2008). In contrast, several genotypes in a population of *G. intraradices* from an agricultural site in Switzerland showed molecular patterns consistent with intergenic recombination (Croll and Sanders, 2009). These recombinant patterns could be attributed to several processes, including meiotic recombination in a cryptic sexual process, or gene exchanges mediated by non-self-vegetative hyphal fusions. However, given the uncertainties concerning genetic makeup of Glomeromycota, it cannot be excluded that these patterns originated due to mitotic recombination between duplicate gene copies (Szostak and Wu, 1980) present in the poorly understood genome of *G. intraradices* (Martin *et al.*, 2008).

The knowledge of population-level processes and patterns that they generate in local and global populations of Glomeromycota is only beginning to emerge. Fundamental problems that include individual genetic makeup and the capacity for gene exchange among individuals remain unresolved. It is clear, however, that, similarly to the *Neotyphodium*–grass mutualism, the evolution of arbuscular mycorrhizae cannot be understood without elucidating the factors and processes that affect distribution and frequency of alleles in fungal populations.

Conclusion

Both Glomeromycota and epichloë fungi are symbionts of plants. They can even coexist in pooid grasses (Guo *et al.*, 1992; Müller, 2003; Mack and Rudgers, 2008). Yet these two classes of fungal cooperators are subject to different and often conflicting selective pressures and represent distinct trajectories of mutualism evolution. Inferring factors that contributed to the Devonian origins of arbuscular mycorrhizae is difficult because of the absence of free-living taxa in the Glomeromycota lineage. The expansions of the biological market framework to accommodate coexistence of multiple unrelated mycorrhizal symbionts in one host offered only recently the first insights into the mechanisms that can stabilize such complex associations (Kummel and Salant, 2006). The ability of both mycorrhizal hosts and symbionts to interact with multiple partners creates a pattern of interactions that resemble networks formed by cooperators involved in animal-mediated plant pollination and seed dispersal. It appears, therefore, that applying conceptual tools that emerged from studying plant-animal networks should be a productive approach for unravelling the evolution of arbuscular mycorrhizae. In contrast to arbuscular mycorrhizae, the emergence of *Neotyphodium* endophytes from their parasitic ancestors is a relatively well-understood process. It is likely that new *Neotyphodium* forms continue to arise in modern ecosystems. While the *Neotyphodium*–grass mutualism is generally well explained by the existing models of mutualism evolution, it is not free of surprises. The discovery of the lack of strong mutualistic outcomes in the *Neotyphodium*–Arizona fescue system challenged the traditional understanding of the *Neotyphodium*–grass mutualism and inspired use of novel theoretical frameworks to explain the antagonistic interaction patterns. The integrated application of ecology and population genetics tools to explore the *Neotyphodium*–Arizona fescue system has lead to the identification of patterns of selection and gene flow that are consistent with the predictions of the geographic mosaic theory of coevolution. This lesson is particularly pertinent to the efforts to understand the evolution of arbuscular mycorrhizae, where the knowledge gaps encompass basic life history traits including the genetic make-up of fungal individuals as well as the fundamentals of transmission and population genetics. Without these data, the processes that contribute to local reciprocal selection in mycorrhizal partners and remixing of coevolving traits on a geographic scale cannot be understood.

Acknowledgements

I am grateful to Chris Schardl and Michael Milgroom for critical comments on the manuscript and to the members of the Pawlowska lab: Dean Castillo, Dan Moebius-Clune, Stephen Mondo, and Nick VanKuren as well as students of the course 'Symbiotic Associations: Evolution and Ecology' for discussions that inspired this chapter.

References

Arnheim, N., Krystal, M., Schmickel, R., Wilson, G., Ryder, O., and Zimmer, E. (1980). Molecular evidence for genetic exchanges among ribosomal genes on nonhomologous chromosomes in man and apes. P. Natl. Acad. Sci. USA 77, 7323–7327.

Arroyo García, R., Martínez Zapater, J.M., García Criado, B., and Zabalgogeazcoa, I. (2002). Genetic structure of natural populations of the grass endophyte *Epichloë festucae* in semiarid grasslands. Mol. Ecol. *11*, 355–364.

Augé, R.M. (2004). Arbuscular mycorrhizae and soil/plant water relations. Can. J. Soil. Sci. *84*, 373–381.

Axelrod, R., and Hamilton, W.D. (1981). The evolution of cooperation. Science *211*, 1390–1396.

Baker, H.G. (1961). The adaptation of flowering plants to nocturnal and crepuscular pollinators. Q. Rev. Biol. *36*, 64–73.

Bascompte, J., and Jordano, P. (2007). Plant-animal mutualistic networks: The architecture of biodiversity. Annu. Rev. Ecol. Evol. S. *38*, 567–593.

Bascompte, J., Jordano, P., Melian, C.J., and Olesen, J.M. (2003). The nested assembly of plant-animal mutualistic networks. P. Natl. Acad. Sci. USA *100*, 9383–9387.

Bécard, G., and Pfeffer, P.E. (1993). Status of nuclear division in arbuscular mycorrhizal fungi during *in vitro* development. Protoplasma *174*, 62–68.

Bever, J.D., and Morton, J. (1999). Heritable variation and mechanisms of inheritance of spore shape within a population of *Scutellospora pellucida*, an arbuscular mycorrhizal fungus. Am. J. Bot. *86*, 1209–1216.

Bever, J.D., and Wang, M. (2005). Arbuscular mycorrhizal fungi – Hyphal fusion and multigenomic structure. Nature *433*, E3–E4.

Borowicz, V.A. (2001). Do arbuscular mycorrhizal fungi alter plant–pathogen relations? Ecology *82*, 3057–3068.

Brem, D., and Leuchtmann, A. (2003). Molecular evidence for host-adapted races of the fungal endophyte *Epichloë bromicola* after presumed host shifts. Evolution *57*, 37–51.

Bucheli, E., and Leuchtmann, A. (1996). Evidence for genetic differentiation between choke-inducing and asymptomatic strains of the *Epichloë* grass endophyte from *Brachypodium sylvaticum*. Evolution *50*, 1879–1887.

Bull, J.J., and Rice, W.R. (1991). Distinguishing mechanisms for the evolution of cooperation. J. Theor. Biol. *149*, 63–74.

Bultman, T.L., and Leuchtmann, A. (2003). A test of host specialization by insect vectors as a mechanism for reproductive isolation among entomophilous fungal species. Oikos *103*, 681–687.

Bultman, T.L., and White, J.F. (1988). Pollination of a fungus by a fly. Oecologia *75*, 317–319.

Chiariello, N., Hickman, J.C., and Mooney, H.A. (1982). Endomycorrhizal role for interspecific transfer of phosphorus in a community of annual plants. Science *217*, 941–943.

Chung, K.R., and Schardl, C.L. (1997). Vegetative compatibility between and within *Epichloë* species. Mycologia *89*, 558–565.

Clay, K., and Schardl, C. (2002). Evolutionary origins and ecological consequences of endophyte symbiosis with grasses. Am. Nat. *160*, S99–S127.

Cooke, J.C., Gemma, J.N., and Koske, R.E. (1987). Observations of nuclei in vesicular-arbuscular mycorrhizal fungi. Mycologia 79, 331–333.

Croll, D., Giovannetti, M., Koch, A.M., Sbrana, C., Ehinger, M., Lammers, P.J., and Sanders, I.R. (2009). Nonself vegetative fusion and genetic exchange in the arbuscular mycorrhizal fungus *Glomus intraradices*. New Phytol. *181*, 924–937.

Croll, D., and Sanders, I.R. (2009). Recombination in *Glomus intraradices*, a supposed ancient asexual arbuscular mycorrhizal fungus. BMC Evol. Biol. 9.

Croll, D., Wille, L., Gamper, H.A., Mathimaran, N., Lammers, P.J., Corradi, N., and Sanders, I.R. (2008). Genetic diversity and host plant preferences revealed by simple sequence repeat and mitochondrial markers in a population of the arbuscular mycorrhizal fungus *Glomus intraradices*. New Phytol. *178*, 672–687.

Doebeli, M., and Knowlton, N. (1998). The evolution of interspecific mutualisms. Proc. Natl. Acad. Sci. USA *95*, 8676–8680.

Dover, G. (1982). Molecular drive: a cohesive mode of species evolution. Nature *299*, 111–117.

Egger, K.N., and Hibbett, D.S. (2004). The evolutionary implications of exploitation in mycorrhizas. Can. J. Bot. *82*, 1110–1121.

Ewald, P.W. (1987). Transmission modes and evolution of the parasitism–mutualism continuum. Ann. NY Acad. Sci. *503*, 295–306.

Faeth, S.H., Bush, L.P., and Sullivan, T.J. (2002). Peramine alkaloid variation in *Neotyphodium*-infected Arizona fescue: Effects of endophyte and host genotype and environment. J. Chem. Ecol. *28*, 1511–1526.

Faeth, S.H., and Hamilton, C.E. (2006). Does an asexual endophyte symbiont alter life stage and long-term survival in a perennial host grass? Microb. Ecol. *52*, 748–755.

Faeth, S.H., Helander, M.L., and Saikkonen, K.T. (2004). Asexual *Neotyphodium* endophytes in a native grass reduce competitive abilities. Ecol. Lett. *7*, 304–313.

Faeth, S.H., and Sullivan, T.J. (2003). Mutualistic asexual endophytes in a native grass are usually parasitic. Am. Nat. *161*, 310–325.

Fleming, T.H., and Muchhala, N. (2008). Nectar-feeding bird and bat niches in two worlds: pantropical comparisons of vertebrate pollination systems. J. Biogeogr. *35*, 764–780.

Foster, K.R., and Kokko, H. (2006). Cheating can stabilize cooperation in mutualisms. Proc. Roy. Soc. B-Biol. Sci. *273*, 2233–2239.

Foster, K.R., and Wenseleers, T. (2006). A general model for the evolution of mutualisms. J. Evol. Biol. *19*, 1283–1293.

Frank, S.A. (1994a). Genetics of mutualism: The evolution of altruism between species. J. Theor. Biol. *170*, 393–400.

Frank, S.A. (1994b). Kin selection and virulence in the evolution of protocells and parasites. Proc. Roy. Soc. Lond. B Biol. *258*, 153–161.

Frank, S.A. (1996a). Host control of symbiont transmission: The separation of symbionts into germ and soma. Am. Nat. *148*, 1113–1124.

Frank, S.A. (1996b). Host-symbiont conflict over the mixing of symbiotic lineages. Proc. Roy. Soc. Lond. B Biol. *263*, 339–344.

Frank, S.A. (1996c). Models of parasite virulence. Q. Rev. Biol. *71*, 37–78.

Frank, S.A. (1997). Models of symbiosis. Am. Nat. *150*, S80–S99.

Freeman, P.W. (1995). Nectarivorous feeding mechanisms in bats. Biol. J. Linn. Soc. *56*, 439–463.

Gabriel, W., Lynch, M., and Burger, R. (1993). Muller's ratchet and mutational meltdowns. Evolution *47*, 1744–1757.

Gandolfi, A., Sanders, I.R., Rossi, V., and Menozzi, P. (2003). Evidence of recombination in putative ancient asexuals. Mol. Biol. Evol. *20*, 754–761.

Gentile, A., Rossi, M.S., Cabral, D., Craven, K.D., and Schardl, C.L. (2005). Origin, divergence, and phylogeny of epichloë endophytes of native Argentine grasses. Mol. Phylogenet. Evol. 35, 196–208.

Giovannetti, M., Azzolini, D., and Citernesi, A.S. (1999). Anastomosis formation and nuclear and protoplasmic exchange in arbuscular mycorrhizal fungi. Appl. Environ. Microb. *65*, 5571–5575.

Giovannetti, M., Sbrana, C., Strani, P., Agnolucci, M., Rinaudo, V., and Avio, L. (2003). Genetic diversity of isolates of *Glomus mosseae* from different geographic areas detected by vegetative compatibility testing and biochemical and molecular analysis. Appl. Environ. Microb. *69*, 616–624.

Gomulkiewicz, R., Drown, D.M., Dybdahl, M.F., Godsoe, W., Nuismer, S.L., Pepin, K.M., Ridenhour, B.J., Smith, C.I., and Yoder, J.B. (2007). Dos and don'ts of testing the geographic mosaic theory of coevolution. Heredity *98*, 249–258.

Gomulkiewicz, R., Nuismer, S.L., and Thompson, J.N. (2003). Coevolution in variable mutualisms. Am. Nat. *162*, S80–S93.

Gomulkiewicz, R., Thompson, J.N., Holt, R.D., Nuismer, S.L., and Hochberg, M.E. (2000). Hot spots, cold spots, and the geographic mosaic theory of coevolution. Am. Nat. *156*, 156–174.

Grishkan, I., Korol, A.B., Nevo, E., and Wasser, S.P. (2003). Ecological stress and sex evolution in soil microfungi. Proc. Roy. Soc. Lond. B Biol. *270*, 13–18.

Guo, B.Z., Hendrix, J.W., An, Z.Q., and Ferriss, R.S. (1992). Role of *Acremonium* endophyte of fescue on inhibition of colonization and reproduction of mycorrhizal fungi. Mycologia *84*, 882–885.

Herre, E.A., Knowlton, N., Mueller, U.G., and Rehner, S.A. (1999). The evolution of mutualisms: exploring the paths between conflict and cooperation. Trends Ecol. Evol. *14*, 49–53.

Hijri, M., and Sanders, I.R. (2005). Low gene copy number shows that arbuscular mycorrhizal fungi inherit genetically different nuclei. Nature *433*, 160–163.

Hoeksema, J.D., and Schwartz, M.W. (2003). Expanding comparative-advantage biological market models: contingency of mutualism on partners' resource requirements and acquisition trade-offs. Proc. Roy. Soc. Lond. B Biol. *270*, 913–919.

Husband, R., Herre, E.A., Turner, S.L., Gallery, R., and Young, J.P. W. (2002a). Molecular diversity of arbuscular mycorrhizal fungi and patterns of host association over time and space in a tropical forest. Mol. Ecol. *11*, 2669–2678.

Husband, R., Herre, E.A., and Young, J.P. W. (2002b). Temporal variation in the arbuscular mycorrhizal communities colonising seedlings in a tropical forest. FEMS Microbiol. Ecol. *42*, 131–136.

Jansa, J., Smith, F.A., and Smith, S.E. (2008). Are there benefits of simultaneous root colonization by different arbuscular mycorrhizal fungi? New Phytol. *177*, 779–789.

Jones, M.D., and Smith, S.E. (2004). Exploring functional definitions of mycorrhizas: Are mycorrhizas always mutualisms? Can. J. Bot. *82*, 1089–1109.

Kiers, E.T., and van der Heijden, M.G. A. (2006). Mutualistic stability in the arbuscular mycorrhizal symbiosis: Exploring hypotheses of evolutionary cooperation. Ecology *87*, 1627–1636.

Koch, A.M., Croll, D., and Sanders, I.R. (2006). Genetic variability in a population of arbuscular mycorrhizal fungi causes variation in plant growth. Ecol. Lett. *9*, 103–110.

Koch, A.M., Kuhn, G., Fontanillas, P., Fumagalli, L., Goudet, I., and Sanders, I.R. (2004). High genetic variability and low local diversity in a population of arbuscular mycorrhizal fungi. P. Natl. Acad. Sci. USA *101*, 2369–2374.

Kondrashov, A.S. (1984). Deleterious mutations as an evolutionary factor. 1. The advantage of recombination. Genet. Res. *44*, 199–217.

Kondrashov, A.S. (1993). Classification of hypotheses on the advantage af amphimixis. J. Hered. *84*, 372–387.

Kuhn, G., Hijri, M., and Sanders, I.R. (2001). Evidence for the evolution of multiple genomes in arbuscular mycorrhizal fungi. Nature *414*, 745–748.

Kummel, M., and Salant, S.W. (2006). The economics of mutualisms: optimal utilization of mycorrhizal mutualistic partners by plants. Ecology *87*, 892–902.

Lehtonen, P.T., Helander, M., Siddiqui, S.A., Lehto, K., and Saikkonen, K. (2006). Endophytic fungus decreases plant virus infections in meadow ryegrass (*Lolium pratense*). Biol. Lett. *2*, 620–623.

Leuchtmann, A. (2007). *Botanophila* flies on *Epichloë* host species in Europe and North America: no evidence for co-evolution. Entomol. Exp. Appl. *123*, 13–23.

Leuchtmann, A., Schardl, C.L., and Siegel, M.R. (1994). Sexual compatibility and taxonomy of a new species of *Epichloë* symbiotic with fine fescue grasses. Mycologia *86*, 802–812.

Lu, B.C. (1974). Genetic recombination in *Coprinus*. IV. Kinetic study of temperature effect on recombination frequency. Genetics *78*, 661–677.

Lynch, M., and Gabriel, W. (1990). Mutation load and the survival of small populations. Evolution *44*, 1725–1737.

Mack, K.M. L., and Rudgers, J.A. (2008). Balancing multiple mutualists: asymmetric interactions among plants, arbuscular mycorrhizal fungi, and fungal endophytes. Oikos *117*, 310–320.

Malinowski, D.P., Alloush, G.A., and Belesky, D.P. (2000). Leaf endophyte *Neotyphodium coenophialum* modifies mineral uptake in tall fescue. Plant Soil *227*, 115–126.

Malinowski, D.P., and Belesky, D.P. (1999). *Neotyphodium coenophialum*-endophyte infection affects the ability of tall fescue to use sparingly available phosphorus. J. Plant. Nutr. *22*, 835–853.

Malinowski, D.P., and Belesky, D.P. (2000). Adaptations of endophyte-infected cool-season grasses to environmental stresses: Mechanisms of drought and mineral stress tolerance. Crop Sci. *40*, 923–940.

Malinowski, D.P., Belesky, D.P., and Fedders, J.M. (1999). Endophyte infection may affect the competitive ability of tall fescue grown with red clover. J. Agronom. Crop Sci. *183*, 91–101.

Martin, F., Gianinazzi-Pearson, V., Hijri, M., Lammers, P., Requena, N., Sanders, I.R., Shachar-Hill, Y., Shapiro, H., Tuskan, G.A., and Young, J.P. W. (2008). The long hard road to a completed *Glomus intraradices* genome. New Phytol. *180*, 747–750.

Maynard Smith, J., Smith, N.H., Orourke, M., and Spratt, B.G. (1993). How clonal are bacteria? P. Natl. Acad. Sci. USA *90*, 4384–4388.

Meijer, G., and Leuchtmann, A. (1999). Multistrain infections of the grass *Brachypodium sylvaticum* by its fungal endophyte *Epichloë sylvatica*. New Phytol. *141*, 355–368.

Meijer, G., and Leuchtmann, A. (2000). The effects of genetic and environmental factors on disease expression (stroma formation) and plant growth in *Brachypodium sylvaticum* infected by *Epichloë sylvatica*. Oikos *91*, 446–458.

Meijer, G., and Leuchtmann, A. (2001). Fungal genotype controls mutualism and sex in *Brachypodium sylvaticum* infected by *Epichloë sylvatica*. Acta. Biol. Hung. *52*, 249–263.

Moon, C.D., Craven, K.D., Leuchtmann, A., Clement, S.L., and Schardl, C.L. (2004). Prevalence of interspecific hybrids amongst asexual fungal endophytes of grasses. Mol. Ecol. *13*, 1455–1467.

Moon, C.D., Guillaumin, J.J., Ravel, C., Li, C.J., Craven, K.D., and Schardl, C.L. (2007). New *Neotyphodium* endophyte species from the grass tribes Stipeae and Meliceae. Mycologia *99*, 895–905.

Moon, C.D., Miles, C.O., Jarlfors, U., and Schardl, C.L. (2002). The evolutionary origins of three new *Neotyphodium* endophyte species from grasses indigenous to the Southern Hemisphere. Mycologia *94*, 694–711.

Moon, C.D., Scott, B., Schardl, C.L., and Christensen, M.J. (2000). The evolutionary origins of *Epichloë* endophytes from annual ryegrasses. Mycologia *92*, 1103–1118.

Moy, M., Belanger, F., Duncan, R., Freehoff, A., Leary, C., Meyer, W., Sullivan, R., and White, J.F. (2000). Identification of epiphyllous mycelial nets on leaves of grasses infected by clavicipitaceous endophytes. Symbiosis *28*, 291–302.

Muller, H.J. (1964). The relation of recombination to mutational advance. Mutat. Res. *1*, 2–9.

Müller, J. (2003). Artificial infection by endophytes affects growth and mycorrhizal colonisation of *Lolium perenne*. Funct. Plant Biol. *30*, 419–424.

Munkvold, L., Kjoller, R., Vestberg, M., Rosendahl, S., and Jakobsen, I. (2004). High functional diversity within species of arbuscular mycorrhizal fungi. New Phytol. *164*, 357–364.

Newman, E.I., Devoy, C.L. N., Easen, N.J., and Fowles, K.J. (1994). Plant species that can be linked by VA mycorrhizal fungi. New Phytol. *126*, 691–693.

Noë, R., and Hammerstein, P. (1994). Biological markets – supply-and-demand determine the effect of partner choice in cooperation, mutualism and mating. Behav. Ecol. Sociobiol. *35*, 1–11.

Noë, R., and Hammerstein, P. (1995). Biological markets. Trends Ecol. Evol. *10*, 336–339.

Normark, B.B., Judson, O.P., and Moran, N.A. (2003). Genomic signatures of ancient asexual lineages. Biol. J. Linn. Soc. *79*, 69–84.

Nuismer, S.L., Thompson, J.N., and Gomulkiewicz, R. (1999). Gene flow and geographically structured coevolution. P. Roy. Soc. Lond. B Bio. *266*, 605–609.

Nuismer, S.L., Thompson, J.N., and Gomulkiewicz, R. (2000). Coevolutionary clines across selection mosaics. Evolution *54*, 1102–1115.

Pawlowska, T.E. (2005). Genetic processes in arbuscular mycorrhizal fungi. FEMS Microbiol. Lett. *251*, 185–192.

Pawlowska, T.E. (2007). How the genome is organized in the Glomeromycota, In Sex in Fungi: Molecular Determination and Evolutionary Implications, Heitman, J., Kronstad, J.W., Taylor, J.W., and Casselton, L.A., eds. (Washington, D.C: ASM Press), pp. 419–430.

Pawlowska, T.E., and Taylor, J.W. (2004). Organization of genetic variation in individuals of arbuscular mycorrhizal fungi. Nature *427*, 733–737.

Pawlowska, T.E., and Taylor, J.W. (2005). Arbuscular mycorrhizal fungi – Hyphal fusion and multigenomic structure – Reply. Nature *433*, E4.

Raab, P.A., Brennwald, A., and Redecker, D. (2005). Mitochondrial large ribosomal subunit sequences are homogeneous within isolates of *Glomus* (arbuscular mycorrhizal fungi, Glomeromycota). Mycol. Res. *109*, 1315–1322.

Redecker, D., Kodner, R., and Graham, L.E. (2000). Glomalean fungi from the Ordovician. Science *289*, 1920–1921.

Remy, W., Taylor, T.N., Hass, H., and Kerp, H. (1994). Four hundred-million-year-old vesicular arbuscular mycorrhizae. Proc. Natl. Acad. Sci. U.S.A. *91*, 11841–11843.

Rodriguez, A., Clapp, J.P., Robinson, L., and Dodd, J.C. (2005). Studies on the diversity of the distinct phylogenetic lineage encompassing *Glomus claroideum* and *Glomus etunicatum*. Mycorrhiza *15*, 33–46.

Rosendahl, S. (2008). Communities, populations and individuals of arbuscular mycorrhizal fungi. New Phytol. *178*, 253–266.

Rosendahl, S., and Matzen, H.B. (2008). Genetic structure of arbuscular mycorrhizal populations in fallow and cultivated soils. New Phytol. *179*, 1154–1161.

Rosendahl, S., and Taylor, J.W. (1997). Development of multiple genetic markers for studies of genetic variation in arbuscular mycorrhizal fungi using AFLP(TM). Mol. Ecol. *6*, 821–829.

Sachs, J.L., Mueller, U.G., Wilcox, T.P., and Bull, J.J. (2004). The evolution of cooperation. Q. Rev. Biol. *79*, 135–160.

Saikkonen, K., Helander, M., Faeth, S.H., Schulthess, F., and Wilson, D. (1999). Endophyte–grass–herbivore interactions: the case of *Neotyphodium* endophytes in Arizona fescue populations. Oecologia *121*, 411–420.

Saikkonen, K., Ion, D., and Gyllenberg, M. (2002). The persistence of vertically transmitted fungi in grass metapopulations. Proc. Roy. Soc. Lond. B Biol. *269*, 1397–1403.

Saikkonen, K., Lehtonen, P., Helander, M., Koricheva, J., and Faeth, S.H. (2006). Model systems in ecology: dissecting the endophyte–grass literature. Trends Plant Sci. *11*, 428–433.

Sanders, I.R., Alt, M., Groppe, K., Boller, T., and Wiemken, A. (1995). Identification of ribosomal DNA polymorphisms among and within spores of the Glomales: application to studies on the genetic diversity of arbuscular mycorrhizal fungal communities. New Phytol. *130*, 419–427.

Schardl, C.L., Leuchtmann, A., and Spiering, M.J. (2004). Symbioses of grasses with seedborne fungal endophytes. Annu. Rev. Plant. Biol. *55*, 315–340.

Schwartz, M.W., and Hoeksema, J.D. (1998). Specialization and resource trade: biological markets as a model of mutualisms. Ecology *79*, 1029–1038.

Selosse, M.A., and Schardl, C.L. (2007). Fungal endophytes of grasses: hybrids rescued by vertical transmission? An evolutionary perspective. New Phytol. *173*, 452–458.

Simms, E.L., and Taylor, D.L. (2002). Partner choice in nitrogen-fixation mutualisms of legumes and rhizobia. Integr. Comp. Biol. *42*, 369–380.

Smith, S.E., and Read, D.J. (2008). Mycorrhizal Symbiosis, 3rd edn (New York: Academic Press).

Stanton, M.L. (2003). Interacting guilds: Moving beyond the pairwise perspective on mutualisms. Am. Nat. *162*, S10–S23.

Steinebrunner, F., Schiestl, F.P., and Leuchtmann, A. (2008a). Variation of insect attracting odor in endophytic *Epichloë* fungi: Phylogenetic constrains versus host influence. J. Chem. Ecol. *34*, 772–782.

Steinebrunner, F., Twele, R., Francke, W., Leuchtmann, A., and Schiestl, F.P. (2008b). Role of odour compounds in the attraction of gamete vectors in endophytic *Epichloë* fungi. New Phytol. *178*, 401–411.

Stukenbrock, E.H., and Rosendahl, S. (2005). Clonal diversity and population genetic structure of arbuscular mycorrhizal fungi (*Glomus* spp.) studied by multilocus genotyping of single spores. Mol. Ecol. *14*, 743–752.

Sullivan, T.J., and Faeth, S.H. (2004). Gene flow in the endophyte *Neotyphodium* and implications for coevolution with *Festuca arizonica*. Mol. Ecol. *13*, 649–656.

Sullivan, T.J., and Faeth, S.H. (2008). Local adaptation in *Festuca arizonica* infected by hybrid and nonhybrid *Neotyphodium* endophytes. Microb. Ecol. *55*, 697–704.

Szostak, J.W., and Wu, R. (1980). Unequal crossing over in the ribosomal DNA of *Saccharomyces cerevisiae*. Nature *284*, 426–430.

Tadych, M., Bergen, M., Dugan, F.M., and White, J.F. (2007). Evaluation of the potential role of water in spread of conidia of the *Neotyphodium* endophyte of *Poa ampla*. Mycol. Res. *111*, 466–472.

Taylor, J.W., Jacobson, D.J., and Fisher, M.C. (1999). The evolution of asexual fungi: Reproduction, speciation and classification. Annual Review of Phytopathology *37*, 197–246.

Thompson, J.N. (2005). The Geographic Mosaic of Coevolution (Chicago: University of Chicago Press).

Tintjer, T., Leuchtmann, A., and Clay, K. (2008). Variation in horizontal and vertical transmission of the endophyte *Epichloë elymi* infecting the grass *Elymus hystrix*. New Phytol. *179*, 236–246.

Towe, A.M., and Stadler, D.R. (1964). Effects of temperature on crossing over in *Neurospora*. Genetics *49*, 577–583.

Tsai, H.F., Liu, J.S., Staben, C., Christensen, M.J., Latch, G.C. M., Siegel, M.R., and Schardl, C.L. (1994). Evolutionary diversification of fungal endophytes of tall fescue grass by hybridization with *Epichloë* species. Proc. Natl. Acad. Sci. USA *91*, 2542–2546.

Vandenkoornhuyse, P., Leyval, C., and Bonnin, I. (2001). High genetic diversity in arbuscular mycorrhizal fungi: evidence for recombination events. Heredity *87*, 243–253.

Wäli, P.R., Ahlholm, J.U., Helander, M., and Saikkonen, K. (2007). Occurrence and genetic structure of the systemic grass endophyte *Epichloë festucae* in fine fescue populations. Microb. Ecol. *53*, 20–29.

Wali, P.R., Helander, M., Nissinen, O., and Saikkonen, K. (2006). Susceptibility of endophyte-infected grasses to winter pathogens (snow molds). Can. J. Bot. *84*, 1043–1051.

Waser, N.M., Chittka, L., Price, M.V., Williams, N.M., and Ollerton, J. (1996). Generalization in pollination systems, and why it matters. Ecology *77*, 1043–1060.

Wilkinson, D.M. (1997). The role of seed dispersal in the evolution of mycorrhizae. Oikos *78*, 394–396.

Wilkinson, D.M., and Sherratt, T.N. (2001). Horizontally acquired mutualisms, an unsolved problem in ecology? Oikos *92*, 377–384.

Yamamura, N., Higashi, M., Behera, N., and Wakano, J.Y. (2004). Evolution of mutualism through spatial effects. J. Theor. Biol. *226*, 421–428.

Yan, Z., and Xu, J. (2005). Fungal mitochondrial inheritance and evolution, In Evolutionary Genetics of Fungi, Xu, J., ed. (Norfolk: Horizon Bioscience), pp. 221–252.

Population Genetics of Pathogenic Fungi in Humans and Other Animals 9

Thomas G. Mitchell

Abstract

This article reviews the more common DNA-based markers that are used to genotype fungi, as well as other eukaryotes, and presents several examples of their application to elucidate the population genetics of mammalian pathogenic fungi. The most common fungal infections are briefly summarized to illustrate the issues that are routinely addressed by population genetics approaches. This exciting area of research is continually growing as the methods become more accessible and medical mycologists recognize the value of studying natural isolates of pathogenic fungi.

Dynamic of pathogenic fungi

Members of the fungal kingdom are eukaryotes. Although there are exceptions, most fungi are chemotrophic, non-photosynthetic, non-motile microbes with a multilayered cell wall composed of characteristic polysaccharides (e.g. chitin, glucan) and glycoproteins (e.g. mannoprotein). Fungi have two fundamental growth forms, as unicellular yeasts or filamentous hyphae, and some fungal species produce both yeast and hyphal growth. The form of vegetative growth manifested by a dimorphic species is triggered by external signals, such as temperature or CO_2 levels. As phenotypes, yeast and hyphal growth evolved multiple times and have no phylogenetic relevance. Approximately 80,000 species of fungi have been described, and the phyla and classes of fungi are depicted in Fig. 9.1. However, less than 400 have been reported as pathogens of humans and other animals, and fewer than 50 species cause more than 90% of the fungal mammalian infections (Mitchell, 2005). For several decades, the overall incidence of infections with fungi (i.e. mycoses) has steadily risen in parallel with the increases in human longevity and medical progress that prolongs the lives of patients with impaired host defences. Concomitantly, the range of etiologic agents has expanded with the emergence of new pathogenic fungi. This phenomenon is driven by several factors:

1 The increase in the susceptible population of humans has precipitated an increase in the prevalence of infections caused by fungi that were rarely pathogenic in past generations.
2 For similar reasons, almost monthly there are new case reports of infections due to fungi that were previously unknown as pathogens. Most of these novel infections occur as opportunists in immunocompromised patients with severe underlying conditions, such as cancer and solid organ and stem cell transplants.
3 Due to advances in the phenotypic and molecular characterization of pathogenic fungi, mycologists have newly discovered cryptic species that were previously unrecognized and misidentified (e.g. *Candida dubliniensis*) (Sullivan and Coleman, 1998).
4 The selection of fungi with resistance to both established and new chemotherapeutic agents has spawned the emergence of infections with drug-resistant strains and species, such as isolates of *Aspergillus terreus* that are resistant to amphotericin B and intrinsic

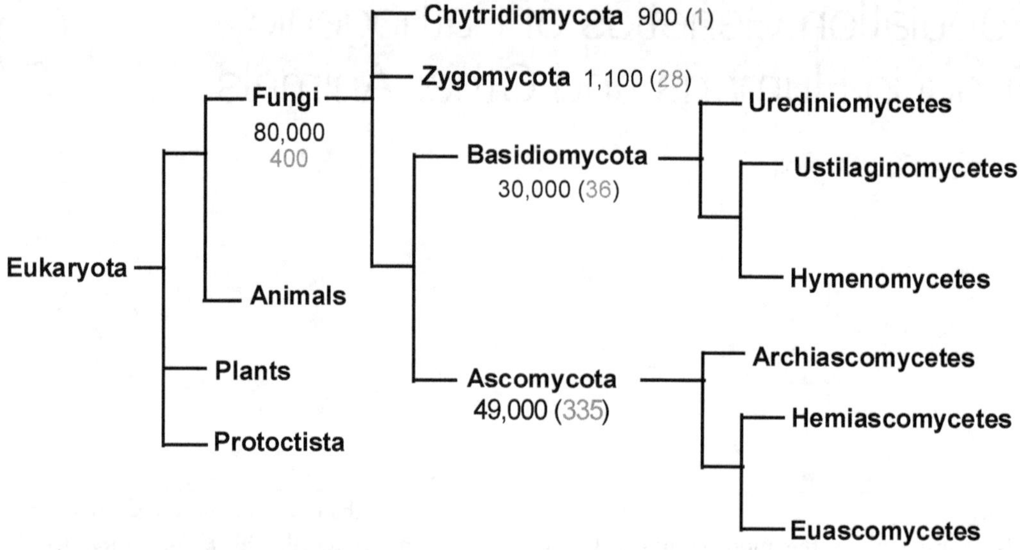

Figure 9.1 Major kingdoms of the Eukaryota, the four phyla of the Fungi, and the three classes of the Basidiomycota and the Ascomycota. Numbers in black indicate the approximate number of described species in each Phylum, and numbers in grey indicate the distribution of the approximately 400 species of mammalian pathogens (Mitchell, 2005).

resistance to fluconazole among isolates of *Candida glabrata* and *Candida krusei* (Pfaller and Diekema, 2004).

This review will discuss only the more common pathogens and those that have been subjected to rigorous population genetics studies. The review is not comprehensive but emphasizes methods that have gained wide acceptance and investigations in which the analysis of population structure has contributed to the resolution of issues regarding the natural history and aetiology of a mycotic infection or created new insights.

Population genetics of pathogenic fungi

As eukaryotes, fungi are complex organisms, and many phenotypes are quantitative traits that are affected by multiple genes. Most of the medical fungi exhibit asexual, mitotic reproduction leading to morphologically distinguishable conidia or sporangiophores. Many other species are also capable of sexual reproduction. Some fungi have complex life cycles involving significant phases during which the cells possess haploid, dikaryotic or diploid nuclei. To fully understand a given pathogen, it is useful to determine its reproduction and population structure because these mechanism(s) propel its evolution. This information has public health significance by predicting the organism's capacity to adapt quickly to changing environments, host species, and antifungal antibiotics. Molecular markers are the attributes used to dissect these mechanisms. In addition, these tools enable individual strains to be identified, providing a system for molecular epidemiology to track mycotic outbreaks.

The molecular markers developed to analyse the population structure of a pathogenic fungus can also be used to investigate its pathogenicity. Most approaches to microbial pathogenesis research entail studies of a few laboratory strains and (i) characterize a single gene (or gene family) to test its hypothetical role in virulence or (ii) focus on the entire genome by comparing the differential expression of genes under conditions that simulate infection. These strategies are problematic in eukaryotic pathogens because virulence is typically governed by hundreds of genes. Alternatively, one can genotype a large number of wild type isolates and use population genetics to identify strains that possess (or lack) the complement of genes required to produce disease, or more commonly, a specific phenotype of clinical importance.

Molecular methods to identify fungi

Traditional methods of identifying fungi rely on phenotypic differences in vegetative and reproductive morphology, metabolism (e.g. patterns of assimilation and fermentation, temperature and pH range of growth), the presence of structural macromolecules (e.g. cell wall polysaccharides, isoprene units of coenzyme Q) and sexual mating. Since most medical fungi are exclusively asexual (i.e. anamorphic), their identification and classification have rested on phenotypic features, which are often difficult to distinguish or highly variable. Consequently, for many pathogens, the species boundaries are loosely defined. For example, any elliptical, fermentative, non-pigmented, asexual yeast producing multilateral buds and pseudohyphae is likely to be placed in the genus *Candida*, which embraces numerous, polyphyletic species (Diezmann *et al.*, 2004).

In contrast, molecular markers are definitive and more stable than most phenotypic features. Different types of molecules may be analysed for the identification of fungi and the diagnosis of mycotic infections. They include antigens, secreted enzymes and proteins, fungal byproducts and secondary metabolites, cell wall mannoproteins and other compounds. However, this review will only focus on methods involving fungal nucleic acids.

Any region of the genome can serve as a marker for genotyping. Some markers are better at discriminating individual strains, discrete species or higher taxonomic groups. For certain studies, it is desirable to use markers in specific genes, and for others, markers in non-coding, usually anonymous portions of the genome are preferable because they are assumed to be neutral or unaffected by selective pressure. The more common DNA-based methods and their applications to fungi have been compared in recent reviews (Mitchell and Xu, 2003; Taylor and Fisher, 2003; Taylor *et al.*, 1999; Mitchell, 2005, 2008; Xu and Mitchell, 2002a,b, 2003).

To genotype individuals for comparison with other isolates of the same species, the most reliable DNA markers identify differences among their nucleotide sequences. Typical DNA markers recognize single nucleotide polymorphisms (SNPs), insertions or deletions (indels), and length variations in tandem oligonucleotide repeats (i.e. microsatellites). Such markers can be detected by a range of molecular techniques, such as PCR amplicon length polymorphisms, differences in the lengths of endonuclease restriction fragments (RFLP), randomly amplified polymorphism DNA (RAPD), amplified fragment length polymorphisms (AFLP), hybridization of nucleotide probes to electrophoretic blots of digested genomic DNA, and direct DNA sequencing of homologous genomic regions of the strains or species of interest (MLST or multilocus sequence typing). There are numerous applications of these methods to pathogenic fungi (Taylor and Fisher, 2003; Mitchell and Xu, 2003; Taylor *et al.*, 2000; Berbee and Taylor, 1999). As noted in Table 9.1, some methods are more suitable for population and phylogenetic studies of closely related taxa and strains of a species (e.g. AFLP and MLST) (Xu and Mitchell, 2003; Mitchell and Xu, 2003; Soll, 2000; Mitchell, 2008).

Other considerations for these methods are their cost and reproducibility. For example, AFLP genotyping has the advantages of being relatively inexpensive, rapid, and capable of generating numerous polymorphic bands. However, AFLP has the disadvantages of being laboratory-specific and somewhat subjective in scoring PCR bands. In addition, the linkage relationships among the AFLP loci cannot be determined. In contrast, MLST markers reflect DNA polymorphisms, usually SNPs, and they are scored by direct sequencing. The only disadvantage is the expense of sequencing multiple genes and strains. However, the resulting data are much more useful. In one recent study, a subset of AFLP-typed strains was also genotyped and analysed by MLST. There was congruency among the gene genealogies and both datasets produced similar Index of Association values (Litvintseva *et al.*, 2003; Litvintseva *et al.*, 2006).

With regard to microarray technology, the criteria for the oligonucleotide probes vary depending upon the application. High-throughput fungal microarrays have been developed for expression analysis, as well as comparative genome

Table 9.1 General DNA-based methods in medical mycology

Method	Application: Species identification	Strain identification, molecular epidemiology, population genetics	Phylogenetics and systematics
Electrophoretic karyotype	✓	✓	
RFLP	✓	✓	
Southern hybridization	✓	✓	
RAPD, PCR fingerprint, AFLP	✓	✓	
Microsatellites	✓	✓	
Microarrays	✓	✓	
SNP, MLST, DNA sequencing	✓	✓	✓

RFLP, restriction fragment length polymorphism; RAPD, random amplified polymorphic DNA; AFLP, amplified fragment length polymorphism; SNP, single nucleotide polymorphism; MLST, multilocus sequence typing (Bruns *et al.*, 1991; Mitchell and Xu, 2003; Mitchell, 2005; Xu and Mitchell, 2002b).

hybridization (Ronald *et al.*, 2005; Selmecki *et al.*, 2005). More recently, microarrays based on species-specific signature sequences within the ribosomal RNA (rDNA) motif have been developed to detect and identify common pathogenic fungi in both cultures (Leinberger *et al.*, 2005; Campa *et al.*, 2008) and clinical specimens (Spiess *et al.*, 2007; Huang *et al.*, 2006). However, few studies to date have used microarrays to investigate fungal populations (Forche *et al.*, 2004; Townsend and Taylor, 2005).

Certain markers can be used for phylogenetics to develop a classification of taxa that reflects their evolutionary relationships. Molecular methods include comparative analyses of the sequences of RNA, DNA and amino acids (Hillis *et al.*, 1996). The most comprehensive phylogenies across broad taxonomic groups have been constructed from comparisons of highly conserved nucleotide sequences, and therefore, it is imperative to select gene(s) or genomic sequences that are relatively conserved across the taxa of interest. Ribosomal RNA, especially alignable regions of the small (18S) and large (26S) subunits (SSU and LSU) of the rDNA genes have proven to be exceptionally useful in phylogenetic studies of fungi as well as other eukaryotes (Iwen *et al.*, 2002).

As with other organisms, both phylogenetic and population genetic methods are used to analyse the DNA-marker-based genotypes to determine the structure of a fungal population. The usual evidence supporting a clonal population structure are (i) over-representation of one or more genotypes; (ii) non-random segregation at individual loci (e.g. deviation of genotypic frequencies from Hardy–Weinberg expectations in diploids); and (iii) absence of recombination between loci (linkage disequilibrium). Phylogenetic criteria for clonality include evidence of phylogenetic structure among the individuals and congruence among multiple trees constructed from different datasets.

The population genetics of pathogenic fungi is a rapidly growing and significant area of research. Advances have been made in developing DNA markers, improving the analytical software, and applying these methods to the ever expanding spectrum of pathogenic fungi. These tools have enabled medical mycologists to (i) resolve taxonomic uncertainties, (ii) investigate the transmission of strains in local outbreaks, (iii) accurately identify fungi in clinical specimens, (iv) recognize strains with clinically important phenotypes (e.g. virulence factors, resistance to antifungal drugs), (v) trace the evolution of mycotic species, and (vi) validate the clinical relevance of strains used in basic research, comparative genomics, and the development of new diagnostic tests and antifungal antibiotics.

Fungal DNA markers

Since the DNA markers used in mycology are similar to those developed for other organisms and described in other chapters, only the most common medical mycological markers will be noted here.

PCR methods

In RAPD analysis, genomic or template DNA is amplified at a low annealing temperature (30–38°C) with a single short oligonucleotide (ca. 10 bases) in the PCR. Multiple PCR products of different electrophoretic mobilities are typically generated, and in comparing species or strains, each isolate will yield several bands of different sizes. RAPD analysis detects variations in the length between two primer binding sites, or sequence length polymorphisms in the fragments between PCR priming regions (Welsh and McClelland, 1990; Mitchell and Xu, 2003). Nucleotide substitutions in the region of PCR primer binding, particularly at the 3′ ends, can prevent binding of the primer to the DNA template and subsequent PCR amplification, and a band will be missing. Similarities in banding profiles among strains (i.e. the number and mobility, but not the density of the bands) can be calculated and used to infer epidemiological relationships. When multiple primers are screened, RAPD analysis is sensitive enough to detect variation among isolates that cannot be observed using other methods, although a combination of methods often provides optimal discrimination (Boekhout *et al.*, 1997; Brandt *et al.*, 1995; Mondon *et al.*, 1997). Despite being technically fast and simple, there are some disadvantages to RAPD. The major drawback is reproducibility. RAPDs can detect minute variation among strains, but as noted, even a single nucleotide substitution in the priming region may affect the production of a characteristic band. Small differences in PCR conditions may also affect binding of the primer. This problem is minimized when the protocol and reagents are carefully standardized. For this reason, the RAPD banding patterns of individual strains are usually specific to each laboratory. RAPDs can also be problematic because bands with the same electrophoretic mobility may not share the same sequence. Another concern with the interpretation of RAPD profiles is the problem of dominant and null alleles. In haploid species, both the dominant (presence) and null (absence) alleles can be scored, but in diploid organisms, it is not possible to distinguish genotypes that are homozygous for the dominant allele from those that are heterozygous. Therefore, RAPD data are generally not ideal for studies of population genetics (Clark and Lanigan, 1993). Nevertheless, for comparing the similarities among strains and developing fingerprints for molecular epidemiology, RAPD analyses have been widely applied to a number of medical fungi.

PCR fingerprinting is similar to RAPD, except that the primers are longer (>15 bases) and the annealing temperatures are more stringent. Most PCR fingerprinting primers are designed from repetitive DNA sequences, microsatellites (see below), or somewhat longer minisatellites (Meyer and Mitchell, 1995; Meyer *et al.*, 1993). Common primers include M13, which is derived from the core sequence of phage M13, T3B, which originates from internal sequences of tRNA genes, and TELO1, which is based on fungal telomere repeat sequences. Due to the increased stringency, PCR fingerprinting is generally more reproducible than RAPD. Although it suffers the same problems of interpretation as RAPDs, under standardized conditions, PCR fingerprinting has proven quite reliable for the identification of species and strains (Gräser *et al.*, 1998; Meyer and Mitchell, 1995; Meyer *et al.*, 1997; Thanos *et al.*, 1996).

The amplification of microsatellite repeats exploits the hypervariability of DNA regions composed of 10 or more tandem repeats of di-, tri- or multiple nucleotides. This hypervariability can be caused by either strand slippage during DNA replication or unequal crossing-over during meiosis. Useful microsatellites can be discovered by probing a genomic library or searching databases of gene sequences (Marra *et al.*, 2004). PCR primers flanking these repeat regions can be designed, and PCR products are typically run on electrophoretic gels to detect size differences in a single microsatellite. Because multiple alleles are typically found at a single locus, the relationships among alleles can be difficult to determine. Other concerns relate to the occasional instability of some microsatellites and PCR artefacts (de Valk *et al.*, 2007b).

Detection of amplified fragment length polymorphisms (AFLPs) is a powerful method for fingerprinting genomic DNA, as well as generating a large number of dominant markers for genotypic analysis (Vos *et al.*, 1995; Vos and Kuiper, 1997). With AFLP, genomic DNA is first digested with two endonucleases (usually a frequent and a rare cutting enzyme), and dsDNA adapters are ligated to the ends of the DNA fragments to create template DNA for the PCR. AFLP adapters consist of a core sequence and an enzyme-specific sequence. AFLP amplification primers consist of the core sequence, the enzyme-specific sequence and a selective extension of one to three nucleotides, which will amplify a subset of the restriction fragments. AFLP usually involves two PCR steps: The preamplification uses unlabelled primers with a single selective nucleotide. After this PCR, the reaction mixtures are diluted ten-fold and used as templates for the second or selective PCR, which uses a longer extension and labelled primers. AFLP has several advantages over other methods. Many more fragments can be generated and analysed. By varying the restriction enzymes and the extension nucleotides, 30 to >100 amplicons can be produced. In addition, these bands are stable and highly reproducible since they are amplified with two specific primers under stringent conditions. Like RAPD markers, AFLP markers are amplified using arbitrary sequences but with greater reproducibility and fidelity. However, AFLP data are also laboratory specific.

Another attractive application of PCR is the use of real-time quantitative PCR (qPCR) to detect fungal DNA in clinical specimens for the direct molecular diagnosis of mycotic infection. Most of the targets are fungal species-specific sequences found within the multi-copy rRNA motif. These methods generally offer superb specificity but less than ideal sensitivity, even with the presence of multiple copies. Despite a number of reports, a variety of protocols, and a range of fungal targets, DNA-detection by qPCR has not become a routine method in clinical laboratories. However, the methods continue to improve, and it is only a matter of time before they are widely accepted by clinical microbiology laboratories (White *et al.*, 2006; Perlin and Zhao, 2008).

RFLP methods

Restriction endonucleases have been used to discriminate species and strains by digesting genomic DNA with a four-base or other frequently cutting enzyme and examining the resultant bands on an electrophoretic gel (Clemons *et al.*, 1997). Although it may be possible to compare digests of genomic DNA for different banding patterns, indicating DNA sequence polymorphisms, this approach can only detect differences in DNA molecules that are present in high copy numbers. More commonly, an amplified gene or other PCR product can be digested and analysed (Xu *et al.*, 1999a). This approach was used to develop multilocus genotypes of *C. albicans* (Xu *et al.*, 1999a; Xu *et al.*, 1999b). In a diploid organism, co-dominant genetic information is obtained, and unlike RAPDs or PCR fingerprinting, both alleles can be scored, and this technique is highly reproducible.

The most widely used RFLP method is a DNA–DNA hybridization technique that involves digesting genomic DNA with restriction endonuclease(s) followed by electrophoretic separation of the DNA fragments. RFLPs are detected by Southern hybridization with labelled probes targeted to single copy markers or repetitive DNA. Several probes have been well characterized for *C. albicans*, such as Ca3, CARE2, 27A, and the RPS family (Lockhart *et al.*, 2001; Soll, 2000). These probes may be as large as several kb. Other established probes are specific for other species of *Candida*, and other pathogens (e.g. *A. fumigatus* and *C. neoformans*). RFLP bands generated by hybridization with probes to multi-copy loci may be difficult to interpret because of the large number of bands and/or the conditions of electrophoresis. These banding patterns can often recognize species or strains, but it is difficult for them to identify the allelic status of individual loci. Conversely, RFLPs detected by hybridization with single-copy probes are uncomplicated, and the data can be used for multiple purposes.

DNA sequencing

The most precise method to identify differences at the DNA level is the direct sequencing of PCR products, cloned genes, or whole genomes. This approach provides the most accurate data for phylogenetic and population genetic analyses.

As with other procedures, several investigations have analysed medical fungi by direct DNA sequencing. For comparing related taxa, the most robust approaches are whole genome sequencing and comparative genomics (Galagan *et al.*, 2005). The more amenable strategy, especially for molecular epidemiology and population genetics, is multilocus sequence typing (MLST), whereby sequences of several genes are compared across members of the species (Mitchell, 2008; Odds and Jacobsen, 2008; Taylor and Fisher, 2003). MLST data are reproducible and the protocols can be implemented by any laboratory with access to DNA sequencing facility. Matthew C. Fisher and colleagues have created a web site that catalogues protocols and data for many pathogenic bacteria and fungi, including *C. albicans*, *C. glabrata*, *C. krusei*, *C. tropicalis*, and *C. neoformans* var. *grubii* (http://www.mlst.net/databases/default.asp). (See also Table 9.4.)

Method selection

There is no single or ideal molecular method for every application. As mentioned above, different typing methods are appropriate for different purposes. For example, rDNA sequences are highly conserved at the species level, and therefore, they provide poor markers for strain identification, but they are excellent for differentiating higher-level taxonomic groups. In contrast, stable mutations are more common in non-coding sequences and also among the third bases of the codons of protein encoding genes. Consequently, for comparisons among strains of a species, MLST using protein encoding genes or non-functional DNA are usually more informative than rDNA. Thus, depending upon the goal, molecular methods can be selected to permit the identification of species, molecular epidemiology, detection of key phenotypic differences (e.g. resistance to antifungals), or analysis of genetic diversity and evolution.

Endemic mycoses

The natural histories of the endemic systemic mycoses – coccidioidomycosis, histoplasmosis, blastomycosis, paracoccidioidomycosis, and penicilliosis – share several features. All are caused by environmental moulds, *Coccidioides* species, *Histoplasma capsulatum*, *Blastomyces dermatitidis*, *Paracoccidioides brasiliensis* and *Penicillium marneffei*, respectively (Table 9.2). Infection is acquired primarily by inhalation of airborne fungal cells or asexual spores (conidia). These infections are not contagious or transmissible among humans or other animals. The aetiology of each mycosis and the primary infection are geographically restricted to specific areas of endemicity. The natural reservoirs of *Coccidioides* and *Histoplasma* are dry soil and soil mixed with guano, respectively. *Blastomyces dermatitidis* and *Paracoccidioides brasiliensis* are presumed to reside in nature, but their ecological niches have not been clearly defined. *P. marneffei* is associated with soil and wild rodent habitats (Mitchell, 2004).

Each of these mycoses is caused by a thermally dimorphic fungus. In nature, they grow as moulds producing septate vegetative hyphae, but in the host, or when cultivated in the laboratory on rich medium at 37°C, all except *Coccidioides* species convert to the yeast form of growth and reproduce by a budding process. Following their inhalation, the conidia (viz., arthroconidia) of *Coccidioides* transform into large spherules that replicate by producing internal endospores, which are released when the mature spherule fractures and restart this cycle. In their respective endemic areas, the rates of infection among humans and other animals is very high for coccidioidomycosis and histoplasmosis. However, with all of these moulds, up to 95% of the infections are asymptomatic. Skin test surveys have confirmed that many healthy people in the respective endemic areas have been exposed to *Coccidioides*, *Histoplasma* and *Paracoccidioides*. The few latent infections that progress to disease may be chronic or progressive, and they frequently involve dissemination from the lungs to other organs. Disease is triggered by a variety of host factors, such as malnutrition, debilitation due to other chronic diseases, or depressed cell-mediated immunity (Mitchell, 2004).

Coccidioides immitis and *C. posadasii*

These two aetiological agents of coccidioidomycosis are phenotypically indistinguishable soil moulds. Decades ago, Charles E. Smith and others recognized the circumscribed ecology of *Coccidioides* and the endemic areas of coccidioidomycosis (Smith *et al.*, 1956; Stevens, 1980). The infection is endemic in humans and other

Table 9.2 Selected dimorphic pathogenic fungi

Fungi	Mycosis	Ecology	Geographic distribution	Host range	Population genetics (references)
Histoplasma capsulatum	Histoplasmosis	Bat and avian habitats; soil	Global, but endemic in central USA, specific regions of NA, SA, Africa and Asia	Humans and other mammals	Carter *et al.* (2001), Kasuga *et al.* (2003), Muniz *et al.* (2001)
Coccidioides posadasii C. immitis	Coccidioidomycosis	Soil, desert rodents	Semiarid regions of southwestern USA, Mexico, Central and South America	Humans and other mammals	Burt *et al.* (2000), Fisher *et al.*, 2001, 2002b), Koufopanou *et al.* (2001)
Blastomyces dermatitidis	Blastomycosis	Unknown (beavers, riverbanks?)	Endemic in USA and Canada; regions bordering Mississippi, Ohio, and St. Lawrence Rivers, and south-eastern USA	Humans and dogs; rarely, other mammals	McCullough *et al.* (2000)
Paracoccidioides brasiliensis	Paracoccidioidomycosis	Unknown	Central and South America	Humans and rarely other mammals	Matute *et al.*, (2006a,b), Nascimento *et al.* (2004)
Penicillium marneffei	Penicilliosis	Soil, bamboo rat habitats	Southeast Asia, India, China	Humans, rodents	Lasker (2006), Vanittanakom *et al.* (2006), Woo *et al.* (2007)
Sporothrix schenckii	Sporotrichosis	Vegetation and soil	Global, but endemic in Mexico, Central America, Brazil, South Africa	Humans and other mammals	Marimon *et al.* (2006), Mesa-Arango *et al.* (2002)

mammals in semiarid regions of the southwestern USA, Central America, and South America. Cases reported outside these areas either have a history of travel to the endemic regions or can be attributed to transmission and exposure to the infectious arthroconidia by fomites. The haploid, asexual species *C. immitis* was thought to be the sole cause of coccidioidomycosis. In an elegant series of pioneering studies from the laboratory of John W. Taylor, PCR and SSCP markers were utilized in both population genetic and phylogenetic approaches to investigate the population structure of isolates of *C. immitis*, and their results provided solid evidence of recombination (Burt *et al.*, 1996). Subsequent analyses of five gene genealogies determined that isolates from California were reproductively isolated from isolates obtained from other endemic areas (Koufopanou *et al.*, 1997), and this finding led to the recognition of the cryptic species, *C. posadasii*, which is responsible for most cases outside California (Fisher *et al.*, 2002a; Tintelnot *et al.*, 2007).

Histoplasma capsulatum

Clinical and environmental isolates of *H. capsulatum* are haploid and normally proliferate by the production of asexual macro- and microconidia. However, like many other pathogenic fungi, *H. capsulatum* was discovered to be heterothallic (i.e. mating occurs between compatible strains with opposite mating type alleles). Strains possess either of two mating types (+ and –) and are capable of sexual reproduction in the laboratory, leading to cleistothecia and ascospores, which characterize the teleomorph, *Ajellomyces capsulatus*. Although the mating types are equally prevalent among environmental isolates, many more infections are caused by the '–' type, suggesting a linkage of this allele with pathogenicity. Classical genetic studies are problematic because after prolonged laboratory cultivation, isolates tend to lose the ability to reproduce asexually or sexually. Another variety also occurs in Africa, *H. capsulatum* var. *duboisii*, which is clinically distinguished from the predominant variety (*H. capsulatum* var. *capsulatum*) by an increase in skin and bone lesions, diminished pulmonary involvement, pronounced giant cell formation, and larger, thick-walled yeast cells in tissue (Mitchell, 2004). *In vitro*, these varieties are indistinguishable morphologically, but they differ in surface antigens and exhibit some ability to mate with each other. A third variety, *H. capsulatum* var. *farciminosum*, is associated with horses and causes cutaneous, subcutaneous and lymphangitic lesions.

The initial molecular studies used a variety of markers to study the population structure of clinical isolates of *H. capsulatum*, which identified individual strains and provided evidence for recombination (Carter *et al.*, 1996, 1997). Another investigation examined isolates from different geographic regions (Africa, and North, Central and South America) and attempted to clarify the relationships among the varieties. They used five gene sequences and analysed 46 isolates from different geographical areas, including *H. capsulatum* var. *duboisii* and *H. capsulatum* var. *farciminosum* (Kasuga *et al.*, 1999). The results indicated strong support for six clades and the possible existence of multiple cryptic species. To resolve these issues, an expanded collection of global isolates was analysed, which defined eight clades or lineages (Kasuga *et al.*, 2003). There was no phylogenetic support for the three varieties, but the lineages were associated with geographic regions and represented phylogenetic species. Based on the genetic diversity and the average DNA substitution rates of four protein-coding genes, these authors estimated that the lineages of *H. capsulatum* originated in Latin America between 3 and 13 MYA (Kasuga *et al.*, 2003).

Blastomyces dermatitidis

Similar to *H. capsulatum*, *B. dermatitidis* is a haploid, heterothallic member of the same teleomorphic genus, *Ajellomyces dermatitidis* (McDonough and Lewis, 1967). The two mating type alleles, '+' and '–', are generated in equal numbers and appear to have comparable pathogenicity (McDonough *et al.*, 1973). *B. dermatitidis* causes pulmonary and disseminated infections in two natural hosts, humans and dogs. More than 95% of these cases occur in the midwestern and southeastern United States and southeastern Canada (Mitchell, 2004). However, there have been rare reports of autochthonous cases in other continents and other mammalian species. Little is know about the ecology of *B. dermatitidis* because there have been only a few isolations from environmental samples (e.g. river banks, beaver habitat) and fewer than 20 reported outbreaks involving three or more cases developing in the same location and time period. In one large molecular study, 59 global isolates of *B. dermatitidis* were typed using digested genomic DNA hybridized with rDNA probes, PCR-RFLP markers, and DNA fingerprinting from which three major groups and multiple genotypes were identified (McCullough *et al.*, 2000). There was marginal correlation between the genotype and origin of the isolates. For example, all 16 isolates from the upper midwest US and southern Canada belonged to the same molecular group, but this group also contained isolates from other regions of North America; furthermore, all three isolates from Africa (Congo, South Africa, Zimbabwe) were in different groups (McCullough *et al.*, 2000).

Paracoccidioides brasiliensis

This haploid, asexual dimorphic pathogen is confined to Central and South America. After pulmonary infection, the period of latency typically extends for many years before active disease manifests, if at all. An acute or subacute form of the disease occurs in a minority of patients with pronounced cellular immunodeficiency (Franco *et al.*, 1994). Paracoccidioidomycosis is one of the most prevalent mycoses in Latin America. Little is known about its ecology because it has only rarely been recovered from the environment, most cases are isolated, and animal infections are rare. Primary hosts include humans and the nine-banded armadillo. An investigation of 33 clinical isolates used five primers to generate RAPD markers, and the similarity dendrogram grouped most of the isolates according to their origin – Argentina, Brazil, Colombia, Peru and Venezuela (Calcagno *et al.*, 1998). The correlation between genotype and country of origin was confirmed by another report that used RFLP markers (Nino-Vega *et al.*, 2000). A more recent population genetics study applied microsatellite markers to 64 isolates and delineated three phylogenetic species of *P. brasiliensis* (Matute *et al.*, 2006b). When the routine laboratory strain (Pb01) that has been the subject of numerous studies of the pathogenicity and immune responses to *P. brasiliensis* was genotyped, it differed considerably from the three phylogenetic species and appeared to be a different species (Carrero *et al.*, 2008).

Penicillium marneffei

Penicilliosis is a systemic infection of predominantly HIV-infected patients in South-East Asia, who develop pulmonary, systemic and cutaneous lesions. The cause is *P. marneffei*, a dimorphic species long associated with bamboo rats and other rodents in the endemic areas. A robust multilocus microsatellite typing method has been used to analyse the population genetics of *P. marneffei* (Fisher *et al.*, 2004), identifying a predominantly clonal population structure with no convincing evidence of recombination. The utility of microsatellite markers was confirmed in another study, which indicated that isolates from China and Thailand were genetically isolated populations (Lasker and Ran, 2004). Studies of the population genetics, molecular epidemiology and ecology of isolates from the endemic areas of Manipur, India, China and Thailand, have determined that indigenous rodents and/or soil are the likely reservoir for infections in patients with AIDS (Vanittanakom *et al.*, 2006).

Opportunistic mycoses

Patients with compromised host defences are susceptible to ubiquitous fungi to which healthy people are exposed but usually resistant. The type of pathogenic fungus and the natural history of the infection are often determined by the predisposing condition of the patient. As members of the normal microbial flora, *Candida* and related yeasts are endogenous opportunists. Most other opportunistic mycoses are caused by exogenous yeasts and moulds, which are globally present in soil, water, and air. For example, in patients with AIDS, susceptibility to opportunistic mycoses is inversely related to the patient's CD4+ lymphocyte count, and they are at risk for mycoses caused by the pathogens they most frequently encounter, namely, species of *Candida* and *Aspergillus*, *Cryptococcus*, endemic pathogens in the appropriate geographical sectors, and other environmental fungi (Table 9.3). All of these opportunistic mycoses have a relatively high prevalence, morbidity and mortality. Population geneticists have long been interested in whether host susceptibility is the crucial determinant of who becomes infected. Strains of most pathogenic fungi have been compared for their virulence in experimental animal models, and strains of most species can

Table 9.3 Selected global pathogenic fungi of humans and other mammals

Fungi (ploidy)	Ecology	Selected population genetics references
Cryptococcosis		
Cryptococcus neoformans (*n*)	Avian faeces, trees, soil	Hanafy *et al.* (2008), Hiremath *et al.* (2008), Litvintseva *et al.* (2006), Mitchell (2008)
C. gattii (*n*)	Trees	Bovers *et al.* (2008a,b), Byrnes *et al.* (2008), Trilles *et al.* (2008)
Candidiasis		
Candida albicans (2*n*) *C. glabrata* (*n*) *C. parapsilosis* (2*n*) *C. tropicalis* (2*n*) *C. dubliniensis* (*n*)	Mammalian mucosal microbiota	Bougnoux *et al.* (2004), Chowdhary *et al.* (2006), Dassanayake *et al.* (2006), Garcia-Hermoso *et al.* (2007), Jacobsen *et al.* (2007), Lockhart *et al.* (2005), Odds *et al.* (2007), Robles *et al.* (2004)
Aspergillosis		
Aspergillus fumigatus (*n*) *A. flavus* (*n*) *A. terreus* (*n*)	Soil, water, air	Alvarez-Perez *et al.* (2008), Balajee *et al.* (2008), de Valk *et al.* (2007a,b), Rydholm *et al.* (2006)
Dermatophytosis		
Trichophyton mentagrophytes (*n*)	Humans and other mammals	Fréalle *et al.* (2007)
T. rubrum (*n*) *T. violaceum* (*n*)	Humans	Chang *et al.* (2008), Gaedigk *et al.* (2003), Gräser *et al.* (1999), Ohst *et al.* (2004)

vary greatly in their virulence and lethality. Consequently, mycologists interested in clinical mycology or fungal pathogenesis are keenly interested in any correlation between genotypes and clinically relevant phenotypes. To what extent is the severity of an invasive opportunistic mycosis due to the condition of the host or the virulence of the infecting strain? This answer to this question may provide predictive information of public health relevance for the prevention, prognosis and epidemiology of a mycotic infection. In addition, if genetic markers can identify strains (or clades) with enhanced or diminished pathogenicity, basic researchers can compare them and use this information to identify genetic determinants of virulence.

This review includes only the most prevalent opportunistic mycoses and global pathogenic fungi whose population structures have been examined most extensively. The excluded fungi include zygomycetes (e.g. *Rhizopus*), dematiaceous ascomyctes (e.g. *Alternaria, Bipolaris, Curvularia*) and non-pigmented ascomycetous (e.g. *Fusarium, Paecilomyces*) moulds.

Candida species

Several species of the yeast genus *Candida* are capable of causing candidiasis, which can entail a myriad of clinical conditions that range from skin and nail infections to invasion of the blood stream and dissemination to multiple organs (Calderone, 2002). The pathogenic species of this large genus are usually members of the normal flora of the skin, mucous membranes, and gastrointestinal tract. *Candida* species colonize the mucosal surfaces of all humans during or soon after birth, and the risk of endogenous infection is ever present. Candidiasis is the most common systemic mycosis, and the most prevalent agents are *C. albicans, C. tropicalis, C. parapsilosis, C. glabrata, C. guilliermondii, C. kefyr,* and *C. dubliniensis*. The widespread use of fluconazole has precipitated the emergence of more resistant species, such as *C. glabrata, C. krusei* and *C. lusitaniae*. *C. albicans* and other *Candida* species are the most investigated of all medical fungi, and there are numerous reports of the population structure and molecular typing of these pathogens (Xu and Mitchell, 2002a,b, 2003; Mitchell and Xu, 2003; Bougnoux *et al.*, 2004; Lockhart *et al.*, 2005).

Some *Candida* species are haploid and others are diploid, and some have well-defined sexual as well as asexual modes of reproduction. The genus is definitely not monophyletic (Diezmann *et al.*, 2004). *C. albicans* is diploid and reproduces asexually in nature. For a decade, the population structure of *C. albicans* has been recognized to be predominantly clonal with some level of historical recombination (Gräser *et al.*, 1996; Xu *et al.*, 1999a). Virtually every type of molecular marker has been applied to *C. albicans* and other common species, as tools for molecular epidemiological investigations of outbreaks. Other studies have attempted to correlate populations with a variety of clinical phenotypes, such as body site, predisposing illness, geographical location, and resistance to antifungal drugs. The most robust methods in current use are standardized MLST markers (Bougnoux *et al.*, 2004; Robles *et al.*, 2004; Odds *et al.*, 2007).

These analyses of the population genetics of *C. albicans* have produced several practical dividends. Several studies using a variety of markers have demonstrated that isolates from patients are less diverse than isolates from healthy people (Xu *et al.*, 1999a; Schönian *et al.*, 1993). *C. dubliniensis* was confirmed to be a cryptic species of the phenotypically similar *C. albicans* by molecular typing studies that employed Southern hybridization probes (microsatellites and the *C. albicans*-specific probe 27A), karyotyping, RAPD markers, and rDNA sequencing (Sullivan *et al.*, 1995). In another study, AFLP markers determined that a sample of phenotypically atypical strains of *C. albicans* did not represent a cryptic species (Forche *et al.*, 1999). In a particularly intriguing report, RAPD genotypes identified a subset of strains of oral isolates with an apparent predilection for patients with AIDS (Xu *et al.*, 1999b).

Aspergillus species

There are many clinical manifestations of aspergillosis – a spectrum of diseases that may be caused by any of a number of *Aspergillus* species (Richardson, 2005). These species are ubiquitous saprobes in nature, and aspergillosis occurs worldwide. *A. fumigatus* is the most common pathogenic species, but many others, including *A flavus*, *A niger*, and *A terreus*, may also cause devastating disease. These moulds produce abundant small conidia that are easily aerosolized. Following their inhalation, atopic individuals often develop severe allergic reactions to the conidial antigens. In immunocompromised patients – especially those with leukaemia, bone marrow transplant patients, and individuals on prolonged regimens of corticosteroids –the conidia may germinate to produce hyphae that invade the blood vessels of the lungs and other organs. This disease, acute invasive aspergillosis, has a significant prevalence and mortality in high risk patients. Dissemination compounds the damage, and lesions in the brain are usually fatal regardless of the immediacy of the diagnosis and initiation of antifungal therapy.

Isolates of *A. fumigatus* are haploid. The population genetics of a global sample of 70 *A. fumigatus* isolates was analysed using intergenic sequence data, and the results revealed a largely clonal structure of relatively low diversity and lacking in geographical partitioning (Rydholm *et al.*, 2006). However, there was genetic evidence of historical recombination (Rydholm *et al.*, 2006; Varga and Tóth, 2003). Although many species of *Aspergillus* engage in sexual reproduction, which is usually heterothallic, *A. fumigatus* was thought to be asexual. However, strains were shown to possess the genes for undergoing sexual reproduction, and its sexual cycle was recently discovered (Paoletti *et al.*, 2005; O'Gorman *et al.*, 2008). Other useful molecular typing methods for *A. fumigatus* include microsatellites and AFLP (de Valk *et al.*, 2007a,b). A recent report using RAPD and microsatellite markers demonstrated that patients may be infected with multiple strains (Alvarez-Perez *et al.*, 2008). In addition, genic RFLP and SNP data are beginning to discriminate subpopulations, if not cryptic species of *A. fumigatus* (Balajee *et al.*, 2007).

Cryptococcus species

Cryptococcosis is caused by *Cryptococcus neoformans* or *Cryptococcus gattii*, and these yeasts can be isolated from avian guano and a variety of trees throughout the world. The infection is acquired by inhalation and may become latent or progress to disease, often with dissemination from the lungs to the reticuloendothelial system, skin, and other sites, but preferentially to the central nervous system (Mitchell and Perfect, 1995; Casadevall and Perfect, 1999). Like other systemic mycoses, cryptococcosis is not contagious, and many more people become infected than develop symptoms, which range from asymptomatic pulmonary infection to life-threatening meningoencephalitis. *C. gattii* is less common globally and usually infects immunocompetent persons, and *C. neoformans* is much more prevalent. Clinical risk factors for *C. neoformans* include HIV/AIDS, treatment with systemic corticosteroids, haematological malignancies, and immunosuppression. As noted above, the immune status of the patient is undoubtedly crucial, but host factors alone do not dictate whether infection leads to disease. Indeed, strains vary greatly in their ability to cause meningitis in experimental infections. The sibling species diverged approximately 37 MYA (Xu *et al.*, 2000). Based on genetic and phenotypic differences, two varieties of *C. neoformans* are recognized as well as hybrid strains. Unlike the other pathogens described here, *Cryptococcus* species are basidiomycetes (Fig. 9.1). Clinical and environmental isolates are invariably haploid but they possess one of two mating type alleles (α or a) and are often capable of sexual reproduction in the laboratory.

Over the past two decades, there have been manifold studies of the phylogenetics and population structure of these species, and a plethora of DNA markers have been employed (Mitchell, 2008). Quite recently, several laboratories involved in this research have developed standard MLST markers for genotyping strains of *C. neoformans* var. *grubii* (Table 9.4), and a similar consensus is being developed for *C. gattii*. The summary results of this collective effort is that there are several distinct genetic subpopulations

Table 9.4 Selected internet resources for population genetics and phylogenetics of fungal mammalian pathogens

Fungal taxa	Resources	Web site
Aspergillus fumigatus, *Candida tropicalis*, *C. krusei*	MLST markers, protocols	http://pubmlst.org/
Candida Genome Database	Genes, DNA sequences	http://www.candidagenome.org/
Candida Database	Genes, DNA sequences	http://genolist.pasteur.fr/CandidaDB/
Candida albicans, *C. glabrata*, *C. krusei*, *C. tropicalis*, *Cryptococcus neoformans* var. *grubii*	MLST markers, protocols	http://www.mlst.net/databases/default.asp
Multiple mammalian and plant pathogens	Broad Institute: Genome sequence projects	http://www.broad.mit.edu/node/568
Aspergillus fumigatus, *Coccidioides posadasii*, *Cryptococcus neoformans* var. *neoformans*	TIGR: Genome sequence projects	http://www.tigr.org/tdb/fungal/
Multiple saprobic and pathogenic fungi	NCBI: Links to genome sequence projects	http://www.ncbi.nlm.nih.gov/genomes/leuks.cgi?p1=10%3A0&p3=11:Fungi|12%3A&taxgroup=11:Fungi|12%3A&
Fungal protocols	DNA markers	http://www.isham.org/Protocol/
Aspergillus species	Genome sequence projects	http://www.aspergillus.org.uk/indexhome.htm?secure/sequence_info/index.php~main
Multiple yeasts	Links to multiple yeast genome sequences	http://cbi.labri.fr/Genolevures/yeastgenomes.php#

of *C. neoformans* and *C. gattii*, designated VNI-VNIV, VNB and VGI-VGIV, respectively (Campbell *et al.*, 2005; Meyer *et al.*, 2003; Kidd *et al.*, 2005; Xu *et al.*, 2000; Kwon-Chung and Varma, 2006; Litvintseva *et al.*, 2006; Bovers *et al.*, 2008b). These methods are resolving the phylogeny of this species-complex. They have discovered that the most diverse isolates of *C. neoformans* var. *grubii* exist in sub-Saharan Africa (Litvintseva *et al.*, 2003; Litvintseva *et al.*, 2006; Litvintseva *et al.*, 2007). In recent years, human and veterinary cases of cryptococcosis due to *C. gattii* emerged on Vancouver Island and have spread to mainland British Columbia and the northwestern USA (Kidd *et al.*, 2004). This major, ongoing outbreak has been tracked by genotyping environmental, clinical and veterinary strains, and MLST markers are being used to determine the origin and evolution of the responsible strains (Kidd *et al.*, 2004; Kidd *et al.*, 2005; Byrnes, III *et al.*, 2008). These extensive analyses of the population genetics of *Cryptococcus* species have posed many provocative questions: What is the origin of the expanding outbreak of rare genotypes of *C. gattii* in British Columbia and the northwestern US? How closely are these isolates related to sibling strains in Australia and South America?

Subcutaneous and cutaneous mycoses

The fungi that cause subcutaneous mycoses normally reside in soil or on vegetation. They enter the skin or subcutaneous tissue by traumatic inoculation with contaminated material. In general, the lesions become granulomatous and expand slowly from the area of implantation. Extension via the lymphatics draining the lesion is slow except in sporotrichosis. These mycoses are usually confined to the subcutaneous tissues, but in rare cases they become systemic and produce life-threatening disease. These infections occur most often in immunocompetent humans or other mammals. The major subcutaneous mycoses are sporotrichosis, chromoblastomycosis, and mycetoma. The last two are caused by several species of ascomycetous moulds, whereas sporotrichosis is caused by a single pathogen, *Sporothrix schenckii*.

S. schenckii is associated with a variety of plants, including grasses, trees, sphagnum moss, rose bushes, and other horticultural plants, and is closely related to several phytopathogens. It is a haploid, asexual thermal dimorphic. At ambient temperatures, it grows as a mould, producing branching, septate hyphae and conidia; and in tissue or *in vitro* at 35–37°C, as a small budding yeast. Most cases follow traumatic introduction of *S. schenckii* into the skin and present as a chronic granulomatous lesion with secondary spread to the draining lymphatics and lymph nodes. Sporotrichosis is endemic in tropical areas of Central America, South America and Africa, and sporadic cases occur in Europe and North America. An outbreak of sporotrichosis among arborists at several nurseries in Florida was traced to contaminated sphagnum moss by genotyping the isolates with RFLP markers (Hajjeh *et al.*, 1997). A recent three-gene MLST analysis of 60 global isolates of S. schenckii identified three major clades dominated by strains from (i) Europe, (ii) Brazil, and (iii) South America and Africa, and a total of six clades represented possible phylogenetic species (Marimon *et al.*, 2006).

With superficial mycoses, hyphae infect the stratum corneum but are restricted from invading deeper tissue. Species of *Malassezia*, which are lipophilic, yeast-like basidiomycetes, routinely colonize the human skin and frequently cause pityriasis versicolor, a chronic, non-inflammatory macular infection of the stratum corneum, as well as seborrhoeic dermatitis (e.g. dandruff). As many as half the world's human population may develop a *Malassezia* infection, and most are caused by *Malassezia globosa, M. restricta, M. sympodialis* and other members of the *M. furfur* complex. Some individuals develop folliculitis due to *Malassezia*. Because these species are difficult to identify by their laboratory phenotypes, SSCP polymorphisms in the ITS1 sequences were used to study 109 isolates from patients with pityriasis versicolor or seborrhoeic dermatitis, and they identified a major clade of *M. globosa* that was significantly correlated with these clinical conditions (Gaitanis *et al.*, 2006).

The dermatophytes cause tinea or ringworm infections, and these keratinophilic moulds are very difficult to differentiate on phenotype alone. The three genera, *Microsporum, Trichophyton,* and *Epidermophyton,* typically lose the ability to undergo sexual reproduction or produce micro- and/or macroconidia as they evolve from geophilic (saprobic existence) to zoophilic (adaptation to specific animal hosts) to anthropophilic (almost exclusive transmission among humans) ecologies. Genotyping and population genetic studies have clarified several epidemiological and evolutionary issues. For example, ITS rDNA and MnSOD gene sequences were used to define two major clades of *T. mentagrophytes*, but both clades included isolates from humans and other animals (Fréalle *et al.*, 2007). A global sample of 96 isolates of *T. rubrum*, which is a cosmopolitan anthropophilic species, well-adapted to human skin, and perhaps the most prevalent agent of human dermatophytosis, was analysed with PCR fingerprinting, RFLP, and RAPD markers that identified 57 loci, but no polymorphisms were observed, indicating clonality and a lack of diversity (Gräser *et al.*, 1999). Similar analyses are gradually describing the population structure and true identification of many of these complex and taxonomically challenging pathogenic fungi (White *et al.*, 2008).

Future trends

The rapid advances in DNA sequencing will continue to generate more MLST and other robust markers for pathogenic fungi. As described in brief here, there are numerous questions about the aetiology, ecology, species boundaries and phenotypes of important fungal human pathogens that can be resolved by understanding their population structure. With continuing progress in the development of high-throughput methods of sequencing and genotyping, the use of population genetics will become more popular and their clinical value will become more evident.

Acknowledgements

This work was supported by a grant from the US Public Health Service, National Institutes of Health, R01 AI 25783.

References

Alvarez-Perez, S., Garcia, M.E., Bouza, E., Peláez, T., and Blanco, J.L. (2009). Characterization of multiple isolates of *Aspergillus fumigatus* from patients: genotype, mating type and invasiveness. Med. Mycol. *47*, 601–608.

Balajee, S.A., de Valk, H.A., Lasker, B.A., Meis, J.F.G.M., and Klaassen, C.H.W. (2008). Utility of a microsatellite assay for identifying clonally related outbreak isolates of *Aspergillus fumigatus*. J. Microbiol. Meth. *73*, 252–256.

Balajee, S.A., Tay, S.T., Lasker, B.A., Hurst, S.F., and Rooney, A.P. (2007). Characterization of a novel gene for strain typing reveals substructuring of *Aspergillus fumigatus* across North America. Eukaryot. Cell *6*, 1392–1399.

Berbee, M.L. and Taylor, J.W. (1999). Fungal phylogeny. In Molecular Fungal Biology, Oliver, R.P. and Schweizer, M., eds. (Cambridge: Cambridge University Press), pp. 21–77.

Boekhout, T., van Belkum, A., Leenders, A.C.A.P., Verbrugh, H.A., Mukamurangwa, P., Swinne, D., and Scheffers, W.A. (1997). Molecular typing of *Cryptococcus neoformans*: taxonomic and epidemiological aspects. Int. J. Syst. Bacteriol. *47*, 432–442.

Bougnoux, M.-E., Aanensen, D.M., Morand, S., Theraud, M., Spratt, B.G., and d'Enfert, C. (2004). Multilocus sequence typing of *Candida albicans*: strategies, data exchange and applications. Infect. Genet. Evol. *4*, 243–252.

Bovers, M., Hagen, F., and Boekhout, T. (2008a). Diversity of the *Cryptococcus neoformans-Cryptococcus gattii* species complex. Rev. Iberoam. Micol. *25*, S4–S12.

Bovers, M., Hagen, F., Kuramae, E.E., and Boekhout, T. (2008b). Six monophyletic lineages identified within *Cryptococcus neoformans* and *Cryptococcus gattii* by multi-locus sequence typing. Fungal Genet. Biol. *45*, 400–421.

Brandt, M.E., Hutwagner, L.C., Kuykendall, R.J., Pinner, R.W., and Cryptococcal Disease Active Surveillance Group (1995). Comparison of multilocus enzyme electrophoresis and random amplified polymorphic DNA analysis for molecular subtyping of *Cryptococcus neoformans*. J. Clin. Microbiol. 33, 1890–1895.

Bruns, T.D., White, T.J., and Taylor, J.W. (1991). Fungal molecular systematics. Ann. Rev. Ecol. Syst. *22*, 525–564.

Burt, A., Carter, D.A., Koenig, G.L., White, T.J., and Taylor, J.W. (1996). Molecular markers reveal cryptic sex in the human pathogen *Coccidioides immitis*. Proc. Natl. Acad. Sci. U.S.A. *93*, 770–773.

Burt, A., Koufopanou, V., and Taylor, J.W. (2000). Population genetics of human pathogenic fungi. In Molecular Epidemiology of Infectious Diseases, Thompson, R.C.A., ed. (London: Arnold Publishers), pp. 229–244.

Byrnes, E.J., I.I.I., Bildfell, R., Frank, S.A., Mitchell, T.G., Marr, K.A., and Heitman, J. (2009). Molecular evidence that the Vancouver Island *Cryptococcus gattii* outbreak has expended into the Pacific Northwest United States. J. Infect. Dis. *199*, 1081–1086.

Calcagno, A.M., Nino-Vega, G.A., San-Blas, F., and San-Blas, G. (1998). Geographic discrimination of *Paracoccidioides brasiliensis* strains by randomly amplified polymorphic DNA analysis. J. Clin. Microbiol. *36*, 1733–1736.

Calderone, R. (2002). Candida *and Candidiasis*. (Washington, D.C.: ASM Press).

Campa, D., Tavanti, A., Gemignani, F., Mogavero, C.S., Bellini, I., Bottari, F., Barale, R., Landi, S., and Senesi,

S. (2008). A DNA microarray based on Arrayed-Primer Extension (APEX) technique for identification of pathogenic fungi responsible for invasive and superficial mycoses. J. Clin. Microbiol. *46*, 909–915.

Campbell, L.T., Currie, B.J., Krockenberger, M.B., Malik, R., Meyer, W., Heitman, J., and Carter, D.A. (2005). Clonality and recombination in genetically differentiated subgroups of *Cryptococcus gattii*. Eukaryot. Cell *4*, 1403–1409.

Carrero, L.L., Nino-Vega, G.A., Teixeira, M.M., Carvalho, M.J., Soares, C.M.A., Pereira, M., Jesuino, R.S., McEwen, J.G., Mendoza, L., Taylor, J.W., Felipe, M.S.S., and San-Blas, G. (2008). New *Paracoccidioides brasiliensis* isolate reveals unexpected genomic variability in this human pathogen. Fungal Genet. Biol. *45*, 605–612.

Carter, D.A., Burt, A., Taylor, J.W., Koenig, G.L., Dechairo, B.M., and White, T.J. (1997). A set of electrophoretic molecular markers for strain typing and population genetic studies of *Histoplasma capsulatum*. Electrophor. *18*, 1047–1053.

Carter, D.A., Burt, A., Taylor, J.W., Koenig, G.L., and White, T.J. (1996). Clinical isolates of *Histoplasma capsulatum* from Indianapolis, Indiana, have a recombining population structure. J. Clin. Microbiol. *34*, 2577–2584.

Carter, D.A., Taylor, J.W., Dechairo, B.M., Burt, A., Koenig, G.L., and White, T.J. (2001). Amplified single-nucleotide polymorphisms and a $(GA)_n$ microsatellite marker reveal genetic differentiation between populations of *Histoplasma capsulatum* from the Americas. Fungal Genet. Biol. *34*, 37–48.

Casadevall, A. and Perfect, J.R. (1999). *Cryptococcus neoformans*. (Washington, D.C.: ASM Press).

Chang, J.-C., Hsu, M.M.L., Barton, R.C., and Jackson, C.J. (2008). High-frequency intragenomic heterogeneity of the ribosomal DNA intergenic spacer region in *Trichophyton violaceum*. Eukaryot. Cell *7*, 721–726.

Chowdhary, A., Lee-Yang, W., Lasker, B.A., Brandt, M.E., Warnock, D.W., and Arthington-Skaggs, B.A. (2006). Comparison of multilocus sequence typing and Ca3 fingerprinting for molecular subtyping epidemiologically-related clinical isolates of *Candida albicans*. Med. Mycol. *44*, 405–417.

Clark, A.G. and Lanigan, C.M.S. (1993). Prospects for estimating nucleotide divergence with RAPDs. Mol. Biol. Evol. *10*, 1096–1111.

Clemons, K.V., Feroze, F., Holmberg, K., and Stevens, D.A. (1997). Comparative analysis of genetic variability among *Candida albicans* isolates from different geographic locales by three genotypic methods. J. Clin. Microbiol. 35, 1332–1336.

Dassanayake, R.S., Samaranayake, Y.H., Yau, J., and Samaranayake, L.P. (2006). DNA fingerprinting elicited evolutionary trend of oral *Candida tropicalis* isolates from diverse geographic locales. Indian J. Med. Microbiol. *24*, 186–194.

de Valk, H.A., Meis, J.F.G.M., de Pauw, B.E., Donnelly, P.J., and Klaassen, C.H.W. (2007a). Comparison of two highly discriminatory molecular fingerprinting assays for analysis of multiple *Aspergillus fumigatus* isolates from patients with invasive aspergillosis. J. Clin. Microbiol. *45*, 1415–1419.

de Valk, H.A., Meis, J.F.G.M., and Klaassen, C.H.W. (2007b). Microsatellite based typing of *Aspergillus fumigatus*: strengths, pitfalls and solutions. J. Microbiol. Meth. *69*, 268–272.

Diezmann, S., Cox, C., Schönian, G., Vilgalys, R.J., and Mitchell, T.G. (2004). Phylogeny and evoluation of medical species of *Candida* and related taxa: a multigenic analysis. Clin, J. Microbiol. *42*, 5624–5635.

Fisher, M.C., Aanensen, D.M., de Hoog, G.S., and Vanittanakom, N. (2004). Multilocus microsatellite typing system for *Penicillium marneffei* reveals spatially structured populations. J. Clin. Microbiol. *42*, 5065–5069.

Fisher, M.C., Koenig, G.L., White, T.J., San-Blas, G., Negroni, R., Alvarez, I.G., Wanke, B., and Taylor, J.W. (2001). Biogeographic range expansion into South America by *Coccidioides immitis* mirrors New World patterns of human migration. Proc. Natl. Acad. Sci. U.S.A. *98*, 4558–4562.

Fisher, M.C., Koenig, G.L., White, T.J., and Taylor, J.W. (2002a). Molecular and phenotypic description of *Coccidioides posadasii* sp. nov., previously recognized as the non-California population of *Coccidioides immitis*. Mycologia *94*, 73–84.

Fisher, M.C., Rannala, B., Chaturvedi, V.P., and Taylor, J.W. (2002b). Disease surveillance in recombining pathogens: multilocus genotypes identify sources of human *Coccidioides* infections. Proc. Natl. Acad. Sci. U.S.A. *99*, 9067–9071.

Forche, A., Magee, P.T., Magee, B.B., and May, G. (2004). Genome-wide single-nucleotide polymorphism map for *Candida albicans*. Eukaryot. Cell 3, 705–714.

Forche, A., Schönian, G., Gräser, Y., Vilgalys, R.J., and Mitchell, T.G. (1999). Genetic structure of typical and atypical populations of *Candida albicans* from Africa. Fungal Genet. Biol. *28*, 107–125.

Franco, M.F., Lacaz, C.S., Restrepo, A., and Del Negro, G.M.B. (1994). *Paracoccidioidomycosis*. (Boca Raton: CRC Press).

Fréalle, E., Rodrigue, M., Gantois, N., Aliouat, C.-M., Delaporte, E., Camus, D., Dei-Cas, E., Kauffmann-Lacroix, C., Guillot, J., and Delhaes, L. (2007). Phylogenetic analysis of *Trichophyton mentagrophytes* human and animal isolates based on MnSOD and ITS sequence comparison. Microbiol. *153*, 3466–3477.

Gaedigk, A., Gaedigk, R., and Abdel-Rahman, S.M. (2003). Genetic heterogeneity in the rRNA gene locus of *Trichophyton tonsurans*. J. Clin. Microbiol. *41*, 5478–5487.

Gaitanis, G., Velegraki, A., Alexopoulos, E.C., Chasapi, V., Tsigonia, A., and Katsambas, A. (2006). Distribution of *Malassezia* species in pityriasis versicolor and seborrhoeic dermatitis in Greece. Typing of the major pityriasis versicolor isolate *M. globosa*. Br. J. Dermatol. *154*, 854–859.

Galagan, J.E., Calvo, S.E., Cuomo, C., Ma, L.-J., Wortman, J.R., Batzoglou, S., Lee, S.-I., Bastürkmen, M., Spevak, C.C., Clutterbuck, J., *et al.* (2005). Sequencing of *Aspergillus nidulans* and comparative analysis with *A. fumigatus* and *A. oryzae*. Nature *438*, 1105–1115.

Garcia-Hermoso, D., Cabaret, O., Lecellier, G., Desnos-Ollivier, M., Hoinard, D., Raoux, D., Costa, J.-M., Dromer, F., and Bretagne, S. (2007). Comparison of microsatellite length polymorphism and multilocus sequence typing for DNA-based typing of *Candida albicans*. J. Clin. Microbiol. *45*, 3958–3963.

Gräser, Y., el Fari, M., Presber, W., Sterry, W., and Tietz, H.-J. (1998). Identification of common dermatophytes (*Trichophyton, Microsporum, Epidermophyton*) using polymerase chain reactions. Br. J. Dermatol. *138*, 576–582.

Gräser, Y., Kuhnisch, J., and Presber, W. (1999). Molecular markers reveal exclusively clonal reproduction in *Trichophyton rubrum*. J. Clin. Microbiol. *37*, 3713–3717.

Gräser, Y., Volovsek, M., Arrington, J., Schönian, G., Presber, W., Mitchell, T.G., and Vilgalys, R.J. (1996). Molecular markers reveal that population structure of the human pathogen *Candida albicans* exhibits both clonality and recombination. Proc. Natl. Acad. Sci. U.S.A. *93*, 12473–12477.

Hajjeh, R.A., McDonnell, S., Reef, S.E., Licitra, C., Hankins, M., Tóth, B., Padhye, A.A., Kaufman, L., Pasarell, L., Cooper, C.R., Jr., Hutwagner, L.C., Hopkins, R.J., and McNeil, M.M. (1997). Outbreak of sporotrichosis among tree nursery workers. J. Infect. Dis. *176*, 499–504.

Hanafy, A.M., Kaocharoen, S., Jover-Botella, A., Katsu, M., Iida, S., Kogure, T., Gonoi, T., Mikami, Y., and Meyer, W. (2008). Multilocus microsatellite typing for *Cryptococcus neoformans* var. *grubii*. Med. Mycol. *46*, 685–696.

Hillis, D.M., Moritz, C., Mable, B.K., and (eds) (1996). Molecular Systematics (Sunderland, MA: Sinauer Associates, Inc.).

Hiremath, S.S., Chowdhary, A., Kowshik, T., Randhawa, H.S., Sun, S., and Xu, J. (2008). Long-distance dispersal and recombination in environmental populations of *Cryptococcus neoformans* var. *grubii* from India. Microbiol. *154*, 1513–1524.

Huang, A., Li, J.-W., Shen, Z.-Q., Wang, X.-W., and Jin, M. (2006). High-throughput identification of clinical pathogenic fungi by hybridization to an oligonucleotide microarray. J. Clin. Microbiol. *44*, 3299–3305.

Iwen, P.C., Hinrichs, S.H., and Rupp, M.E. (2002). Utilization of the internal transcribed spacer regions as molecular targets to detect and identify human fungal pathogens. Med. Mycol. *40*, 87–109.

Jacobsen, M.D., Gow, N.A.R., Maiden, M.C.J., Shaw, D.J., and Odds, F.C. (2007). Strain typing and determination of population structure of *Candida krusei* by multilocus sequence typing. J. Clin. Microbiol. *45*, 317–323.

Kasuga, T., Taylor, J.W., and White, T.J. (1999). Phylogenetic relationships of varieties and geographical groups of the human pathogenic fungus *Histoplasma capsulatum* Darling. J. Clin. Microbiol. *37*, 653–663.

Kasuga, T., White, T.J., Koenig, G.L., McEwen, J.G., Restrepo, A., Castañeda, E., da Silva Lacaz, C., Heins-Vaccari, E.M., De Freitas, R.S., Zancopé-Oliveira, R.M., *et al.* (2003). Phylogeography of the fungal pathogen *Histoplasma capsulatum*. Mol. Ecol. *12*, 3383–3401.

Kidd, S.E., Guo, H., Bartlett, K.H., Xu, J., and Kronstad, J.W. (2005). Comparative gene genealogies indicate that two clonal lineages of *Cryptococcus gattii* in British Columbia resemble strains from other geographical areas. Eukaryot. Cell *4*, 1629–1638.

Kidd, S.E., Hagen, F., Tscharke, R.L., Huynh, M., Bartlett, K.H., Fyfe, M., MacDougall, L., Boekhout, T., Kwon-Chung, K.J., and Meyer, W. (2004). A rare genotype of *Cryptococcus gattii* caused the cryptococcosis outbreak on Vancouver Island (British Columbia, Canada). Proc. Natl. Acad. Sci. U.S.A. *101*, 17258–17263.

Koufopanou, V., Burt, A., Szaro, T.M., and Taylor, J.W. (2001). Gene genealogies, cryptic species, and molecular evolution in the human pathogen *Coccidioides immitis* and relatives (Ascomycota, Onygenales). Mol. Biol. Evol. *18*, 1246–1258.

Koufopanou, V., Burt, A., and Taylor, J.W. (1997). Concordance of gene genealogies reveals reproductive isolation in the pathogenic fungus *Coccidioides immitis*. Proc. Natl. Acad. Sci. U.S.A. *94*, 5478–5482.

Kwon-Chung, K.J. and Varma, A. (2006). Do major species concepts support one, two or more species within *Cryptococcus neoformans*? FEMS Yeast Res. *6*, 574–587.

Lasker, B.A. (2006). Nucleotide sequence-based analysis for determining the molecular epidemiology of *Penicillium marneffei*. J. Clin. Microbiol. *44*, 3145–3153.

Lasker, B.A. and Ran, Y. (2004). Analysis of polymorphic microsatellite markers for typing *Penicillium marneffei* isolates. J. Clin. Microbiol. *42*, 1483–1490.

Leinberger, D.M., Schumacher, U., Autenrieth, I.B., and Bachmann, T.T. (2005). Development of a DNA microarray for detection and identification of fungal pathogens involved in invasive mycoses. J. Clin. Microbiol. *43*, 4943–4953.

Litvintseva, A.P., Lin, X., Templeton, I., Heitman, J., and Mitchell, T.G. (2007). Many globally isolated AD hybrid strains of *Cryptococcus neoformans* originated in Africa. PLoS Pathog. 3, e114.

Litvintseva, A.P., Marra, R.E., Nielsen, K., Heitman, J., Vilgalys, R.J., and Mitchell, T.G. (2003). Evidence of sexual recombination among *Cryptococcus neoformans* serotype A isolates in sub-Saharan Africa. Eukaryot. Cell 2, 1162–1168.

Litvintseva, A.P., Thakur, R., Vilgalys, R.J., and Mitchell, T.G. (2006). Multilocus sequence typing reveals three genetic subpopulations of *Cryptococcus neoformans* var. *grubii* (serotype A), including a unique population in Botswana. Genetics *172*, 2223–2238.

Lockhart, S.R., Pujol, C., Dodgson, A.R., and Soll, D.R. (2005). Deoxyribonucleic acid fingerprinting methods for *Candida* species. Meth. Mol. Med. *118*, 15–26.

Lockhart, S.R., Pujol, C., Joly, S., and Soll, D.R. (2001). Development and use of complex probes for DNA fingerprinting the infectious fungi. Med. Mycol. *39*, 1–8.

Marimon, R., Gené, J., Cano, J.F., Trilles, L., Dos Santos Lazéra, M., and Guarro, J. (2006). Molecular phylogeny of *Sporothrix schenckii*. J. Clin. Microbiol. *44*, 3251–3256.

Marra, R.E., Huang, J.C., Fung, E., Nielsen, K., Heitman, J., Vilgalys, R.J., and Mitchell, T.G. (2004). A genetic linkage map of *Cryptococcus neoformans* variety *neoformans* serotype D (*Filobasidiella neoformans*). Genetics *167*, 619–631.

Matute, D.R., McEwen, J.G., Puccia, R., Montes, B.A., San-Blas, G., Bagagli, E., Rauscher, J.T., Restrepo, A., Morais, F.V., Nino-Vega, G.A., and Taylor, J.W. (2006a). Cryptic speciation and recombination in the fungus *Paracoccidioides brasiliensis* as revealed by gene genealogies. Mol. Biol. Evol. *23*, 65–73.

Matute, D.R., Sepulveda, V.E., Quesada, L.M., Goldman, G.H., Taylor, J.W., Restrepo, A., and McEwen, J.G. (2006b). Microsatellite analysis of three phylogenetic species of *Paracoccidioides brasiliensis*. J. Clin. Microbiol. *44*, 2153–2157.

McCullough, M.J., Di Salvo, A.F., Clemons, K.V., Park, P., and Stevens, D.A. (2000). Molecular epidemiology of *Blastomyces dermatitidis*. Clin. Infect. Dis. *30*, 328–335.

McDonough, E.S. and Lewis, A.L. (1967). *Blastomyces dermatitidis*: production of the sexual stage. Science *156*, 528–529.

McDonough, E.S., McNamara, W.J., Chan, D.M., and Wallenfang, M.A. (1973). Geographic distribution of '+' and '-' isolates of *Blastomyces (Ajellomyces) dermatitidis* in North America. Am. J. Epidemiol. *98*, 63–67.

Mesa-Arango, A.C., Reyes-Montes, M.R., Pérez-Mejía, A., Navarro-Barranco, H., Souza, V., and Toriello, C. (2002). Phenotyping and genotyping of *Sporothrix schenckii* isolates according to geographic origin and clinical form of sporotrichosis. J. Clin. Microbiol. *40*, 3004–3011.

Meyer, W., Castañeda, A., Jackson, S., Huynh, M., and Castañeda, E. (2003). Molecular typing of IberoAmerican *Cryptococcus neoformans* isolates. Emerg. Infect. Dis. *9*, 189–195.

Meyer, W., Latouche, G.N., Daniel, H.-M., Thanos, M., Mitchell, T.G., Yarrow, D., Schönian, G., and Sorrell, T.C. (1997). Identification of pathogenic yeasts of the imperfect genus *Candida* by polymerase chain reaction fingerprinting. Electrophor. *18*, 1548–1559.

Meyer, W. and Mitchell, T.G. (1995). PCR fingerprinting in fungi using single primers specific to minisatellites and simple repetitive DNA sequences: strain variation in *Cryptococcus neoformans*. Electrophoresis *16*, 1648–1656.

Meyer, W., Mitchell, T.G., Freedman, E.Z., and Vilgalys, R.J. (1993). Hybridization probes for conventional DNA fingerprinting used as single primers in the polymerase chain reaction to distinguish strains of *Cryptococcus neoformans*. J. Clin. Microbiol. *31*, 2274–2280.

Mitchell, T.G. (2004). Systemic fungi. In Infectious Diseases, Cohen, J. and Powderly, W.G., eds. (London: Mosby), pp. 2363–2381.

Mitchell, T.G. (2005). Kingdom fungi: fungal phylogeny and systematics. In Topley & Wilson's Microbiology & Microbial Infections. Medical Mycology, Merz, W.G. and Hay, R.J., eds. (London: Hodder Arnold), pp. 43–68.

Mitchell, T.G. (2008). Genomic approaches to investigate the pathogenicity of *Cryptococcus neoformans*. Curr. Fungal Infect. Rep. 2, 172–179.

Mitchell, T.G. and Perfect, J.R. (1995). Cryptococcosis in the era of AIDS–100 years after the discovery of *Cryptococcus neoformans*. Clin. Microbiol. Rev. *8*, 515–548.

Mitchell, T.G. and Xu, J. (2003). Molecular methods to identify pathogenic fungi. In Pathogenic Fungi in Humans and Animals, Howard, D.H., ed. (New York: Marcel Dekker), pp. 677–702.

Mondon, P., Brenier, M.P., Symoens, F., Rodriguez, E.R., Coursange, E., Chaib, F., Lebeau, B., Piens, M.A., Tortorano, A.M., Mallié, M., Chapuis, F., Carlotti, A., Villard, J., Viviani, M.A., Nolard, N., Bastide, J.-M., Ambroise-Thomas, P., and Grillot, R. (1997). Molecular typing of *Aspergillus fumigatus* strains by sequence-specific DNA primer (SSDP) analysis. FEMS Immunol. Med. Microbiol. *17*, 95–102.

Muniz, M.M., Pizzini, C.V., Peralta, J.M., Reiss, E., and Zancopé-Oliveira, R.M. (2001). Genetic diversity of *Histoplasma capsulatum* strains isolated from soil, animals, and clinical specimens in Rio de Janeiro state, Brazil, by a PCR-based random amplified polymorphic DNA assay. J. Clin. Microbiol. *39*, 4487–4494.

Nascimento, E., Martinez, R., Rodrigues Lopes, A., de Souza Bernardes, L.A., Pomponio Barco, C., Goldman, M.H.S., Taylor, J.W., McEwen, J.G., Nóbrega, M.P., Nóbrega, F.G., and Goldman, G.H. (2004). Detection and selection of microsatellites in the genome of *Paracoccidioides brasiliensis* as molecular markers for clinical and epidemiological studies. J. Clin. Microbiol. *42*, 5007–5014.

Nino-Vega, G.A., Calcagno, A.M., San-Blas, G., San-Blas, F., Gooday, G.W., and Gow, N.A.R. (2000). RFLP analysis reveals marked geographical isolation between strains of *Paracoccidioides brasiliensis*. Med. Mycol. 38, 437–441.

O'Gorman, C.M., Fuller, H.T., and Dyer, P.S. (2008). Discovery of a sexual cycle in the opportunistic fungal pathogen *Aspergillus fumigatus*. Nature 457, 471–474.

Odds, F.C., Bougnoux, M.-E., Shaw, D.J., Bain, J.M., Davidson, A.D., Diogo, D., Jacobsen, M.D., Lecomte, M., Li, S.-Y., Tavanti, A., Maiden, M.C.J., Gow, N.A.R., and d'Enfert, C. (2007). Molecular phylogenetics of *Candida albicans*. Eukaryot. Cell *6*, 1041–1052.

Odds, F.C. and Jacobsen, M.D. (2008). Multi-locus sequence typing of pathogenic *Candida* species. Eukaryot. Cell 7, 1075–1084.

Ohst, T., de Hoog, G.S., Presber, W., Stavrakieva, V., and Gräser, Y. (2004). Origins of microsatellite diversity in the *Trichophyton rubrum-T. violaceum* clade (dermatophytes). J. Clin. Microbiol. *42*, 4444–4448.

Paoletti, M., Rydholm, C., Schwier, E.U., Anderson, M.J., Szakacs, G., Lutzoni, F.M., Debeaupuis, J.-P., Latgé, J.-P., Denning, D.W., and Dyer, P.S. (2005). Evidence for sexuality in the opportunistic fungal pathogen *Aspergillus fumigatus*. Curr. Biol. *15*, 1242–1248.

Perlin, D.S. and Zhao, Y. (2008). Molecular diagnostic platforms for detecting *Aspergillus*. Med. Mycol. *47*, S223–S232.

Pfaller, M.A. and Diekema, D.J. (2004). Rare and emerging opportunistic fungal pathogens: concern for resistance beyond *Candida albicans* and *Aspergillus fumigatus*. J. Clin. Microbiol. *42*, 4419–4431.

Richardson, M.D. (2005). Aspergillosis. In *Topley & Wilson's* Microbiology *& Microbial Infections. Medical Mycology*, Merz, W.G. and Hay, R.J., eds. (London: Hodder Arnold), pp. 687–738.

Robles, J.C., Koreen, L., Park, S., and Perlin, D.S. (2004). Multilocus sequence typing is a reliable alternative method to DNA fingerprinting for discriminating among strains of *Candida albicans*. J. Clin. Microbiol. *42*, 2480–2488.

Ronald, J., Akey, J.M., Whittle, J., Smith, E.N., Yvert, G., and Kruglyak, L. (2005). Simultaneous genotyping, gene-expression measurement, and detection of allele-specific expression with oligonucleotide arrays. Genome Res. *15*, 284–291.

Rydholm, C., Szakacs, G., and Lutzoni, F.M. (2006). Low genetic variation and no detectable population structure in *Aspergillus fumigatus* compared to closely related *Neosartorya* species. Eukaryot. Cell *5*, 650–657.

Schönian, G., Meusel, O., Tietz, H.-J., Meyer, W., Gräser, Y., Tausch, I., Presber, W., and Mitchell, T.G. (1993). Identification of clinical strains of *Candida albicans* by DNA fingerprinting with the polymerase chain reaction. Mycoses *36*, 171–179.

Selmecki, A., Bergmann, S., and Berman, J.G. (2005). Comparative genome hybridization reveals widespread aneuploidy in *Candida albicans* laboratory strains. Mol. Microbiol. *55*, 1553–1565.

Smith, C.E., Saito, M.T., and Simons, S.A. (1956). Pattern of 39, 500 serologic tests in coccidioidomycosis. J. Am. Med. Assoc. *160*, 546–552.

Soll, D.R. (2000). The ins and outs of DNA fingerprinting the infectious fungi. Clin. Microbiol. Rev. *13*, 332–370.

Spiess, B., Seifarth, W., Hummel, M., Frank, O., Fabarius, A., Zheng, C., Mörz, H., Hehlmann, R., and Buchheidt, D. (2007). DNA microarray-based detection and identification of fungal pathogens in clinical samples from neutropenic patients. J. Clin. Microbiol. *45*, 3743–3753.

Stevens, D. A. (1980). Coccidioidomycosis. (Plenum Press: New York).

Sullivan, D.J. and Coleman, D.C. (1998). *Candida dubliniensis*: characteristics and identification. J. Clin. Microbiol. *36*, 329–334.

Sullivan, D.J., Westerneng, T.J., Haynes, K.A., Bennett, D.E., and Coleman, D.C. (1995). *Candida dubliniensis* sp. nov.: phenotypic and molecular characterization of a novel species associated with oral candidosis in HIV-infected individuals. Microbiol. *141*, 1507–1521.

Taylor, J.W. and Fisher, M.C. (2003). Fungal multilocus sequence typing – it's not just for bacteria. Curr. Opin. Microbiol. *6*, 351–356.

Taylor, J.W., Geiser, D.M., Burt, A., and Koufopanou, V. (1999). The evolutionary biology and population genetics underlying fungal strain typing. Clin. Microbiol. Rev. *12*, 126–146.

Taylor, J.W., Jacobson, D.J., Kroken, S., Kasuga, T., Geiser, D.M., Hibbett, D.S., and Fisher, M.C. (2000). Phylogenetic species recognition and species concepts in fungi. Fungal Genet. Biol. *31*, 21–32.

Thanos, M., Schönian, G., Meyer, W., Schweynoch, C., Gräser, Y., Mitchell, T.G., Presber, W., and Tietz, H.-J. (1996). Rapid identification of *Candida* species by DNA fingerprinting with PCR. J. Clin. Microbiol. *34*, 615–621.

Tintelnot, K., de Hoog, G.S., Antweiler, E., Losert, H., Seibold, M., Brandt, M.A., Gerrits van den Ende, A.H.G., and Fisher, M.C. (2007). Taxonomic and diagnostic markers for identification of *Coccidioides immitis* and *Coccidioides posadasii*. Med. Mycol. *45*, 385–393.

Townsend, J.P. and Taylor, J.W. (2005). Designing experiments using spotted microarrays to detect gene regulation differences within and among species. Meth. Enz. *395*, 597–617.

Trilles, L., Lazéra, M.S., Wanke, B., Oliveira, R.V., Barbosa, G.G., Nishikawa, M.M., Morales, B.P., and Meyer, W. (2008). Regional pattern of the molecular types of *Cryptococcus neoformans* and *Cryptococcus gattii* in Brazil. Mem. Inst. Oswaldo Cruz *103*, 455–462.

Vanittanakom, N., Cooper, C.R., Jr., Fisher, M.C., and Sirisanthana, T. (2006). *Penicillium marneffei* infection and recent advances in the epidemiology and molecular biology aspects. Clin. Microbiol. Rev. *19*, 95–110.

Varga, J. and Tóth, B. (2003). Genetic variability and reproductive mode of *Aspergillus fumigatus*. Infect. Genet. Evol. *3*, 3–17.

Vos, P., Hogers, R., Bleeker, M., Reijans, M., van de Lee, T., Hornes, M., Freijters, A., Pot, J., Peleman, J., Kuiper, M., and Zabeau, M. (1995). AFLP: a new technique for DNA fingerprinting. Nucleic Acids Res. *23*, 4407–4414.

Vos, P. and Kuiper, M. (1997). AFLP analysis. In DNA Markers. Protocols, Applications, and Overviews, Caetano-Anollës, G. and Gresshoff, P.M., eds. (New York: Wiley-Liss, Inc.), pp. 115–132.

Welsh, J. and McClelland, M. (1990). Fingerprinting genomes using PCR with arbitrary primers. Nucleic Acids Res. *18*, 7213–7218.

White, P.L., Barton, R.C., Guiver, M., Linton, C.J., Wilson, S., Smith, M., Gomez, B.L., Carr, M.J., Kimmitt, P.T., Seaton, S., Rajakumar, K., Holyoake, T., Kibbler, C.C., Johnson, E.M., Hobson, R.P., Jones, B., and Barnes, R.A. (2006). A consensus on fungal polymerase chain reaction diagnosis?: a United Kingdom-Ireland evaluation of polymerase chain reaction methods for detection of systemic fungal infections. J. Mol. Diagn. *8*, 376–384.

White, T.C., Oliver, B.G., Gräser, Y., and Henn, M.R. (2008). Generating and testing molecular hypotheses in the dermatophytes. Eukaryot. Cell *7*, 1238–1245.

Woo, P.C.Y., Lau, C.C.Y., Chong, K.T.K., Tse, H., Tsang, D.N.C., Lee, R.A., Tse, C.W.S., Que, T.-L., Chung, L.M.W., Ngan, A.H.Y., Hui, W.-T., Wong, S.S.Y., Lau, S.K.P., and Yuen, K.-Y. (2007). MP*1* homologue-based multilocus sequence system for typing the pathogenic fungus *Penicillium marneffei*: a novel approach using lineage-specific genes. J. Clin. Microbiol. *45*, 3647–3654.

Xu, J. and Mitchell, T.G. (2002a). Population genetic analyses of *Candida*. In *Candida* and Candidiasis, Calderone, R.A., ed. (Washington, D.C.: ASM Press), pp. 55–64.

Xu, J. and Mitchell, T.G. (2002b). Strain variation and clonality in *Candida* spp. and *Cryptococcus neoformans*. In Fungal Pathogenesis: Principles and Clinical Applications, Calderone, R.A. and Cihlar, R.L., eds. (New York: Marcel Dekker), pp. 739–749.

Xu, J. and Mitchell, T.G. (2003). Population genetic analyses of medically important fungi. In Pathogenic Fungi in Humans and Animals, Howard, D.H., ed. (New York: Marcel Dekker), pp. 703–722.

Xu, J., Mitchell, T.G., and Vilgalys, R.J. (1999a). PCR-restriction fragment length polymorphism (RFLP) analyses reveal both extensive clonality and local genetic differences in *Candida albicans*. Mol. Ecol. *8*, 59–73.

Xu, J., Vilgalys, R.J., and Mitchell, T.G. (1999b). Lack of genetic differentiation between two geographically diverse samples of *Candida albicans* isolated from patients infected with human immunodeficiency virus. J. Bacteriol. *181*, 1369–1373.

Xu, J., Vilgalys, R.J., and Mitchell, T.G. (2000). Multiple gene genealogies reveal recent dispersion and hybridization in the human pathogenic fungus *Cryptococcus neoformans*. Mol. Ecol. *9*, 1471–1481.

Population Genetics of Human Malaria Parasites

10

Deirdre A. Joy

Abstract

Genetic diversity, population structure, and evolutionary history of malaria parasites are among the factors that influence our ability to identify genes contributing to drug resistance, parasite development, and disease pathogenesis. These factors also have an impact on vaccine and drug development, parasite source tracking, and the formulation of other disease prevention and control measures. For example, a highly polymorphic parasite population will contain ample genetic diversity capable of generating drug resistance genotypes at an accelerated rate; while the presence of homogeneous parasite populations should aid in the development of an effective malaria vaccine. Malaria research has entered the post-genomic era, and population genomics, the study of sequence variation across the genome, has begun to shed light on the relative importance of mutation, recombination, natural selection and genetic drift in shaping genetic variation in this important human parasite.

Introduction

Malaria parasites belong to the genus *Plasmodium* and infect a broad range of vertebrate hosts, from lizards, to birds, to mammals, including several species of non-human primates, and humans. The number of described species of *Plasmodium* and related parasite numbers greater than 500 (Levine, 1988). Among the five *Plasmodium* species that infect humans, *Plasmodium falciparum* is by far the most deadly, while *Plasmodium vivax* is the most geographically widespread. These two parasites are found throughout the tropical regions of the world, with the exception of Africa, where *P. vivax* is nearly absent. According to the 2008 World Health Organization (WHO) World Malaria Report, malaria causes nearly 1 million deaths worldwide each year. Children under 5 years of age account for about 85% of all deaths and the vast majority are from Africa. Other high-risk groups include pregnant women and immunocompromised travellers.

Malaria parasites have a complex life cycle, requiring both a vertebrate host and a mosquito vector. They are haploid during most of their life cycle, which greatly simplifies many population genetics analyses. A short-lived sexual diploid stage occurs in the mosquito midgut, where haploid gametes in the blood meal fuse to form a diploid zygote. This brief sexual stage provides an opportunity for recombination between genetically distinct parasite isolates, termed outcrossing, and generates genetic diversity. Selfing occurs when only one parasite genotype is present in the blood meal. A recent genetic analysis of oocysts, a mosquito stage of the parasite, confirmed that outcrossing between genetically distinct parasite isolates occurs frequently in areas of moderate to high malaria transmission in Africa (Mzilahowa *et al.*, 2007). In regions with relatively low transmission, such as South and Central America, outcrossing appears to be much less common and therefore less important for generating genetic diversity (Anderson *et al.*, 2000).

In order to fully understand the population genetics of a protozoan parasite, one must also take into account the population genetics of the host and, as is the case with malaria, the vector that transmits the parasite between hosts. Human, mosquito, and malaria parasite population dynamics interact with and influence each other. For example, vector ecology and genetic mutations in humans both have an impact on parasite population genetics.

One classic example of the inter-relationship of host and pathogen population histories is that of *P. vivax* malaria and human Duffy blood-group genotypes in sub-Saharan Africa (Miller *et al.*, 1976). Duffy negativity results from a single nucleotide polymorphism (SNP) in the promoter region of the human Duffy antigen/receptor for chemokines (DARC) gene. Duffy negative individuals are homozygous for the DARC allele FY*B^{null} and as a result do not express the DARC protein on the red cell surface. Consequently, they are completely protected from *P. vivax*. In western and central Africa, regions where *P. vivax* malaria is extremely rare, the Duffy negative phenotype is near fixation (Culleton *et al.*, 2008). The high frequency of Duffy negativity has effectively restricted *P. vivax* from these regions of Africa. Whether or not *P. vivax* was the selective agent that drove the FY*B^{null} allele to fixation in Africa is not known definitively. An alternative scenario espoused by Livingstone (1984) is that the FY*B allele was fixed in Africa prior to the arrival of *P. vivax* by an unknown selective agent, thus preventing the establishment of *P. vivax* in the first place. He reasoned that a non-lethal pathogen was not likely to be responsible for such an extensive selective sweep. Additionally, phylogenetic and population evidence suggests an Asian rather than an African origin for *P. vivax* (Escalante *et al.*, 2005; Jongwutiwes *et al.*, 2005; Mu *et al.*, 2005b). The resolution of this question will probably require more extensive global sampling of *P. vivax*.

Molecular markers

The earliest population genetics studies in *P. falciparum* were limited to a small number of loci which included the merozoite surface proteins MSP-1 and MSP-2, the circumsporozoite surface protein (CSP), and the S-antigen locus (Babiker *et al.*, 1991, 1997; Daubersies *et al.*, 1996; Paul *et al.*, 1995). Because all of these genes are assumed to be under host immune selection, they are likely to produce biased and inaccurate estimates of population genetics parameters. For example, if we observe elevated levels of non-synonymous substitutions this may be due to immune pressure for antigenic variants (diversifying selection). Observed linkage disequilibrium (LD) based on genes under selection may not be an accurate reflection of actual recombination rates, but rather may be due to a selective sweeps. In general, selection pressure is likely to result in an overestimation of genetic diversity within, and differentiation among, populations.

Microsatellites (MS) are an attractive tool for population genetics studies in malaria for several reasons. MS are extremely abundant in *P. falciparum*, in part due to the very high AT content of the genome. This is reflected in the fact that the majority of MS in *P. falciparum* are $(TA)_n$, or $(TAA)_n$. The frequency of MS in the *P. falciparum* genome is about one in every 1–2 kb genome-wide (Gardner *et al.*, 1998; Su and Wellems, 1996). Although the *P. vivax* genome lacks the extreme AT-bias of the *P. falciparum* genome, MS also appear to be wide spread. Another attractive feature of MS as a tool for malaria population genetics is the high mutation rate. An estimate of the *P. falciparum* MS mutation rate based on a genetic cross with 35 progeny genotyped for more than 800 MS makers puts the mutation rate between 6.98×10^{-5} and 3.7×10^{-4}, with dinucleotide repeat loci evolving slightly faster than trinucleotide repeats (Anderson *et al.*, 2000). The advantage of a high mutation rate for population studies is that MS markers often exhibit multiple alleles in population surveys, making them potentially very informative. Lastly, MS are generally presumed to be selectively neutral, and thus avoid many of the pitfalls associated with protein coding makers.

Evolutionary history

The two most important human malaria parasites in terms of mortality and morbidity, *P. falciparum* and *P. vivax*, differ markedly from each other with respect to evolutionary history and origin as human parasites. Phylogenetic evidence points to *P. reichenowi*, a chimpanzee parasite, as the closest known relative to *P. falciparum* (Escalante and Ayala, 1994; Martinsen *et al.*, 2008; Perkins and Schall, 2002). Mitochondrial DNA further indicates that *P. falciparum* underwent a rapid population expansion and subsequent migration from Africa ~10,000 years ago (Joy *et al.*, 2003). The timing of malaria population growth overlaps with the rise of human agricultural societies and a concomitant increase in African mosquito transmission (Coluzzi, 1999; Coluzzi *et al.*, 2002). If the dates prove to be correct, this would provide an excellent example of the linkage between the population dynamics of the host, vector, and parasite.

In contrast to *P. falciparum*, and as mentioned earlier, the weight of the phylogenetic evidence indicates an Asia origin for *P. vivax* (Escalante and Ayala, 1994; Escalante *et al.*, 1998). This has been corroborated by population analyses showing Asia as the area of highest *P. vivax* diversity and the oldest time to the most recent ancestor (Cornejo and Escalante, 2006; Jongwutiwes *et al.*, 2005). Based on this model *P. vivax* originated as a human parasite via a host switch from non-human primates in Asia (Escalante *et al.*, 2005; Mu *et al.*, 2005b). However, as mentioned earlier, this view is not entirely reconciled with the very high frequency of Duffy negativity among sub-Saharan Africans, which in itself could imply a long co-evolutionary history between malaria parasite and Africans.

Population structure

Knowledge of the genetic structure of malaria populations is crucial from a control standpoint, because it informs predictions about how the parasite will evolve in response to interventions such as drugs, vaccines, and vector control measures. Over the years, the genetic structure of *P. falciparum* has been alternatively described as effectively clonal, epidemic, and panmictic. In an effectively clonal population, sexual recombination between parasites with different genotypes may occur, but is extremely rare. In an epidemic population, periods of relatively low incidence of disease are interspersed with large outbreaks, resulting in the clonal expansion of a few parasite genotypes. Both of these population types have low genetic diversity. In contrast, a panmictic population is one in which genetic diversity is high and sexual recombination is frequent. One of the measures of sexual recombination in natural microbial populations is the rapid decay in linkage disequilibrium (LD) with physical distance along a chromosome.

Anderson *et al* (2001) were the first to use MS loci for multi-locus genotyping in *Plasmodium*. They genotyped *P. falciparum* from three continents using 12 MSs, and found that malaria parasites from Africa had the highest genetic diversity followed by Southeast Asia, with South America being the least diverse (Anderson *et al.*, 2000). *P. falciparum* was shown to be panmictic in Africa with frequent recombination. In contrast, high levels of multi-locus linkage disequilibrium were associated with low diversity in low endemic regions outside of Africa, where either clonal or epidemic structures are assumed to be common.

Since this first paper, several others have explored the population genetics of *P. falciparum* on both global and local scales, most recently with genome wide markers. Certain trends have emerged, several of which were identified in the original Anderson *et al.*, paper. A survey of SNPs across chromosome 3 has confirmed that *P. falciparum* populations primarily cluster by continent (Mu *et al.*, 2005a). The authors also found that recombination rates varied widely, both among populations and across the chromosome; and that despite this variability, recombination hotspots at the ends of chromosome 3 were conserved. *P. falciparum* chromosome ends, or telomeres, contain a high proportion of genes implicated in host interactions and are likely to be under strong immune-mediated diversifying selection. High recombination rates are one means of maintaining allelic diversity. A separate study of 372 SNPs among 54 parasites from Africa, Asia, and Latin America further confirmed the hypothesis of continental subdivision of *P. falciparum* populations (Volkman *et al.*, 2007). In agreement with the findings based on data from chromosome 3, the authors found that LD extended over much shorter distances in African parasite populations compared with Asian parasite populations.

More recently, as markers became available, investigators have explored the population structure of *P. vivax*. One of the first population studies of *P. vivax* showed extremely low levels of diversity among eight MS markers, suggestive of a severe and recent population bottleneck (Leclerc *et al.*, 2004). However this was later shown to be an artefact of marker choice. The average number of repeats in the MS chosen for this study was biased downward and consequently the MS loci lacked variability (Imwong *et al.*, 2006). Subsequent studies employing the complete mitochondrial genome (Jongwutiwes *et al.*, 2005; Mu *et al.*, 2005b) and MS loci (Imwong *et al.*, 2007; Joy *et al.*, 2008; Karunaweera *et al.*, 2008) have in fact found that the genetic diversity of *P. vivax* is at least equal to that of *P. falciparum*, if not greater.

The global population structure of *P. vivax* is still being ascertained. However, current results suggest that it probably differs from that of *P. falciparum*. Given their divergent histories and the significant evolutionary distance that separates them, such differences are not surprising. In contrast with *P. falciparum*, although there are exceptions (Imwong *et al.*, 2007), most studies have not detected strong continental structuring in *P. vivax* (Cornejo and Escalante, 2006; Karunaweera *et al.*, 2008; Mu *et al.*, 2005b). The general lack of global continental structure in *P. vivax* may be related to the phenomenon of relapsing malaria. Unlike *P. falciparum*, *P. vivax* has a hypnozoite stage, which lies dormant in the liver and can cause a relapse of the illness long after the primary blood stage episode has been treated and cleared. The length of the relapse period differs among geographical regions and is probably related to temperature. Shorter latent periods (3–5 weeks) and more frequent relapse episodes are characteristic of tropical regions, whereas longer latent periods (5–10 months) and fewer incidences of relapse are more characteristic of *P. vivax* from temperate regions. The dormant liver stage, especially where it is long lasting, may act to break down geographic associations as infected humans move between geographical regions (Karunaweera *et al.*, 2008).

The local population structure of *P. vivax* mirrors some of what has already been observed in *P. falciparum*, although other patterns have also emerged (Imwong *et al.*, 2007). In low transmission areas *P. falciparum* typically exhibits low genetic diversity, high LD and infrequent multi-clone infections. Surprisingly, *P. vivax* in low transmission areas of Sri Lanka, Columbia, and Brazil exhibits both high genetic diversity and high multi-locus LD (Ferreira *et al.*, 2007; Imwong *et al.*, 2007; Karunaweera *et al.*, 2008). It appears as though low-level transmission does not severely constrain *P. vivax* genetic diversity. One possible explanation for this pattern is the presence of more than one genetically isolated *P. vivax* population in a single geographic area. This was in fact shown to be the case in southern Mexico, where the different *P. vivax* populations were preferentially transmitted by different local mosquito vectors (Joy *et al.*, 2008). An earlier study also identified two distinct *P. vivax* populations, designated Old World and New World types, and demonstrated that they too differed markedly with respect to mosquito vector compatibility (Li *et al.*, 2001). Vector-parasite compatibilities may play an important role in generating population structure in *P. vivax*.

Selection

Microsatellites have proven quite useful for studying selection, particularly drug-mediated selection, in malaria parasites. By measuring the genetic variation of MS that flank genes of interest, and identifying regions of unexpectedly low heterozygosity, researchers have been able to identify loci that had not been previously linked to drug resistance. Additionally, when drug resistance genes are known, the flanking MS haplotypes can be compared between drug sensitive and drug resistance malaria parasites. This information can be used to measure the strength of selection, deduce evolutionary relationships, and infer past demographic processes.

Based on their analysis of MS variation in the region flanking *pfcrt*, a known drug resistance gene, researchers concluded that drug resistant forms have evolved more than once and are rapidly spreading around the globe (Wootton *et al.*, 2002). Due to high and increasing levels of chloroquine resistance, chloroquine has been replaced by sulphadoxine pyrimethamine (SP) in many regions, including much of Africa as a first line treatment for malaria. Unfortunately, SP resistance has become prevalent in many parts of the world as well. The recent spread of SP resistance, both regionally and globally, has been documented by a number of researchers using MS variation in the region flanking the dihydrofolate reductase (*dhfr*) gene (Cortese *et al.*, 2002; Mita *et al.*, 2007; Ndiaye *et al.*, 2006; Pearce *et al.*, 2005; Roper *et al.*, 2004).

Future directions

In recent years, a wealth of genomic data has created unprecedented opportunities for in-depth evolutionary and population genetics studies of malaria parasite. It is now possible to bring whole genome datasets to bear on important evolutionary questions related to disease dynamics. In 2007, a genome wide catalogue of SNPs, MS, and indels (insertion/deletions) for a large number of malaria isolates was generated by three separate research groups to produce a detailed map of genetic variation across the *P. falciparum* genome (Jeffares *et al.*, 2007; Su *et al.*, 2007; Volkman *et al.*, 2007). Such a high-density haplotype map is essential for linking phenotype with genotype via association studies, and represents a potential wealth of data for population genomic studies of malaria. Since the release of the *P. falciparum* genome in 2002, the number of population studies in malaria that employ genome wide marker sets has increased rapidly. Now that the *P. vivax* genome sequence has also been released, we expect to see a similar trend of increasing population genetic studies for this parasite.

Increasingly genome-wide assays are being developed for use with large numbers of field isolates. One example is the molecular 'bar-coding' assays developed for genotyping *P. falciparum* (Daniels *et al.*, 2008). This assay uses 24 SNPs with high minimum allele frequencies to produce a unique genetic signature for each malaria parasite thus far tested. Such an assay could prove very useful for field-based population studies. Higher throughput SNP genotyping platforms are currently being developed for *P. falciparum* as well. For example, a survey of 372 genome-wide SNPs from 54 isolates separated parasite populations based on continent (Volkman, 2007). This same research group employed an array with 3000 SNP markers to illustrate the strong influence of natural selection on the genomic diversity of *P. falciparum* (Neafsey *et al.*, 2008). Other genome-wide genotyping methods such as the 'tiling' genotyping array are able to detect SNPs, indels, and MS, as well as copy number variation (CNV) (Jiang *et al.*, 2008). CNV can occur when regions of the genome differ in size among individuals due to gene amplification or gene loss events. High density comparative genome hybridization (CGH) is another method for surveying the entire genome for CNV. CGH has been applied to *P. falciparum*, for which CNV appears to be widespread across the genome (Ribacke *et al.*, 2007). Recently CGH has been used to study the emergence and spread of drug resistance in *P. falciparum* (Nair *et al.*, 2007, 2008). It is thought that gene amplification of drug resistance genes is adaptive though the exact mechanism is not yet known.

Mixed infections, consisting of more than one malaria parasite genotype, can greatly complicate population analyses. Typically mixed infections have been excluded from some analyses (Anderson *et al.*, 2000). It would be useful to accurately estimate haplotypes in mixed infections, for example before and after large-scale vaccine trials, or before and after the introduction of new anti-malaria drugs. Pyrosequencing has recently been used to accurately quantify alleles in mixed malaria infections containing up to three haplotypes (Takala *et al.*, 2006). Advantages are that it is fast and relatively inexpensive.

The complete genomes of several *Plasmodium* species – including two human malarias, three rodent malarias, and one primate malaria – have been sequenced (Carlton *et al.*, 2002; Gardner *et al.*, 2002; Hall *et al.*, 2005; Pain *et al.*, 2008). Comparison of *Plasmodium* genomes has revealed a high level of conservation of gene order, known as synteny (Carlton *et al.*, 2008); and the degree of synteny is positively correlated with evolutionary relatedness. Comparative genomics of human and closely related non-human malaria parasites may help to decipher what characteristics are associated with successful human pathogens. Increasingly, population genetics is giving way to population genomics in malaria research. Genome sequencing of large numbers of parasites from the field is becoming increasingly practical and cost effective. An understanding of the genomic distribution of sequence variation at the population level promises to yield new insights into parasite biology and evolutionary history.

References

Anderson, T.J. C., Haubold, B., Williams, J.T., Estrada-Franco, J.G., Richardson, L., Mollinedo, R., Bockarie, M., Mokili, J., Mharakurwa, S., French, N., *et al.* (2000). Microsatellite markers reveal a spectrum of population structures in the malaria parasite *Plasmodium falciparum*. Mol. Biol. Evol. *17*, 1467–1482.

Babiker, H.A., Creasey, A.M., Fenton, B., Bayoumi, R.A., Arnot, D.E., and Walliker, D. (1991). Genetic diversity of *Plasmodium falciparum* in a village in eastern Sudan. 1. Diversity of enzymes, 2D-PAGE proteins and antigens. Trans. R. Soc. Trop. Med. Hyg. *85*, 572–577.

Babiker, H.A., Lines, J., Hill, W.G., and Walliker, D. (1997). Population structure of *Plasmodium falciparum* in villages with different malaria endemicity in east Africa. Am J Trop Med Hyg *56*, 141–147.

Carlton, J.M., Adams, J.H., Silva, J.C., Bidwell, S.L., Lorenzi, H., Caler, E., Crabtree, J., Angiuoli, S.V., Merino, E.F., Amedeo, P., *et al.* (2008). Comparative genomics of the neglected human malaria parasite *Plasmodium vivax*. Nature *455*, 757–763.

Carlton, J.M., Angiuoli, S.V., Suh, B.B., Kooij, T.W., Pertea, M., Silva, J.C., Ermolaeva, M.D., Allen, J.E., Selengut, J.D., Koo, H.L., *et al.* (2002). Genome sequence and comparative analysis of the model rodent malaria parasite *Plasmodium yoelii yoelii*. Nature *419*, 512–519.

Coluzzi, M. (1999). The clay feet of the malaria giant and its African roots: hypotheses and inferences about origin, spread and control of Plasmodium falciparum. Parassitologia *41*, 277–283.

Coluzzi, M., Sabatini, A., Torre, A. d., Deco, M.A. D., and Petrarca, V. (2002). A polytene chromosome analysis of the *Anopheles gambiae* species complex. Science *298*, 1415–1418.

Cornejo, O.E., and Escalante, A.A. (2006). The origin and age of *Plasmodium vivax*. Trends Parasitol. *22*, 558–563.

Cortese, J.F., Caraballo, A., Contreras, C.E., and Plowe, C.V. (2002). Origin and dissemination of *Plasmodium falciparum* drug-resistance mutations in South America. J. Infect. Dis. *186*, 999–1006.

Culleton, R., Mita, T., Ndounga, M., Unger, H., Cravo, P., Paganotti, G., Takahashi, N., Kaneko, A., Eto, H., Tinto, H., *et al.* (2008). Failure to detect *Plasmodium vivax* in West and Central Africa by PCR species typing. Malar. J. *7*, 174.

Daniels, R., Volkman, S., Milner, D., Mahesh, N., Neafsey, D., Park, D., Rosen, D., Angelino, E., Sabeti, P., Wirth, D., and Wiegand, R. (2008). A general SNP-based molecular barcode for *Plasmodium falciparum* identification and tracking. Malar. J. *7*, 223.

Daubersies, P., Sallenave-Saies, S., Magne, S., Trape, J.-F., Contamin, H., Fandeur, T., Rogier, C., Mercereau-Puijalon, O., and Druilhe, P. (1996). Rapid turnover on *Plasmodium falciparum* populations in asymptomatic individuals living in a high transmission area.. Am. J. Trop. Med. Hyg. *54*, 18–26.

Escalante, A.A., and Ayala, F.J. (1994). Phylogeny of the malarial genus *Plasmodium*, derived from rRNA gene sequences. Proc. Natl. Acad. Sci. U.S.A. *91*, 11373–11377.

Escalante, A.A., Cornejo, O.E., Freeland, D.E., Poe, A.C., Durrego, E., Collins, W.E., and Lal, A.A. (2005). A monkey's tale: The origin of *Plasmodium vivax* as a human malaria parasite. Proc. Natl. Acad. Sci. U.S.A. *102*, 1980–1985.

Escalante, A.A., Freeland, D.E., Collins, W.E., and Lal, A.A. (1998). The evolution of primate malaria parasites based on the gene encoding cytochrome b from the linear mitochondrial genome. Proc. Natl. Acad. Sci. U.S.A. *95*, 8124–8129.

Ferreira, M., Karunaweera, N., Silva-Nunes, M., Silva, N.D., Wirth, D., and Hartl, D. (2007). Population structure and transmission dynamics of *Plasmodium vivax* in rural Amazonia. J. Infect. Dis. *195*, 1218–1226.

Gardner, M.J., Hall, N., Fung, E., White, O., Berriman, M., Hyman, R.W., Carlton, J.M., Pain, A., Nelson, K.E., Bowman, S., *et al.* (2002). Genome sequence of the human malaria parasite Plasmodium falciparum. Nature *419*, 498–511.

Gardner, M.J., Tettelin, H., Carucci, D.J., Cummings, L.M., Aravind, L., Koonin, E.V., Shallom, S., Mason, T., Yu, K., Fujii, C., and Pederson, J. (1998). Chromosome 2 sequence of the human malaria parasite Plasmodium falciparum. Science *282*, 1126–1132.

Hall, N., Karras, M., Raine, J.D., Carlton, J.M., Kooij, T.W. A., Berriman, M., Florens, L., Janssen, C.S., Pain, A., Christophides, G.K., *et al.* (2005). A comprehensive survey of the *Plasmodium* life cycle by genomic, transcriptomic, and proteomic analyses. Science *307*, 82–86.

Imwong, M., Nair, S., Pukrittayakamee, S., Sudimack, D., Williams, J.T., Mayxay, M., Newton, P.N., Kim, J.R., Nandy, A., Osorio, L., *et al.* (2007). Contrasting genetic structure in *Plasmodium vivax* populations from Asia and South America. Int. J. Parasitol. *37*, 1013–1022.

Imwong, M., Sudimack, D., Pukrittayakamee, S., Osorio, L., Carlton, J.M., Day, N.P. J., White, N.J., and Anderson, T.J. C. (2006). Microsatellite variation, repeat array length, and population history of *Plasmodium vivax*. Mol. Biol. Evol. 23, 1016–1018.

Jeffares, D.C., Pain, A., Berry, A., Cox, A.V., Stalker, J., Ingle, C.E., Thomas, A., Quail, M.A., Siebenthall, K., Uhlemann, A.-C., *et al.* (2007). Genome variation and evolution of the malaria parasite *Plasmodium falciparum*. Nat. Genet. *39*, 120–125.

Jiang, H., Yi, M., Mu, J., Zhang, L., Ivens, A., Klimczak, L., Huyen, Y., Stephens, R., and Su, X.-z. (2008). Detection of genome-wide polymorphisms in the AT-rich *Plasmodium falciparum* genome using a high-density microarray. BMC Genomics *9*, 398.

Jongwutiwes, S., Putaporntip, C., Iwasaki, T., Ferreira, M.U., Kanbara, H., and Hughes, A.L. (2005). Mitochondrial genome sequences support ancient population expansion in *Plasmodium vivax*. Mol. Biol. Evol. *22*, 1733–1739.

Joy, D.A., Feng, X., Mu, J., Furuya, T., Chotivanich, K., Krettli, A.U., Ho, M., Wang, A., White, N.J., Suh, E., *et al.* (2003). Early origin and recent expansion of *Plasmodium falciparum*. Science *300*, 318–321.

Joy, D.A., Gonzalez-Ceron, L., Carlton, J.M., Gueye, A., Fay, M., McCutchan, T.F., and Su, X.-z. (2008). Local adaptation and vector-mediated population structure in *Plasmodium vivax* malaria. Mol. Biol. Evol. *25*, 1245–1252.

Karunaweera, N.D., Ferreira, M.U., Munasinghe, A., Barnwell, J.W., Collins, W.E., King, C.L., Kawamoto, F., Hartl, D.L., and Wirth, D.F. (2008). Extensive microsatellite diversity in the human malaria parasite *Plasmodium vivax*. Gene *410*, 105–112.

Leclerc, M.C., Durand, P., Gauthier, C., Patot, S., Billotte, N., Menegon, M., Severini, C., Ayala, F.J., and Renaud, F. (2004). Meager genetic variability of the human malaria agent *Plasmodium vivax*. Proc. Natl. Acad. Sci. U.S.A. *101*, 14455–14460.

Levine, N. (1988). The Protozoan Phylum Apicomplexa. (Boca Raton, FL: CRC Press).

Li, J., Collins, W.E., Wirtz, R.A., Rathore, D., Lal, A., and McCutchan, T.F. (2001). Geographic subdivision of the range of the malaria parasite, *Plasmodium vivax*. Emerging Infect. Dis. *7*, 35–42.

Livingstone, F.B. (1984). The Duffy blood groups, vivax malaria, and malaria selection in human populations: a review. Hum Biol *56*, 413–425..

Martinsen, E.S., Perkins, S.L., and Schall, J.J. (2008). A three-genome phylogeny of malaria parasites (Plasmodium and closely related genera): Evolution of life-history traits and host switches. Mol. Phyl. Evol. *47*, 261–273.

Miller, L., Mason, S., Clyde, D., and McGuiness, M. (1976). The resistance factor to *Plasmodium vivax* in blacks. The Duffy-blood-group genotype, FyFy. N. Engl. J. Med. *295*, 302–304.

Mita, T., Tanabe, K., Takahashi, N., Tsukahara, T., Eto, H., Dysoley, L., Ohmae, H., Kita, K., Krudsood, S., Looareesuwan, S., *et al.* (2007). Independent evolution of pyrimethamine resistance in *Plasmodium falciparum* isolates in Melanesia. Antimicrob. Agents Chemother. *51*, 1071–1077.

Mu, J., Awadalla, P., Duan, J., McGee, K.M., Joy, D.A., McVean, G.A. T., and Su, X.-z. (2005a). Recombination hotspots and population structure in *Plasmodium falciparum*. PLoS Biol. *3*, e335.

Mu, J., Joy, D.A., Duan, J., Huang, Y., Carlton, J., Walker, J., Barnwell, J., Beerli, P., Charleston, M.A., Pybus, O.G., and Su, X.-z. (2005b). Host switch leads to emergence of *Plasmodium vivax* malaria in humans. Mol. Biol. Evol. *22*, 1686–1693.

Mzilahowa, T., McCall, P.J., and Hastings, I.M. (2007). 'Sexual' population structure and genetics of the malaria agent *P. falciparum*. PLoS ONE *2*, e613.

Nair, S., Miller, B., Barends, M., Jaidee, A., Patel, J., Mayxay, M., Newton, P., Nosten, F. o., Ferdig, M.T., and Anderson, T.J. C. (2008). Adaptive copy number evolution in malaria parasites. PLoS Genet. *4*, e1000243.

Nair, S., Nash, D., Sudimack, D., Jaidee, A., Barends, M., Uhlemann, A.-C., Krishna, S., Nosten, F., and Anderson, T.J. C. (2007). Recurrent gene amplification and soft selective sweeps during evolution of multidrug resistance in malaria parasites. Mol. Biol. Evol. *24*, 562–573.

Ndiaye, D., Daily, J.P., Sarr, O., Ndir, O., Gaye, O., Mboup, S., Roper, C., and Wirth, D.F. (2006). Defining the origin of *Plasmodium falciparum* resistant *dhfr* isolates in Senegal. Acta Tropica *99*, 106–111.

Neafsey, D., Schaffner, S., Volkman, S., Park, D., Montgomery, P., Milner, D., Lukens, A., Rosen, D., Daniels, R., Houde, N., *et al.* (2008). Genome-wide SNP genotyping highlights the role of natural selection in *Plasmodium falciparum* population divergence. Genome Biol. *9*, R171.

Pain, A., Bohme, U., Berry, A.E., Mungall, K., Finn, R.D., Jackson, A.P., Mourier, T., Mistry, J., Pasini, E.M., Aslett, M.A., *et al.* (2008). The genome of the simian and human malaria parasite *Plasmodium knowlesi*. Nature *455*, 799–803.

Paul, R.E., Packer, M.J., Walmsley, M., Lagog, M., Ranford-Cartwright, L.C., Paru, R., and Day, K.P. (1995). Mating patterns in malaria parasite populations of Papua New Guinea. Science *269*, 1709–1711.

Pearce, R., Malisa, A., Kachur, S.P., Barnes, K., Sharp, B., and Roper, C. (2005). Reduced variation around drug-resistant *dhfr* alleles in African *Plasmodium falciparum*. Mol. Biol. Evol. *22*, 1834–1844.

Perkins, S., and Schall, J. (2002). A molecular phylogeny of malarial parasites recovered from cytochrome b gene sequences. J. Parasitol. *88*, 972–978.

Ribacke, U., Mok, B.W., Wirta, V., Normark, J., Lundeberg, J., Kironde, F., Egwang, T.G., Nilsson, P., and Wahlgren, M. (2007). Genome wide gene amplifications and deletions in Plasmodium falciparum. Mol. Biochem. Parasitol. *155*, 33–44.

Roper, C., Pearce, R., Nair, S., Sharp, B., Nosten, F., and Anderson, T. (2004). Intercontinental spread of pyrimethamine-resistant malaria. Science *305*, 1124.

Su, X.-z., and Wellems, T.E. (1996). Toward a high-resolution *Plasmodium falciparum* linkage map: polymorphic markers from hundreds of simple sequence repeats. Genomics *33*, 430–444.

Su, X., Hayton, K., and Wellems, T.E. (2007). Genetic linkage and association analyses for trait mapping in *Plasmodium falciparum*. Nat. Rev. Genet. *8*, 497–506.

Takala, S.L., Smith, D.L., Stine, O.C., Coulibaly, D., Thera, M.A., Doumbo, O.K., and Plowe, C.V. (2006). A high-throughput method for quantifying alleles and haplotypes of the malaria vaccine candidate *Plasmodium falciparum* merozoite surface protein-1, 19 kDa. Malar. J. *5*, 31.

Volkman, S.K., Sabeti, P.C., DeCaprio, D., Neafsey, D.E., Schaffner, S.F., Milner, D.A., Daily, J.P., Sarr, O., Ndiaye, D., Ndir, O., *et al.* (2007). A genome-wide map of diversity in *Plasmodium falciparum*. Nat. Genet. *39*, 113–119.

Wootton, J.C., Feng, X., Ferdig, M.T., Cooper, R.A., Mu, J., Baruch, D.I., Magill, A.J., and Su, X.-z. (2002). Genetic diversity and chloroquine selective sweeps in *Plasmodium falciparum*. Nature *418*, 320–323.

Population Genetics and Epidemiology of Human Viral Pathogens

11

Fernando González Candelas and Rafael Sanjuán

Abstract

Many viral pathogens, especially those with an RNA genome, are characterized by their high mutation rates and large population sizes. These features are responsible for the high levels of genetic variation usually found in viral populations and for their rapid response to different selective challenges encountered during their infection and transmission processes. They are quantitatively and qualitatively so different from most other organisms that special models and concepts, such as the quasispecies model, have been developed to better describe the evolutionary dynamics of viral populations. Here, we review these and other salient features of viral pathogens, with a special emphasis on RNA viruses. We describe how population genetics theory provides an adequate frame-work for analysing and interpreting genetic variation in viral populations, for understanding their dynamics, and to study adaptive processes occurring therein. However, not all the evolutionary changes observed are due to the action of positive selection and this is not an all-mighty agent of evolutionary change. We finish this review by introducing a recently developed framework for integrating the evolutionary and epidemic behaviour of infectious organisms, known as phylodynamics, which is especially well-suited for fast evolving organisms such as viruses.

Introduction

Viruses represent one of the most important groups of pathogens for humans and, in general, for all other living organisms (Domingo and Holland, 1997). From minor nuisance or even harmless particles to deadly pathogens, the effects of viral infections span a full range of possibilities. Despite being denoted under a single name, viruses encompass very different organisms with only a few common features. Among these, their obligate intracellular way of life as parasites and the pervading effects of genetic variation in their populations probably are the most relevant. They do not share a common genetic material nor genome organization or structure. This is the result of their different origins, for which many hypotheses have been postulated and which we will not cover in this chapter (see reviews and opposing views in Bamford, 2003; Forterre, 2006; Burnett, 2006; Claverie, 2006; Koonin, 2006). From a population genetics point of view, variation is a key component of viral biology, not only by its effects on human health but also by its close links with the mechanisms of replication and transmission, the nature of their genetic material, or genome organization.

From an evolutionary biology perspective, viruses constitute a fascinating group of organisms in which evolutionary change can be observed in real time. This has led to an increasing interest in using viruses as model organisms for basic evolutionary studies (Elena and Sanjuán, 2007; Bull and Molineux, 2008; Forde *et al.*, 2008) which exploit many of their intrinsic features for analysing complex processes under simple and controlled conditions. The two most general properties of viruses for offering such an appropriate model system for evolution (large population sizes, short generation times) also play a crucial role in their interaction with the human host. Two additional important features of some viruses are their high level of genetic variability and highly compact genomes with a few genes and virtually no dispensable information even in their non-coding regions.

One of the basic features on which the current classification for viruses (Fauquet *et al.*, 2005) (International Committee on Taxonomy of Viruses, http://www.ictvonline.org/) is based is the nature and organization of their genetic material (Table 11.1). Among them, those containing RNA constitute the most important group of human pathogens both by the number of diseases, the number of people infected and the death-toll they take every year (WHO statistics on causes of death, http://www.who.org). They include old acquaintances of humanity such as Influenza A virus (FLUVA) or Rabies virus as well as new or recently known foes like Hepatitis C virus (HCV) or HIV. Their diverse evolutionary origins, genome organizations, life-cycles, modes of transmission, mechanisms for replication or cell-entry, etc., make it extremely difficult to derive generalizations but about a few aspects of viral biology. Our intention in this chapter is to delineate some of their more general features, many of which derive from their extraordinary potential to generate genetic variation, and how

Table 11.1 Main groups of viruses classified according to the nature and organization of their genetic material

Genome organization	Family	Species	Disease
dsDNA	Poxviridae	Variola virus	Smallpox
	Herpesviridae	Human herpes simplex virus (1–2)	Shingles, chickenpox
	Papillomaviridae	Human papillomavirus	Cervical cancer
ssDNA	Parvoviridae	Human parvovirus	Fifth disease
dsRNA	Reoviridae	Rotavirus A	Diarrhoea
ssRNA (+ polarity)	Picornaviridae	Hepatitis A virus	Hepatitis A
	Coronaviridae	SARS-coronavirus	SARS
	Enteroviridae	Poliovirus	Poliomyelitis
	Flaviviridae	Hepatitis C virus (HCV)	Hepatitis C
	Togaviridae	Rubella virus	Rubella
ssRNA (– polarity)	Rhabdoviridae	Rabies virus	Rabies
	Filoviridae	Marburg virus	Marburg haemorrhagic fever
	Paramyxoviridae	Measles virus	Measles
	Orthomyxoviridae	Influenza virus A (FLUVA)	Influenza
	Bunyaviridae	Hantaan virus	Haemorrhagic fever with renal syndrome
dsDNA-RT viruses	Hepadnaviridae	Hepatitis B virus	Hepatitis B
ssRNA-RT	Retroviridae	Human immunodeficiency virus (HIV)	AIDS

Some representatives of each group and the diseases they produce in humans are shown.

Genomes of (+) polarity can directly be used as mRNA for translation, whereas those of (–) polarity are complementary to the mRNA.

ds, double-stranded; ss, single-stranded; RT, retro-transcribing.

they affect both their evolution and their interaction with their human hosts.

Generation of variation

Mutation is the ultimate source of generic variation in every organism. However, there are other processes that can affect the levels of variation found in natural populations. Genetic drift and natural selection usually act by diminishing genetic variability whereas migration and recombination usually increase it and will substantially affect the rate of substitution – the rate at which a new variant is fixed (or nearly so) – in a viral population. Single-stranded (ss) RNA viruses, such as HCV, are characterized by very high mutation rates, usually estimated in the range of 10^{-4}–10^{-5} mutations per site and round of replication (Duffy *et al.*, 2008). This is the result of the lack of proofreading capacity of RNA-dependent RNA polymerases. Using a new approach, Cuevas *et al.* (2009a) have estimated the spontaneous mutation rate of HCV to be $1.15 \pm 0.29 \times 10^{-4}$ per site, within the usual range for ssRNA viruses (Belshaw *et al.*, 2008; Drake and Holland, 1999; Duffy *et al.*, 2008). Given the rate of generation of new viral particles during infection, this mutation rate is high enough to generate every single and most double mutants in the genomes of RNA-viruses every single day.

An alternative mechanism generating genetic variation in most RNA viruses is recombination (Rambaut *et al.*, 2004). For some viruses, most notably HIV but, among others, also hepatitis B and C viruses (Fares and Holmes, 2002; Kalinina *et al.*, 2002; Sentandreu *et al.*, 2008); Dengue virus, Japanese encephalitis virus or St. Louis encephalitis virus (Twiddy *et al.*, 2003) and enteroviruses (Lukashev, 2005) recombination plays an important role in the generation of new variation that allows a much faster and thorough exploration of the mutational space. This process is further reinforced in viruses with segmented genomes, such as orthomyxoviruses, including FLUVA, and reoviruses, which include rotaviruses, by reassortment.

For viral recombination to occur, two different viral genomes must be present simultaneously in the same cell, which can reside either in the human or in any other reservoir or vector host (Hon *et al.*, 2008). There are multiple constraints on coinfection of the same cell by two or more viruses. Some are temporal, depending on the duration of infection at the host and the cellular levels, while others depend on the availability of cell receptors enabling viral infection or cell-membrane modifications triggered by viral infection. Although the molecular mechanisms are different from those related to recombination, reassortment of segments in viruses with multipartite genomes, such as FLUVA and picornaviruses (Lai, 1992), also requires simultaneous infection of the same cell to happen. Hence, the same restrictions indicated above apply for this form of generating variability of some viruses.

The relevance of recombination and reassortment in the evolution and epidemiology of RNA viruses such as HIV-1 and FLUVA virus has been amply documented (Bollyky *et al.*, 1996; Chan and Abubakar, 2006; Chao, 2007; Charpentier *et al.*, 2006; Costa-Mattioli *et al.*, 2003; Gibbs *et al.*, 2001; Holmes *et al.*, 1999; Lukashev, 2005; Rambaut *et al.*, 2004; Shriner *et al.*, 2004a). In fact, high rates of recombination are usually found in positive-stranded RNA viruses (Simmonds, 2006) with HCV representing a notable exception.

Until recently there has been limited evidence of HCV recombination between different genotypes/subtypes (Yun *et al.*, 1996), suggesting that these events are rare *in vivo* and that the resulting recombinants are usually not viable (Smith *et al.*, 1997; Simmonds *et al.*, 1994; Viazov *et al.*, 2000). However, in the last few years, a few natural intergenotypic recombinants of HCV have been identified (RF1_2k/1b, RF2_1a/1b and RF3_2b/1b) and the crossover points have been mapped to the NS2, NS5B and NS3 regions, respectively (Kalinina *et al.*, 2002, 2004; Colina *et al.*, 2004; Kageyama *et al.*, 2006). More recently, a new natural intergenotypic (2/5) recombinant of HCV has been found whose crossover point is located between genes NS2 and NS3 (Legrand-Abravanel *et al.*, 2007). All these reports have described HCV recombination between different genotypes/subtypes but fewer studies have identified intra-subtype recombinants of HCV. One recombinant strain was detected by analysis of NS5A sequences from intra-patient populations belonging to a patient undergoing anti-viral therapy (Moreno *et al.*, 2006).

In a much more extensive study involving >17,000 sequences from 111 patients, Sentandreu *et al.* (Sentandreu *et al.*, 2008) were able to detect intra-subtype recombinant sequences in samples from a variety of patients: before and after antiviral treatment, with and without coinfection with HIV-1. In all cases, the frequency of the recombinant sequences was relatively low (one or two sequences from 25 to 302 sequences in each analysis) which reinforces the idea that HCV-recombinants usually are not advantageous. The difficulties for detecting recombinant strains of HCV (Sentandreu *et al.*, 2008) along with the relative low incidence of dual infections (with different genotypes or subtypes) (Viazov *et al.*, 2000) explain the little attention recombination has received as a factor generating genetic variation in HCV although at least one of these recombinant strains has started to spread (Kurbanov *et al.*, 2008).

Quantification of variation

Genetic variation can be described and quantified at different levels and with different goals. In general, only appropriate statistical measures of variability are valid for making scientifically valid comparisons and deriving conclusions. In certain contexts, the absolute numbers of variants present in a sample or population may be informative, but this is very unlikely to provide a valid parameter to compare different samples or populations, since its value is highly dependent on sample or population sizes. This can be solved easily by computing relative measures of variability such as entropy or genetic diversity. Both can be applied from the most fundamental unit of comparison, a specific nucleotide position, to the most encompassing one, the complete genome, and they are derived from the relative frequencies of the different alternatives for the corresponding unit. The former is derived from information theory and it is very popular in ecology and physical systems, while the latter is more closely connected to population genetics. But both measures consider only the frequencies of the different variants and do not incorporate any information on the nature or extent of the differences among them, although these are critical for making inferences about evolutionary processes.

Nucleotide diversity (π) is the most commonly used parameter to summarize and compare genetic variability in population genetics. It incorporates information on the frequency of the variants and their relatedness and can be customized to include only certain types of variants (e.g. synonymous or non-synonymous). This is one of most important parameters in molecular population genetics theory since it is closely related to q, the population mutation rate, which determines the levels of genetic variability or the expected frequencies of neutral variants in a population at mutation-drift equilibrium.

One advantage of using well-founded statistical parameters to quantify genetic variability is that they can be used to partition and compare the contribution of different levels of biological organization to the total genetic variation observed in a population. For instance, it is possible to quantitatively assign the contribution of within- and between-individuals or within- or between-cells to the total variation in a population or an individual, respectively. This is achieved under an analysis of variance framework (Weir, 1996) which was extended by Excoffier *et al.* (1992) to take genetic differences into account. This method is known as analysis of molecular variance (AMOVA).

The dynamics of variation

Genetic variability in a population does not increase inevitably along time, since it is the result of factors acting in opposite directions: some processes introduce variation in the populations while others remove it. Two main processes deplete variation from viral populations: selection and drift. By increasing the proportion of viral particles that carry particular, high-fitness variants, selection may reduce transitorily genetic variability in populations while the effect of drift is often associated to transmission events and can act more indiscriminately.

There are three main modes of action of natural selection at the molecular level. Selection can act purging new variants arising in the population which diminish some fitness-related property such as replication capacity or encapsidation efficiency. This form is known as purifying or negative selection and it has a moderate effect in reducing genetic variability, since it will mainly eliminate the less fit variants and those that appear associated to them in the same genome. The opposite form of action is known as positive selection, and it consists in the favouring of a new mutant which rises in frequency at the expense of other, less fit variants, which leads to a temporary reduction of genetic variation in the population known as selective sweep. This reduction is not at random: only those variants that incorporate the selected mutations will increase in frequency in the population until they become fixed, an effect that will extend to the other variants linked to the advantageous mutation (a process usually known as hitchhiking). Finally, a third mode of action of natural selection at the molecular level actually promotes and contributes to increased levels of genetic variation. This is known as diversifying selection and corresponds to those cases in which the virus benefits directly from generating new variation at particular genome sites. For instance, in the epitopes presented to major histocompatibility complex class I molecules, the variants at higher frequency in the population will tend to be selectively disfavoured because the immune system will target them preferentially. This particular form of diversifying selection is called frequency-dependent selection.

Genetic drift is the result of the sampling process that occurs in every population in which the total number of individuals is limited. This limit can be very high (billions or trillions, in the case of viral loads of HIV-1 or HCV in plasma) or very low, as when an individual becomes infected by a few viral particles. In the former case, the reduction in genetic variability is almost imperceptible and it is easily compensated by the new genetic variation that is generated continuously. On the contrary, extreme reductions in population size, especially during the transmission from one infected host to a new one, result in a drastic elimination of genetic variability after which only a few of the initially present variants are represented in the newly established population. In this case, the variants that originate the new population may be drawn at random from those initially present and the particular variants included are not necessarily associated with increased fitness. Since most mutations are deleterious, they will tend to be less fit than their progenitors.

Although usually overlooked, if not ignored, in the study of genetic variation in viruses, the neutral theory of molecular evolution (Kimura, 1968, 1983) sustains that most variation found at the molecular level does not have an impact on viral fitness and, as a consequence, is neutral in terms of natural selection. The original proposal was subsequently expanded (Ohta and Kimura, 1971; Ohta, 1992) by incorporating the evolutionary consequences of slightly deleterious mutations whose fate depends on the relative reduction in fitness they produce and the size of the population. Stochastic processes, usually associated with genetic drift, will dominate the fate of these mutations if effective population size is lower than the reciprocal of the corresponding selection coefficients. When population sizes or selection coefficients are larger and the above inequality no longer holds then deterministic processes will dominate and selection will be the main evolutionary force in the population. Given the large population sizes associated with viruses, it is usually considered that genetic drift is not as important as selection in determining evolutionary change in viral populations. However, this is not the case during transmission and also during chronic infection. Estimates of effective population sizes for HIV-1 within individual patients are around 10^3 (Leigh Brown, 1997; Shriner *et al.*, 2004b). These values imply that stochastic factors may have an important role in the evolution of viral populations at this level for slightly deleterious mutations. One additional, often overlooked, aspect of the quasi-neutral theory is that it also applies to slightly favourable mutations. Despite their increased fitness, their dynamics, stochastic or deterministic, will be determined by the relationship between effective population size and selection coefficient: a slightly advantageous mutation may easily disappear from a small population while it will probably increase in frequency in a large one.

The quasispecies paradigm

The term quasispecies was borrowed from chemistry and refers to an ensemble of quasi-identical molecules, as opposed to molecular species, which consist of identical molecules. Indeed, the word is very often used as a mere synonym of 'variable population', but variability is an intrinsic feature of biological populations and hence this usage is rather trivial. However, there is a theory underlying the quasispecies term, originally developed by Manfred Eigen (Eigen, 1971; Eigen *et al.*, 1988) to model the population dynamics of replicons during the early stages of life. Due to its emphasis on short genomes and error-prone replication, it became soon applied to RNA viruses and as increasing amounts of genetic variability data became available, so the term quasispecies became increasingly used and ended up dominating the field of evolutionary virology (Domingo *et al.*, 2006). Its widespread use could give the impression that the quasispecies theory is well-supported by data, but the fact is that it is very rarely used in reference to the underlying theory but instead, in its most trivial meaning.

The basic quasispecies model describes the population dynamics of macromolecular sequences maintained by error-prone replication. Mutations occurring during the copying process can modify the replication rate of the newly arisen sequences, allowing Darwinian selection to operate. Quasispecies theory is thus a population genetics theory with special emphasis on high mutation rates (Wilke, 2005). Because it originated in the field of physical chemistry, its terminology is somewhat different, though conceptually equivalent, to that of used by population geneticists. In general, populations are assumed to be at the mutation-selection balance, and the corresponding stationary sequence distribution consisting of a master sequence and a distribution of mutants is called the quasispecies. The system can be described by a value matrix $\mathbf{W}$ in which diagonal elements W_{kk} equal the number of non-erroneous copies resulting from exact replication of sequence k and off-diagonal elements W_{kj} equal the number of copies of sequence k resulting from erroneous replication of sequence j. At the stationary state, the average biological fitness of the population equals the largest eigenvalue of $\mathbf{W}$ and the abundance of each sequence is given by its associated eigenvector.

Although quasispecies provide a very general theoretical framework, the huge dimension of $\mathbf{W}$ (equal to the genome length) requires the use of simplifications in order to make the system tractable. Assumptions in the original quasispecies models (Eigen, 1971; Eigen *et al.*, 1988) include the following: (i) there is a single fittest sequence; (ii) all other sequences show equal fitness regardless of their distance to the master; (iii) the environment is constant; (iv) the population size is infinite (v) there is no recombination. The situation in which there is a single fittest sequence and all other genotypes have constant fitness corresponds to a two-class single-peak landscape, an assumption that is clearly divorced from reality.

Because quasispecies theory focuses on small, error-prone replicons, the usual set of assumptions adopted by quasispecies models often differs from those used in classical population genetics, sometimes leading to different predictions. For instance, the Haldane-Muller principle states that the average fitness of the population at the mutation-selection balance does not depend on selection coefficients (Haldane, 1937). This holds when multiple and back-mutations are neglected, which is reasonable for large genomes and low mutation rates. However, in small genomes and at high mutation rates, the evolutionary fate of any given mutant genotype depends not only on its fitness, but also on its ability to tolerate secondary mutations (Schuster and Swetina, 1988). As a result, in quasispecies models, the average population fitness depends on the mutation rate, but also on the average selection coefficient, implying that a slow replicator can potentially outgrow a faster competitor by virtue of its higher mutational robustness, a phenomenon called the 'survival of the flattest' (Wilke *et al.*, 2001). This effect can be magnified in small populations, in which genetic drift favours the accumulation of mutations (Krakauer and Plotkin, 2002).

Another well-known prediction of quasispecies models is the existence of an error threshold, defined as the critical error rate beyond which selection cannot further maintain the genetic integrity of the population (Eigen, 1971; Eigen *et al.*, 1988). Below the threshold, the quasispecies is stably localized around one or several sequences, but if the mutation rate exceeds the threshold, the variety of mutants that are in amounts comparable to the fittest types becomes very large and the population enters into error catastrophe. However, the existence of an error threshold is not a necessary prediction of the quasispecies theory (Wilke, 2005) and the original quasispecies model produced an error threshold because all mutants were assumed to have a constant fitness regardless of their mutational distance to the master sequence (Summers and Litwin, 2006). In this case, the error threshold deterministically takes place when $L = \log \sigma_0/\mu$, where L is genome length, σ_0 is the superiority of the master sequence over the mutants, and μ is the per-base mutation rate. It must be noted, though, that even in this case, the error threshold does not produce a sharp transition in average fitness. Indeed, average fitness smoothly decreases with increasing mutation rate until it reaches a plateau (Bull *et al.*, 2007).

In general, whether there is an error threshold depends on the assumed fitness landscape. Some generalizations of the two-class single-peak landscape show error catastrophes and, in general, the outcome depends on the sign of epistasis. Error catastrophe is more likely if epistasis is antagonistic, such that mutations tend to have progressively less fitness effects as they accumulate (notice that the two-class single-peak landscape represents an extreme form of antagonistic epistasis) (Wiehe, 1997). However, recent work has shown that with null epistasis, there can be an error catastrophe if lethal mutations are allowed (Takeuchi and Hogeweg, 2007).

A priori, the quasispecies theory should be relevant to the evolution of small genomes replicating at elevated per-base mutation rates (Domingo, 1985). The smallest genomes described correspond to viroids, plant pathogens with a minimal (<400 nt) non-protein-coding RNA genome (Flores *et al.*, 2005). Moreover, recent work has shown that the mutation rate of some viroids might be as high as 1/400, the highest reported for any biological entity (Gago *et al.*, 2009). The second highest mutation rates (10^{-5}–10^{-4}) correspond to ssRNA viruses (Drake and Holland, 1999; Sniegowski *et al.*, 2000). Moreover, RNA viruses also possess small genomes, ranging 3–30 kb, with most not exceeding 15 kb. Finally, ssDNA viruses show similarly small genome sizes and slightly lower mutation rates (10^{-7}–10^{-5}) (Duffy *et al.*, 2008). Hence, the natural domain of application of quasispecies models should be viroids, RNA viruses and, less likely, ssDNA viruses.

However, most evolutionary properties of RNA viruses are satisfactorily explained by classical population genetics, as for example the high adaptability of large populations, the deleterious effects of mutation accumulation, the relationship between mutation and adaptation, the existence of fitness tradeoffs and epistasis, or the fitness decay of small populations (Elena *et al.*, 2007). Therefore, the extra complexity brought by the quasispecies seems justified only in cases where there is evidence supporting its specific predictions. One such case is the survival of the flattest. The first experimental evidence for this phenomenon was provided with viroids (Codoñer *et al.*, 2006). Two different species were used, one characterized by fast growth and low genetic variability, and another by slower growth and higher variability. Under normal conditions, the faster replicator outcompeted its slower competitor, but when the mutation rate was artificially increased using UVC radiation, the relative fitness of the slower replicator increased, presumably due to its higher mutational robustness. Similar results were obtained with Vesicular stomatitis virus (VSV) (Sanjuán *et al.*, 2007). Differences in robustness between two genotypes were first established based on the analysis of plaque size distributions, genetic variability, and fitness decay under relaxed selection. The more robust genotype replicated more slowly under normal conditions but was able to outcompete its less robust counterpart after the addition of base-analogue mutagens as 5-fluorouracil or 2-azacytidine.

On the other hand, there are no consistent experimental evidences supporting the existence of an error threshold. The consequences of artificially increasing error rates have been explored in cell culture experiments with a variety of RNA viruses, including VSV, HIV-1, Poliovirus-1, Foot-and-mouth disease virus (FMDV), and Lymphochoriomeningitis virus (LCMV) (Anderson *et al.*, 2004). All these studies have shown that mutagens are highly detrimental to viral fitness and in some cases, lead to extinction. For example, in HIV-1, the addition of the base analogue 5-hydroxydeoxycytidine can result in loss of infectivity after 10–20 serial passages (Loeb *et al.*, 1999). This has led to postulate RNA virus-specific mutagens as candidate therapeutic agents (Domingo, 2003). However, lethal mutagenesis tells us nothing about error thresholds. Lethal mutagenesis implies extinction, whereas the error catastrophe is a shift in sequence space without a bearing on the survival of the population. Further, attempts to observe the predicted shift in sequence space have failed (Grande-Pérez *et al.*, 2002). Recently, a theory for lethal mutagenesis of viruses dissociated from error catastrophe has been proposed (Bull *et al.*, 2007). According to this theory, extinction should deterministically occur if $e^{-U}R < 1$, where R is the progeny number per infectious particle, U is the genomic mutation rate, and e^{-U} equals the mean equilibrium fitness of the population.

Finally, current evidence from comparative data does not support or definitely argues against the existence of quasispecies in nature. Many RNA viruses, as for instance HIV or HCV, exhibit high levels of intra-host genetic diversity (Anastassopoulou and Kostrikis, 2006; Pawlotsky, 2006), but this is not necessarily a consequence of high mutation rates. It can also be the result of relaxed selection, positive selection, recombination, or demographic factors such as migration. Finally, there is at present no definitive example from a natural RNA virus population undergoing the survival of the flattest or error catastrophe.

Analysis of adaptive evolution in viruses

One common postulate among virologists is the almost unlimited capacity of viral populations to quickly adapt to any change in environmental conditions. We will consider limitations to this potential below but now we are interested in analysing the source of viral adaptations and the methods used to infer them from the study of genetic variation in viral populations.

Adaptations are the result of natural selection acting on the existing variation in a natural population in which some factor(s) act limiting its growth. As commented above, genetic variation usually arises from mutation but recombination and reassortment are also significant factors contributing to genetic variability in many viruses. Their relevance for adaptive evolution is not intrinsic to the mechanism but it depends on the nature and environment on which the new variants have arisen (van Opijnen *et al.*, 2007). Occasionally, variation arising from recombination has been related directly to adaptation (van der Walt *et al.*, 2009) and the recent spread of a new FLUVA virus subtype resulting from a complex reassortment of viral genome segments provides just another example of the adaptive potential derived from abrupt changes in the immunogenic properties of this virus (Trifonov *et al.*, 2009; Garten *et al.*, 2009; Neumann *et al.*, 2009). However, increased frequency or rapid spread in a population is not synonymous of adaptation and disentangling this from other phenomena is not usually easy.

The first requisite for adaptive evolution is the existence of genetic variation for the trait in question. For viral populations, this is usually not a problem given the tremendous amount of variants generated continuously. However, the mere existence of large amounts of genetic variation does not guarantee the presence or appearance of a specific mutation or a set of these in the population so that they have an increased fitness compared to previous or new arisen variants. Usually, the process is dynamic and continuous, because other factors, both viral and from the host, are involved in determining the fitness of a viral mutation, so that, at least at the individual level, there is usually only a limited period in which the new mutants can appear and be favoured by selection. Otherwise, the viral population may go extinct before they even appear.

An adaptive mutation confers an advantage to its carriers over those which do not possess it. This leads to higher chances for an increase in the population frequency of the former group at the expense of the latter. Since the onset of the process until the fixation of the advantageous variant a series of events occur at different levels. They can be used to infer the existence of ongoing or recent adaptation. However, their fingerprint becomes blurred with time and the signal created can be confounded with that of other, non-adaptive, processes. The three main indications of positive selection at the molecular population genetic level are: (i) a change in the composition of the mutant spectrum at the population level, with a significant deviation from the expected distribution under mutation-drift equilibrium (Ewens, 1972; Ganapathy and Uyenoyama, 2009) in which the favoured variation rises in frequency at the expense of the other variants in the population; (ii) generation of strong linkage disequilibrium (LD) around the favoured site, with decreasing levels of LD as we move away from the site under selection; and (iii) a higher proportion of non-synonymous substitutions than of synonymous ones at the corresponding sites when the mutation favoured by selection falls within a protein coding locus.

Many methods have been developed to identify and link these changes to the action of positive selection (Shapiro *et al.*, 2009). However, each method has a different temporal frame of application and conclusions obtained after an inadequate method has been used can be very misleading. This note of caution is even more relevant for the analysis of virus adaptation, since their fast evolutionary rates often imply a very rapid action of selection at one level which is not necessarily accompanied by a similar change at other levels (see above).

The most popular class of methods for the inference of adaptive evolution in viral populations is based on the identification of positive Darwinian selection at the genetic level, by analysing the relative proportions of non-synonymous changes per non-synonymous positions (dN) to that corresponding to synonymous substitutions (dS)(Nei and Gojobori, 1986). The basic logic of this approach is simple and attractive. Due to its nature, adaptive evolution is expected to produce a change at the phenotypic level which, at least for coding regions – that usually span most of the compacted genome of viruses, depends on a change at the amino acid level. Under the neutral theory of evolution, changes at the molecular level will have no effect on viral fitness and selection will not favour nor disfavour non-synonymous changes over synonymous ones. This results in approximately equal chances for both kinds of mutations of becoming fixed when they appear in a population, since fixation is, under neutrality, a purely stochastic process. The proportion of the two types of mutations ($w = dN/dS$) observed in a phylogeny will be approximately equal to one. This value will also reflect deviations from the purely neutral model: if selection acts purging or removing most of the new mutations that produce a change at the amino acid level (non-synonymous mutations) because they result in deleterious or even lethal effects, then dN/dS will deviate towards values lower than one, since synonymous mutations will be more likely to become fixed in the population than non-synonymous mutations. On the contrary, if selection acts favouring changes at the amino acid level then the ratio will be higher than one, since non-synonymous mutations will prevail over synonymous ones, which will be governed by the stochastic nature of genetic drift.

This reasoning has a clear attractive: if true, it can help to identify the genetic loci on which selection has acted. The hallmark of adaptive selection acting on changes at the protein level is the detection of genes with ratios of non-synonymous to synonymous substitution rates larger than 1. A similar argument can be used for the detection of the opposite mode of selection: loci in which $w < 1$ correspond to instances of purifying or negative selection, in which changes at the amino acid level are penalized and purged from the population.

One immediate problem with this method is that the unit of selection considered is the whole gene, while evidence has accumulated indicating that a more fundamental level corresponds to codons. Hence, several methods were soon developed to analyse the action of natural selection at the codon level (Goldman and Yang, 1994; Nielsen and Yang, 1998; Yang *et al.*, 2000). These methods consider not the average value of dN/dS for the whole set of codons in a gene, but the distribution of these values over individual codons. By applying a maximum likelihood method, these methods first determine whether there is a subset of codons in the alignment which have evolved with an $w = dN/dS > 1$. If the null hypothesis that there are no such codons is rejected, then a Bayesian inference procedure is applied to ascertain which individual codons belong to that set. Several software packages and servers are available to perform these analyses (Yang, 2007; Kosakovsky Pond *et al.*, 2005). The output informs on the mode of selection (value of w) at each codon that best explains the observed evolutionary pattern.

The application of these methods has revealed a very large number of cases of adaptive evolution in viruses. Among many other examples, adaptive evolution has been associated to antigenic 'drift' in FLUVA (Blackburne *et al.*, 2008), virulence in West Nile virus (Brault *et al.*, 2007), immune evasion in HCV (Brown *et al.*, 2005; Cuevas *et al.*, 2009b), escape from antiretroviral therapy in HIV-1 (Frost *et al.*, 2001), and cell-tropism and virus-mediated membrane fusion in Dengue virus (Twiddy *et al.*, 2002). In many cases, the inference of positive selection at a certain codon is reinforced by previous analyses in which the effect(s) of changes at the corresponding amino acid have been ascertained and the evolutionary analysis informs on the pattern of evolution at those sites. In others, there is no such direct information and the identification of putative adaptive selection at a site should be taken more cautiously although it does provide important clues for further functional and mutational analyses. Further improvements to this basic procedure for the inference of adaptive evolution based on d*N*/d*S* have been developed. A recent more detailed review of these methods can be found in Aguileta *et al.* (2009).

One important limitation of these methods is that they only apply to protein-coding genes and to selection acting at the amino acid level. Non-coding portions of the viral genomes cannot be analysed under this framework despite they can be subjected to both purifying and positive selection (Wong and Nielsen, 2004; Hahn, 2007). Besides, there are additional levels at which natural selection can operate. For instance, secondary structure plays an important functional role in RNA-based viruses (Gultyaev *et al.*, 2007; McMullan *et al.*, 2007; Diviney *et al.*, 2008) and the action of natural selection on it cannot be ascertained by calculating w.

An additional problem for analysing adaptive evolution only on the basis of d*N*/d*S* is that these methods were developed for studying the pattern of changes fixed in a phylogeny. A fixed change is assumed to be shared by all the members of a species, or at least by the vast majority of them. However, in many cases the action of adaptive evolution in viruses results in increased levels of polymorphism at the population or the species level, although it may lead to fixation within a host. Furthermore, the rate of evolution within an infected individual can be large enough as to prevent the detection of changes in the codon position(s) undergoing adaptive evolution. Hence, alternative methods are necessary to analyse adaptive evolution in viral populations (Shapiro *et al.*, 2009; Aguileta *et al.*, 2009).

A complementary approach for detecting adaptive evolution to methods based on dN/dS is based on the analysis of polymorphism within populations. The former class of methods is useful for long-term action of natural selection, which can be short in an absolute time-scale for rapidly evolving viruses, while the later is more appropriate for the analysis of recent selection. Methods based on divergence are more robust to changes in the demography and they allow us to analyse coevolution and compensatory changes but methods based on polymorphism are more adequate to study recent changes in selection pressures driven by colonization or migration into a new host or population (Aguileta *et al.*, 2009).

If a newly arisen variant is favoured by natural selection it will increase in frequency at the corresponding level thus provoking a reduction in the frequency of previously existing variants. This change in the spectrum of frequencies can be tested against the expected distribution under neutral evolution. This distribution is given by Ewens' formula (Ewens, 1972) and the corresponding test was derived by Ewens (1972) and Watterson (1978). However, this test simply indicates that the frequency distribution of the different alleles found in a population deviates from the expectation under neutrality, but it does not provide any indication about the causes of such a deviation. The action of selection is one possibility, but there are other population genetics and demographic processes that can produce such a deviation (Nielsen, 2005).

Changes in the spectrum of allele frequencies over the expectation under neutrality underlie many tests of adaptive evolution based on polymorphism. These tests compare estimates of some basic parameter, usually the population rate of mutation ($\theta = 2Nm$), obtained by two different methods which will provide the same result if neutrality holds true but will differ otherwise. This class includes popular tests such as Tajima's (1989), Fu and Li's (1993), and Fay and Wu's (2000). More recent developments include methods to correct for the confounding effects of demography (Jensen *et al.*, 2005; Nielsen *et al.*, 2005).

Tests based on the spectrum of allele frequencies are informative on the action of positive selection at the haplotype level but they fail to reveal which specific positions are responsible for the observed selective process. Positive selection also results in a hitchhiking effect by which those variants closely linked to the target of selection also rise in frequency in the corresponding population. However, genetic recombination tends to break down these associations and the result is a decreasing LD value as we move away from the position under positive selection (Andolfatto *et al.*, 1999; Kim and Nielsen, 2004). In consequence, a number of methods have been developed to detect which positions are evolving, or have recently evolved, under positive selection based on the progressive dissipation of LD around those sites (Andolfatto *et al.*, 1999; Sabeti *et al.*, 2002; Kim and Nielsen, 2004; Voight *et al.*, 2006).

Another consequence of natural selection acting in a population is the promotion of genetic differentiation with respect to other populations. The level of between-populations differentiation is usually evaluated by means of *F*-statistics (Wright, 1931) and their derivatives (Rousset, 2000; Excoffier, 2000), which incorporate information not only on the frequencies of the different genetic variants in different populations but also on their nature and genetic distances. Recently, complex statistical models have been developed to test the levels of genetic differentiation between populations and to disentangle selective and demographic processes (Tenaillon *et al.*, 2004; Beaumont and Balding, 2004; Fagundes *et al.*, 2007). The identification of significant differentiation among viral populations within individual patients (see 'Within-host evolution' below) also provides an opportunity for detection of different selective pressures on viral populations occupying different organs or cell compartments. However, few of these methods are applied to viral populations although they are more appropriate or at least complementary to the usual d*N*/d*S*-based methods previously discussed.

Within-host evolution

The rapid turnover and high mutation rates of most RNA viruses imply that significant evolutionary change can take place within individual hosts, thus making organ or tissue compartmentalization possible. The viral population sampled from plasma is predominantly contributed by viruses released from the major cellular targets of the virus, e.g. hepatocytes in HCV or CD4 cells in HIV-1. However, viruses often infect other cell types, as for instance peripheral blood mononuclear cells (PBMC), which are the precursors of mature macrophages, and HIV-1 can be found at autopsy in samples from all organs. Once viruses from different body sites have been genetically characterized, it is possible to address viral compartmentalization. This has often been done by comparing directly the average distance between pairs of sequences within and between organs, or using phylogenetic analysis. However, a more specific tool is the AMOVA (Excoffier *et al.*, 1992), in which the genetic variance between and within organs is used to calculate the fixation index (F_{ST}), a particular form of *F*-statistic used to assess population structure.

HIV-1 infection typically starts with a genetically homogeneous population as a consequence of the population bottleneck occurring upon transmission, but diversity is rapidly regenerated during the initial acute stage of the disease and steadily increases during the non-symptomatic stage. Viruses from different organs can then diverge and eventually evolve separately. The best established viral compartment is the one infecting the central nervous system, but compartmentalization can also occur between blood and semen, blood and cervical swabs, or blood and lung, among others (McGrath *et al.*, 2001; Petito, 2004). Although some studies have found no compelling evidence for compartmentalization, approximately 90% of patients show statistically significant differences between two or more organs at autopsy, and differences between compartments may increase with disease progression (Sanjuán *et al.*, 2004). HIV-1 genotypes from brain, lung, testicles, and liver tend to be the most differentiated. These conclusions are mainly based on the analysis of sequences from the V3 hypervariable region of the envelope gene, although analysis of *pol* and *gag* genes generally leads to similar conclusions (Bordería *et al.*, 2007).

Compartmentalization has also been studied in HCV. Variants isolated from plasma and

PBMC are genetically differentiated in 10–40% of the patients (Zehender *et al.*, 2005; Roque-Afonso *et al.*, 2005), and PBMC-specific amino acids have been identified in patients coinfected with HIV-1 (Blackard *et al.*, 2007). HCV is also capable of neuroinvasion, which might aggravate the encephalopathy primarily caused by liver failure. Sequences from brain microglia have also been found to be significantly different from those found in plasma and lymphatic nodes (Forton *et al.*, 2004; Wilkinson *et al.*, 2009). Finally, cervix and plasma sequences are also differentiated in many patients (Minosse *et al.*, 2006).

In principle, compartmentalization can be caused by selection or genetic drift. Several selective pressures can lead to the establishment of intra-patient population structure. First, the evolution of viruses with different cell tropisms has been amply demonstrated. HIV-1 infects lymphoid cells (T-tropic) and monocytic cells (M-tropic). Different co-receptors are used for recognition and for entry in these two cell types and co-receptor use is at least partially determined by the V3 region of the *env* gene (Milich *et al.*, 1993). Resident macrophages are the main focus of infection for non-lymphoid organs and hence, differences in cell tropism can account for organ compartmentalization. Lung, colon, testicles and, brain are mainly M-tropic organs whereas lymphatic nodes, spleen, blood and, bone marrow are predominantly T-tropic.

Second, variation in local immune pressures, as for instance reduced immune activity in brain, could potentially contribute to the evolution of organ-specific variants. Third, drug penetration can vary among body sites and may be particularly poor in the central nervous system due to the presence of the haematoencephalic barrier. In this vein, it has been shown that, following the initiation of antiretroviral treatment, HIV-1 isolates from blood become resistant more rapidly than those from brain (Strain *et al.*, 2005).

However, genetic drift can also contribute to organ compartmentalization. For example, physical barriers to virus migration (i.e. the haematoencephalic barrier) or restricted cell trafficking might effectively isolate viral subpopulations. HIV-1 replicates more slowly in M-tropic cells than in T-tropic cells and hence drift might play a role in the former microenvironment (Karlsson *et al.*, 1998). Similarly, HCV samples from brain show specific mutations that lead to reduced translation efficiency and hence limit local spread of the virus (Forton *et al.*, 2004). Also, HCV replicates more slowly and might be under relaxed selective pressure in tumoral hepatocytes compared to normal ones (Sobesky *et al.*, 2007). The relative contributions of selection and drift to organ compartmentalization are sometimes difficult to disentangle, although the analysis of synonymous and non-synonymous variation suggests that there is a significant contribution of organ-specific adaptation in most cases (Sanjuán *et al.*, 2004; Bordería *et al.*, 2007).

Limits to RNA virus adaptability

The extremely high levels of genetic diversity resulting from high mutation rates are generally believed to fuel the adaptive capacity of RNA viruses, although elevated population sizes and short generation times can also accelerate the rate of evolution substantially (Holland *et al.*, 1982). On one hand, high mutations rates increase the number of beneficial mutations that spontaneously appear per time unit but, on the other hand, most mutations with phenotypic effect are deleterious (Sniegowski *et al.*, 2000; Elena and Sanjuán, 2005). The balance between these two opposing factors makes the level of mutations maximizing the rate of adaptive evolution to reside at some intermediate point (Fisher, 1930). Below this optimum, the rate-limiting factor for evolvability is the waiting time until the appearance of beneficial mutations, whereas above the optimum, the limiting factor is genetic contamination with deleterious mutations which interfere with the fixation of beneficial ones. Importantly, the optimal mutation rate depends directly on the selection coefficient, at least for asexual populations (Orr, 2000; Johnson and Barton, 2002). This implies that error-prone replication is an efficient strategy to respond to episodes of strong selection, therefore explaining why RNA viruses show rapid emergence of immune escape variants and resistances to vaccines or treatments. Evidence for the advantage of high mutation rates under immune pressure was provided by a study using Poliovirus-1 in which a high-fidelity variant was showed to have reduced pathogenicity in mice (Pfeiffer and Kirkegaard, 2005). In sum, current data suggest that, owing to their high mutation rates, RNA viruses respond readily to strong selective pressures, but that their evolutionary capacity under weaker selective pressures, as well as their ability to explore vast regions of the fitness landscape might be less remarkable (Belshaw *et al.*, 2008).

Being true that high mutation rates can provide an evolutionary advantage in some scenarios, it does not follow that high mutation rates have evolved because they confer an advantage in terms of evolvability. Given that deleterious mutations are more abundant than beneficial ones, the net effect of increasing the mutation rate is a short-term decrease in mean fitness. Since natural selection acts in the short term and cannot optimize traits for future benefits, genotypes with low mutation rates will be favoured. Transient elevations in mutation rates can result from genetic linkage of mutator genotypes with beneficial mutations (Taddei *et al.*, 1997), but this linkage will be rapidly dissipated by recombination (Drake *et al.*, 1998; Sniegowski *et al.*, 2000). The fact that many RNA viruses show high levels of recombination makes highly unlikely that hitchhiking of mutator genotypes is relevant to error-prone replication in RNA viruses. Further, work with digital organisms has shown that selection fails to maintain high levels of mutation in asexual organisms if the underlying fitness landscape is rugged, a condition that is likely to be met for RNA viruses, in light of experiments showing that epistatic interactions are frequent (Sanjuán *et al.*, 2004). Finally, it must be noted that DNA viruses and many microorganisms have life-styles similar to those of RNA viruses and hence should be under similar selective pressures, but with the exception of ssDNA viruses, the divide between RNA and DNA genomes seems well-established in terms of mutation rates (Duffy *et al.*, 2008). This leads to the hypothesis that high mutation rates in RNA viruses might be inextricably linked to RNA biochemistry, though current evidence indicates that RNA replication fidelity can be increased to some extent (Pfeiffer and Kirkegaard, 2003; Furió *et al.*, 2007). All in all, it seems more appropriate to see high mutation rates in RNA viruses as resulting from evolutionary constraints than as the product of evolutionary optimization. Related to this, it has been suggested that elevated mutation rates might have evolved in response to a trade-off between replication speed and accuracy, such that increased fidelity would pay the cost of slower replication (Furió *et al.*, 2005, 2007; Belshaw *et al.*, 2008).

While the evolutionary explanation of error-prone replication in RNA viruses remains controversial, there is ample consensus that high mutation rates impose a limit to the evolution of RNA viruses (Holmes, 2003; Biebricher and Eigen, 2005; Domingo *et al.*, 2006; Elena and Sanjuán, 2007). Early quasispecies models predicted that the maximum genome length that can be stably maintained without losing the integrity of genetic information is negatively correlated with mutation rate (Eigen, 1971; Eigen *et al.*, 1988). Therefore, error-prone replication limits genome size. In fact, few RNA virus genomes exceed 15,000 bases, with the notable exception of coronaviruses. In turn, such small genomes leave small room for accommodating complex functions and, in particular, for the appearance of proofreading. This leads to an evolutionary dead-end or chicken-egg problem known as the Eigen's paradox (Eigen, 1971), in which low replication fidelity limits genome size, which in turn limits replication fidelity. Interestingly, recent findings indicate that, contrary to other RNA viruses, coronavirus replicases possess 3′ exonuclease proofreading activity (Minskaia *et al.*, 2006) which presumably reduces the overall mutation rate of replication and relaxes the genome size restriction.

A priori, the highest growth rates should correspond to the smallest genomes. However, experimental evidence does not support this expectation in the case of viruses. In an experiment in which RNA and DNA phages with ample differences in genomes size were allowed to evolve under conditions favouring rapid growth, DNA phages with relatively large genomes showed higher replication rates than those with small genomes, RNA viruses being the slowest ones (Bull *et al.*, 2004). This suggests that the ability to replicate fast depends not only genome size, but also on other factors related to genome organization, the viral infection cycle, and host factors (Springman *et al.*, 2005). In this vein, it is important to note that viral genetic information is densely compressed, with little non-coding or regulatory regions, frequent overlapping reading frames, and multifunctional regions, implying that evolutionary constraints must be very strong. In the case of RNA genomes, the abundance of secondary RNA structures adds another significant level at which selection acts and hence increases the level of restriction.

As a result of these restrictions, most mutations in RNA viruses are deleterious rather than neutral. Indirect evidence for such low tolerance to mutation came from early mutation accumulation experiments (Chao, 1990) in which drastic fitness losses were observed. However, the first direct estimation of selection coefficients in RNA viruses came from a study in which a series of single-point mutants of VSV were obtained by site-directed mutagenesis and assayed for fitness in cell cultures, revealing that most random mutations are highly deleterious, up to 40% being lethal for the virus (Elena and Sanjuán, 2007). Very similar results were obtained with Tobacco etch virus *in vivo* (Carrasco *et al.*, 2007), confirming the extremely low tolerance of RNA viruses to mutation. Elevated selection coefficients and high mutation rates imply that RNA viruses are subject to extremely high mutational loads compared to most DNA viruses and microorganisms.

Functional constraint has important consequences for the evolution of RNA viruses. For instance, antagonistic pleiotropy, which takes place when mutations that are beneficial in one environment become deleterious in another environment, should be especially frequent. This issue has been investigated in the context of host radiation. VSV populations serially passaged in new cell types showed fitness increases to the detriment of the fitness in the original host (Turner and Elena, 2000). In another study with VSV, alternating-host passages involving mammalian and insect cells also revealed a fitness trade-off (Zárate and Novella, 2004). Experimental studies with bacteriophage Φ6 have confirmed that mutations conferring a selective advantage in a novel host are generally deleterious in standard hosts (Duffy *et al.*, 2006, 2007; Ferris *et al.*, 2007). Another important consequence of functional constraint is parallel or convergent evolution at the molecular level. Several cases have been identified in laboratory conditions, using phages (Bull *et al.*, 1997; Wichman *et al.*, 1999, 2000; Crill *et al.*, 2000) and animal viruses (Cuevas *et al.*, 2002), as well as in plant viruses *in vivo* (Rico *et al.*, 2006) The parallel or convergent changes consisted not only of amino acid replacements, but also involved synonymous mutations (Wichman *et al.*, 1999; Cuevas *et al.*, 2002; Remold *et al.*, 2008) and deletions (Cunningham *et al.*, 1997). These data suggest that RNA viruses have access to a very limited number of alternative solutions given an evolutionary problem.

Phylodynamics: reconstructing the epidemiological dynamics of viral infections

Nucleotide and amino acid sequences record information on the evolutionary history of individuals, populations and species. Under appropriate theoretical models, this information can be properly retrieved and used to infer relevant events occurred in the past. The rapid pace of evolutionary change which characterizes most viruses, especially RNA viruses as a result of their high mutation rates, large population sizes and short generation times, implies that the time scale over which these analyses can provide reliable information can be very short, even as short as months or years. This possibility has been used to develop a new methodology for studying the evolutionary and epidemiological dynamics of viral pathogens, giving rise to a new integrative concept known as pathogen phylodynamics (Grenfell *et al.*, 2004).

Phylodynamic analysis seeks to unify the study of processes occurring at different scales, from intrapatient to the species level, which are responsible for the spatial-temporal incidence and the phylogenetic patterns observed in viral populations. In consequence, these analyses take into account factors arising from three main sources. Firstly, the fast (acute) or slow (persistent) type of infection produced by the pathogen is a major determinant of its epidemiological spread. This is related to the epidemic period which in the former case can be measured in scales of days or weeks, whereas in the latter can extend for years. The second component has to do with the forces determining the phylogenetic patterns of pathogens. Here, the action of natural selection, genetic drift, and population structure, among other factors, are of primary importance. For instance, natural selection during transmission from one host to another, directly or mediated by vectors, results from the immune effects that the different variant strains of the pathogen exert on a population of susceptible hosts. These hosts are not genetically identical and they can substantially differ in their susceptibility to infection and/or intrinsic or temporary capabilities to prevent the pathogen to establish an infection or to further spread it. As a consequence, non-deterministic factors will also play an important role in the observed phylogenetic patterns. Finally, mutation and, eventually, recombination, as primary sources of variation, are necessary to fuel the continuous evolutionary process in which viral pathogens are engaged.

Combinations of these main properties led Grenfell and colleagues (Grenfell *et al.*, 2004) to propose four main categories of viral phylodynamics:

- *Short infections with strong cross-immunity* This pattern of viral phylodynamics is characterized by a strong epidemic behaviour with complex recurrent epidemics. It represents one end of the spectrum of dynamics, in which bursts of infections spread epidemically in a population of susceptible hosts. The strong cross-immunity and the rapid spread cause a rapid exhaustion of the susceptible population which results in the self-containment of the viral spread until a new opportunity arises. This dynamics is well illustrated by several morbilliviruses, such as measles virus, which show a markedly epidemic dynamics. The establishment of world-wide vaccination programs is aimed at eradicating this disease, a task that should be facilitated by the permanent immunity conferred by the appropriate vaccine. However, recent declines in compliance with the vaccination protocol have reduced the level of coverage below the threshold to prevent new measles epidemics and, in fact, several large outbreaks of measles have been produced in the last few years all over the world. This dynamics does not leave a clear imprint onto the phylogeny, since strong immune response against the different viral strains does not allow strong selective action against one or a few of them. As a result, the dynamics of different strains in time and space are responsible for the phylogenetic patterns that these viruses present.
- *Short infections with partial cross-immunity* This phylodynamics pattern is characterized by a strong selective pressure resulting from partial cross-immunity. This leads to rapid turnover of viral strains which may fluctuate periodically in a seasonal manner. The most representative case is probably FLUVA, although the dynamics of this virus is a bit more complicated due to antigenic shifts caused by reassortment of variants in the neuraminidase (NA) and haemagglutinin (HA) genes. The recent discovery of a subtropical reservoir for the virus has helped to complete the picture of both the regional and global dynamics of influenza. The strong selection drives the within subtype evolution, which follows a typical ladder-like pattern in the reconstructed phylogenies.
- *Infections with immune enhancement* Antibody-dependent enhancement explains a particular pattern of some viral infections in which the symptoms produced in a secondary infection are much more severe than those in the primary one. This is the case, for instance, of dengue virus (DENV), one of the most important emerging viruses and the most common vector-borne virus infecting humans. There are only four circulating serotypes and several subtypes of DENV and their genetic variation have been extensively studied in the last few years (Holmes and Twiddy, 2003). The reconstructed phylogenies have revealed a complex pattern with lineage extinction and replacement within serotypes, generating fluctuations in genetic diversity and influencing the population structuring of the virus. Phylogenetic trees also reveal a distinction between endemic strains, which have circulated in particular areas for prolonged periods of time, and epidemic subtypes, which spread rapidly through many populations.
- *Persistent infections* Persistent viral infections coupled with their high evolutionary rate and capacity for adaptive changes add a new dimension to viral phylogenies. These aspects can now be studied within infected hosts.

These viruses, among which HIV-1 and the HCV occupy relevant positions, do not elicit a strong cross-immunity, since co-infections with different viral subtypes or serotypes are relatively frequent. However, their within-host dynamics is strongly driven by the viral response to continuous and strong selection pressure from neutralizing antibodies or cytotoxic T lymphocytes. These result in an extremely fast viral dynamics, with within-host phylogenies resembling inter-host patterns in other viruses. This dynamics can be further complicated in cases of compartmentalization of the virus, in which the strong differentiation effects of genetic drift and population structure under restricted migration can overlap the previously described effects, thus yielding extremely complex patterns with apparent resuscitation of supposedly extinct lineages which enter the bloodstream (where circulating viruses are usually sampled). At the population level, the dynamics is governed more by epidemiological factors than by population genetic effects at the within-host level.

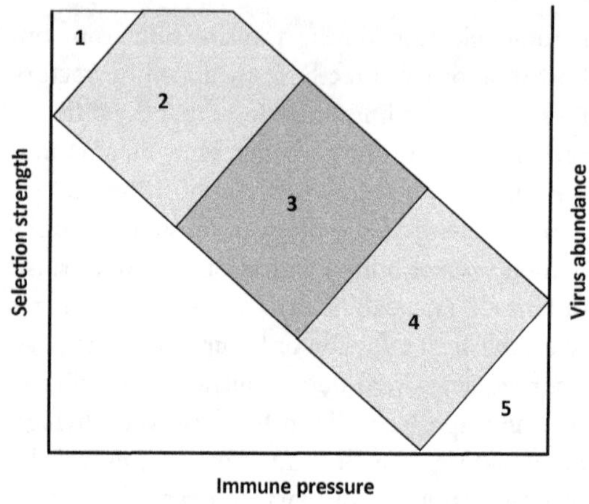

Figure 11.1 Schematic representation of viral adaptation under immune selective pressure. Five different phylodynamic patterns are shown (detailed in the main text). Data from Grenfell *et al.* (2004).

The different phylodynamic categories arise from the combination of three main factors (i.e. mutation, drift and selection), acting at two essential levels (i.e. within- and between hosts) (Mideo *et al.*, 2008). As previously described, these processes have potentially important consequences of medical relevance such as infectivity, disease progression or resistance to treatments.

There are two main stages in which natural selection can act during viral infections: within infected hosts, as the result of immune pressure, and among hosts, during transmission and including intermediates and reservoirs in-between infections. The consideration of the stochastic effects of genetic drift, resulting especially from the bottlenecks during host-to-host transmission, led Grenfell *et al.* (2004) to propose an integrated model in which the different possibilities and values of the parameters governing these processes are combined (Fig. 11.1). This model defines five basic patterns of viral adaptation to the immune pressure from the infected host:

When there is no immune pressure from the host there is no effective adaptive response in the virus, and most variants segregating within individual hosts are neutral. If an advantageous variant arises which is capable of replicating faster and/or more efficiently than the rest it may eventually become fixed in the population. This pattern can be observed in HCV-infected patients after liver transplantation when they receive strong immunosuppressing therapy (López-Labrador *et al.*, 2004, 2006).

Low immune response and low viral adaptation. This pattern can be observed in HCV patients rapidly progressing to chronic infection after the incorporation of a few adaptive mutations to overcome a weak immune response (Farci *et al.*, 2000) or in patients who do not respond to antiviral therapy and with little evidence of adaptive change in their infecting viral populations (Cuevas *et al.*, 2008).

Medium immune response with a fast viral adaptation. This is the result in the coevolutionary arms race between host and pathogen when the immune pressure exerted by the former is not strong enough to fully control the infectious agent and that the viral population is capable of generating enough variation for beneficial mutations to appear and drive its adaptation. The best documented examples probably correspond to changes during the chronic infection phase of HIV-1 before the exhaustion of the immune response by the host (Williamson, 2003, 2005; Frost *et al.*, 2005).

A strong immune response that prevents adaptive responses from the virus by reducing its population size down enough so that beneficial mutations have very little chance to arise. One likely example for this dynamics is given by self-resolving HCV infections in which, after an acute phase, infection does not progress to a chronic state.

An ever stronger immune response that allows an even more rapid control of the viral infection and which leads to persistent immune protection such as that provided to measles and rubella viruses (Grenfell *et al.*, 2001; Grenfell *et al.*, 2004).

These different phylodynamic patterns are partially reflected in the phylogenetic reconstructions derived for the corresponding virus populations, thus allowing the inference of the former from the analysis of intrapatient and population-level phylogenies. Depending on the inferred phylodynamics the questions that can be solved from the analysis of evolutionary patterns incorporated into the phylogenies will be different. For instance, in those cases in which the underlying phylodynamics results from a low or null adaptive response to immune pressure (cases 1 and 2 above), the phylogenies could reveal the spatial and epidemiological processes driving the evolution of the corresponding viruses. On the other hand, if immune selection is a strong component of the evolutionary change in the viral populations, the resulting phylogenies will reflect more of adaptive evolution and will have much less epidemiological and geographical information (Grenfell *et al.*, 2004).

Acknowledgements

This work has been funded by projects BFU2008-03000BMC and BFU2008-03978BMC from Ministerio de Ciencia e Investigación (Spain) and Conselleria de Sanitat de la Generalitat Valenciana.

References

Aguileta, G., Refrégier, G., Yockteng, R., Fournier, E., and Giraud, T. (2009). Rapidly evolving genes in pathogens: Methods for detecting positive selection and examples among fungi, bacteria, viruses and protists. Infect. Genet. Evol. *9*, 656–670.

Anastassopoulou, C.G. and Kostrikis, L.G. (2006). Global genetic variation of HIV-1 infection. Curr. HIV Res. *4*, 365–373.

Anderson, J.P., Daifuku, R., and Loeb, L.A. (2004). Viral error catastrophe by mutagenic nucleosides. Annu. Rev. Microbiol. *58*, 183–205.

Andolfatto, P., Wall, J.D., and Kreitman, M. (1999). Unusual haplotype structure at the proximal breakpoint of In(2L)t in a natural population of *Drosophila melanogaster*. Genetics *153*, 1297–1311.

Bamford, D.H. (2003). Do viruses form lineages across different domains of life? Res. Microbiol. *154*, 231–236.

Beaumont, M.A. and Balding, D.J. (2004). Identifying adaptive genetic divergence among populations from genome scans. Mol Ecol *13*, 969–980.

Belshaw, R., Gardner, A., Rambaut, A., and Pybus, O.G. (2008). Pacing a small cage: mutation and RNA viruses. Trends Ecol. Evol. *23*, 188–193.

Biebricher, C.K. and Eigen, M. (2005). The error threshold. Virus Res. *107*, 117–127.

Blackard, J.T., Hiasa, Y., Smeaton, L., Jamieson, D.J., Rodriguez, I., Mayer, K.H., and Chung, R.T. (2007). Compartmentalization of hepatitis C virus (HCV) during HCV/HIV coinfection. J. Infect. Dis. *195*, 1765–1773.

Blackburne, B.P., Hay, A.J., and Goldstein, R.A. (2008). Changing selective pressure during antigenic changes in human influenza H3. PLoS Pathog. *4*, e1000058.

Bollyky, P.L., Rambaut, A., Harvey, P.H., and Holmes, E.C. (1996). Recombination between sequences of hepatitis B virus from different genotypes. Journal of Molecular Evolution *42*, 97–102.

Bordería, A.V., Codoñer, F.M., and Sanjuán, R. (2007). Selection promotes compartmentalization in HIV-1: evidence from *gag* and *pol* genes. Evolution *61*, 272–279.

Brault, A.C., Huang, C.Y.H., Langevin, S.A., Kinney, R.M., Bowen, R.A., Ramey, W.N., Panella, N.A., Holmes, E.C., Powers, A.M., and Miller, B.R. (2007). A single positively selected West Nile viral mutation confers increased virogenesis in American crows. Nat. Genet. *39*, 1162–1166.

Brown, R.J.P., Juttla, V.S., Tarr, A.W., Finnis, R., Irving, W.L., Hemsley, S., Flower, D.R., Borrow, P., and Ball, J.K. (2005). Evolutionary dynamics of hepatitis C virus envelope genes during chronic infection. J. Gen. Virol. *86*, 1931–1942.

Bull, J.J., Badgett, M.R., Springman, R., and Molineux, I.J. (2004). Genome properties and the limits of adaptation in bacteriophages. Evolution *58*, 692–701.

Bull, J.J., Badgett, M.R., Wichman, H.A., Huelsenbeck, J.P., Hillis, D.M., Gulati, A., Ho, C., and Molineux, I.J. (1997). Exceptional Convergent Evolution in a Virus. Genetics *147*, 1497–1507.

Bull, J.J. and Molineux, I.J. (2008). Predicting evolution from genomics: experimental evolution of bacteriophage T7. Heredity *100*, 453–463.

Bull, J.J., Sanjuán, R., and Wilke, C.O. (2007). Theory of lethal mutagenesis for viruses. J. Virol. *81*, 2930–2939.

Burnett, R.M. (2006). More barrels from the viral tree of life. Proc. Natl. Acad. Sci. U.S.A. *103*, 3–4.

Carrasco, P., de la Iglesia, F., and Elena, S.F. (2007). Distribution of fitness and virulence effects caused by single-nucleotide substitutions in *Tobacco Etch Virus*. J. Virol. *81*, 12979–12984.

Chan, Y.F. and Abubakar, S. (2006). Phylogenetic evidence for inter-typic recombination in the emergence of human enterovirus 71 subgenotypes. BMC Microbiol. *6*, 74.

Chao, L. (1990). Fitness of RNA Virus Decreased by Muller's Ratchet. Nature *348*, 454–455.

Chao, M. (2007). RNA recombination in hepatitis delta virus: Implications regarding the abilities of mammalian RNA polymerases. Virus Research *127*, 208–215.

Charpentier, C., Nora, T., Tenaillon, O., Clavel, F., and Hance, A.J. (2006). Extensive recombination among Human Immunodeficiency Virus type 1 quasispecies makes an important contribution to viral diversity in individual patients. J. Virol. *80*, 2472–2482.

Claverie, J.M. (2006). Viruses take center stage in cellular evolution. Genome Biol. *7*, 110.

Codoñer, F.M., Daròs, J.A., Solé, R., and Elena, S.F. (2006). The fittest versus the flattest: experimental confirmation of the quasispecies effect with subviral pathogens. PLoS Pathog. *2*, e136.

Colina, R., Casane, D., Vasquez, S., Garcia-Aguirre, L., Chunga, A., Romero, H., Khan, B., and Cristina, J. (2004). Evidence of intratypic recombination in natural populations of hepatitis C virus. J. Gen. Virol. *85*, 31–37.

Costa-Mattioli, M., Ferre, V., Casane, D., Perez-Bercoff, R., Coste-Burel, M., Imbert-Marcille, B.M., Andre, E.C.M., Bressollette-Bodin, C., Billaudel, S., and Cristina, J. (2003). Evidence of recombination in natural populations of hepatitis A virus. Virology *311*, 51–59.

Crill, W.D., Wichman, H.A., and Bull, C.M. (2000). Evolutionary reversals during viral adaptation to alternating hosts. Genetics *154*, 27–37.

Cuevas, J.M., Elena, S.F., and Moya, A. (2002). Molecular basis of adaptive convergence in experimental populations of RNA viruses. Genetics *162*, 533–542.

Cuevas, J.M., González-Candelas, F., Moya, A., and Sanjuán, R. (2009a). Effect of ribavirin on the mutation rate and spectrum of Hepatitis C virus *in vivo*. J. Virol. *83*, 5760–5764.

Cuevas, J.M., Torres-Puente, M., Jiménez-Hernández, N., Bracho, M.A., García-Robles, I., Carnicer, F., Olmo, J.d., Ortega, E., González-Candelas, F., and Moya, A. (2009b). Combined therapy of interferon plus ribavirin promotes multiple adaptive solutions in hepatitis C virus. J. Med. Virol. *81*, 650–656.

Cuevas, J.M., Torres-Puente, M., Jiménez-Hernández, N., Bracho, M.A., García-Robles, I., Wróbel, B., Carnicer, F., Olmo, J.d., Ortega, E., Moya, A., and González-Candelas, F. (2008). Genetic variability of Hepatitis C Virus before and after combined therapy of interferon plus ribavirin. PLoS ONE 3, e3058.

Cunningham, C.W., Jeng, K., Husti, J., Badgett, M., Molineux, I.J., Hillis, D.M., and Bull, J.J. (1997). Parallel molecular evolution of deletions and nonsense mutations in bacteriophage T7. Mol. Biol. Evol. *14*, 113–116.

Diviney, S., Tuplin, A., Struthers, M., Armstrong, V., Elliott, R.M., Simmonds, P., and E.V.A.N.S., D.J. (2008). A hepatitis C virus cis-acting replication element forms a long-range RNA–RNA interaction with upstream RNA sequences in NS5B. J. Virol. *82*, 9008–9022.

Domingo, E. (1985). The quasispecies (extremely heterogeneous) nature of viral RNA genome populations: biological relevance – a review. Gene *40*, 1–8.

Domingo, E. (2003). Quasispecies and the development of new antiviral strategies. Prog. Drug Res. *60*, 133–158.

Domingo, E. and Holland, J.J. (1997). RNA virus mutations and fitness for survival. Annu. Rev. Microbiol. *51*, 151–178.

Domingo, E., Martín, V., Perales, C., Grande-Pérez, A., García-Arriaza, J., and Arias, A. (2006). Viruses as quasispecies: biological implications. Curr. Top. Microbiol Immunol. *299*, 51–82.

Drake, J.W., Charlesworth, B., Charlesworth, D., and Crow, J.F. (1998). Rates of spontaneous mutation. Genetics *148*, 1667–1686.

Drake, J.W. and Holland, J.J. (1999). Mutation rates among RNA viruses. Proc. Natl. Acad. Sci. U.S.A. *96*, 13910–13913.

Duffy, S., Burch, C.L., and Turner, P.E. (2007). Evolution of host specificity drives reproductive isolation among RNA viruses. Evolution *61*, 2614–2622.

Duffy, S., Shackelton, L.A., and Holmes, E.C. (2008). Rates of evolutionary change in viruses: patterns and determinants. Nat. Rev. Genet. *9*, 267–276.

Duffy, S., Turner, P.E., and Burch, C.L. (2006). Pleiotropic costs of niche expansion in the RNA bacteriophage f6. Genetics *172*, 751–757.

Eigen, M. (1971). Self-organisation of matter and the evolution of biological macromolecules. Naturwissenschaften *58*, 465.

Eigen, M., McCaskill, J.S., and Schuster, P. (1988). Molecular quasi-species. J. Phys. Chem. *92*, 6881–6891.

Elena, S.F. and Sanjuán, R. (2007). Virus evolution: insights from an experimental approach. Annu. Rev. Ecol. Syst. *38*, 27–52.

Elena, S.F. and Sanjuán, R. (2005). RNA viruses as complex adaptive systems. Biosystems *81*, 31–41.

Elena, S.F., Wilke, C.O., Ofria, C., and Lenski, R.E. (2007). Effects of population size and mutation rate on the evolution of mutational robustness. Evolution *61*, 666–674.

Ewens, W.J. (1972). The sampling theory of selectively neutral alleles. Theor. popul. Biol. *3*, 87–112.

Excoffier, L. (2000). Analysis of population subdivision. In Handbook of Statistical Genetics, Balding, D.J., Bishop, M., and Cannings, C., eds. (Chichester: John Wiley & Sons Ltd.), pp. 271–307.

Excoffier, L., Smouse, P.E., and Quattro, J.M. (1992). Analysis of molecular variance inferred from metric distances among DNA haplotypes: application to human mitochondrial DNA restriction data. Genetics *131*, 479–491.

Fagundes, N.J.R., Ray, N., Beaumont, M., Neuenschwander, S., Salzano, F.M., Bonatto, S.L., and

Excoffier, L. (2007). Statistical evaluation of alternative models of human evolution. Proc. Natl. Acad. Sci. U.S.A. *104*, 17614–17619.

Farci, P., Shimoda, A., Coiana, A., Diaz, G., Peddis, G., Melpolder, J.C., Strazzera, A., Chien, D.Y., Munoz, S.J., Balestrieri, A., Purcell, R.H., and Alter, H.J. (2000). The outcome of acute hepatitis C predicted by the evolution of the viral quasispecies. Science *288*, 339–344.

Fares, M.A. and Holmes, E.C. (2002). A revised evolutionary history of Hepatitis B Virus (HBV). Journal of Molecular Evolution *54*, 807–814.

Fauquet, C.M., Mayo, M.A., Maniloff, J., Desselberger, U., and Ball, L.A. (2005). Virus Taxonomy: VIIIth Report of the International Committee on Taxonomy of Viruses (New York: Academic Press).

Fay, J.C. and Wu, C.I. (2000). Hitchhiking under positive darwinian selection. Genetics *155*, 1405–1413.

Ferris, M.T., Joyce, P., and Burch, C.L. (2007). High frequency of mutations that expand the host range of an RNA virus. Genetics *176*, 1013–1022.

Fisher, R.A. (1930). The Genetical Theory of Natural Selection (Oxford: Oxford Univ. Press).

Flores, R., Hernández, C., Martínez de Alba, A.E., Daròs, J.A., and di Serio, F. (2005). Viroids and viroid–host interactions. Annu. Rev. Phytopathol. *43*, 117–139.

Forde, S.E., Beardmore, R.E., Gudelj, I., Arkin, S.S., Thompson, J.N., and Hurst, L.D. (2008). Understanding the limits to generalizability of experimental evolutionary models. Nature *455*, 220–223.

Forterre, P. (2006). The origin of viruses and their possible roles in major evolutionary transitions. Virus Res. *17*, 5–16.

Forton, D.M., Karayiannis, P., Mahmud, N., Taylor-Robinson, S.D., and Thomas, H.C. (2004). Identification of unique hepatitis C virus quasispecies in the central nervous system and comparative analysis of internal translational efficiency of brain, liver, and serum variants. J. Virol. *78*, 5170–5183.

Frost, S.D., Gunthard, H.F., Wong, J.K., Havlir, D., Richman, D.D., and Leigh Brown, A.J. (2001). Evidence for positive selection driving the evolution of HIV-1 *env* under potent antiviral therapy. Virology *284*, 250–258.

Frost, S.D.W., Wrin, T., Smith, D.M., Pond, S.L.K., Liu, Y., Paxinos, E., Chappey, C., Galovich, J., Beauchaine, J., Petropoulos, C.J., Little, S.J., and Richman, D.D. (2005). Neutralizing antibody responses drive the evolution of human immunodeficiency virus type 1 envelope during recent HIV infection. Proc. Natl. Acad. Sci. U.S.A. *102*, 18514–18519.

Fu, Y.X. and Li, W.H. (1993). Maximum likelihood estimation of population parameters. Genetics *134*, 1261–1270.

Furió, V., Moya, A., and Sanjuán, R. (2007). The cost of replication fidelity in human immunodeficiency virus type 1. Proc. Biol. Sci. *274*, 225–230.

Furió, V., Moya, A., and Sanjuán, R. (2005). The cost of replication fidelity in an RNA virus. Proc. Natl. Acad. Sci. U.S.A. *102*, 10233–10237.

Gago, S., Elena, S.F., Flores, R., and Sanjuán, R. (2009). Extremely high mutation rate of a hammerhead viroid. Science *323*, 1308.

Ganapathy, G. and Uyenoyama, M.K. (2009). Site frequency spectra from genomic SNP surveys. Theoretical Population Biology *75*, 346–354.

Garten, R.J., Davis, C.T., Russell, C.A., Shu, B., Lindstrom, S., Balish, A., Sessions, W.M., Xu, X., Skepner, E., Deyde, V., *et al.* (2009). Antigenic and genetic characteristics of swine-origin 2009 A(H1N1) Influenza viruses circulating in humans. Science 1176225.

Gibbs, M.J., Armstrong, J.S., and Gibbs, A.J. (2001). Recombination in the hemagglutinin gene of the 1918 'Spanish flu'. Science *293*, 1842–1845.

Goldman, N. and Yang, Z. (1994). A codon-based model of nucleotide substitution for protein- coding DNA sequences. Mol. Biol. Evol. *11*, 725–736.

Grande-Pérez, A., Sierra, S., Castro, M.G., Domingo, E., and Lowenstein, P.R. (2002). Molecular indetermination in the transition to error catastrophe: Systematic elimination of lymphocytic choriomeningitis virus through mutagenesis does not correlate linearly with large increases in mutant spectrum complexity. Proc. Natl. Acad. Sci. U.S.A. *99*, 12938–12943.

Grenfell, B.T., Bjornstad, O.N., and Kappey, J. (2001). Travelling waves and spatial hierarchies in measles epidemics. Nature *414*, 716–723.

Grenfell, B.T., Pybus, O.G., Gog, J.R., Wood, J.L.N., Daly, J.M., Mumford, J.A., and Holmes, E.C. (2004). Unifying the epidemiological and evolutionary dynamics of pathogens. Science *303*, 327–332.

Gultyaev, A.P., Heus, H.A., and Olsthoorn, R.C.L. (2007). An RNA conformational shift in recent H5N1 influenza A viruses. Bioinformatics 23, 272–276.

Hahn, M.W. (2007). Detecting natural selection on cis-regulatory DNA. Genetica *129*, 7–18.

Haldane, J.B.S. (1937). The effect of variation on fitness. Am. Nat. *71*, 337–349.

Holland, J., Spindler, K., Horodyski, F., Grabau, E., Nichol, S., and VandePol, S. (1982). Rapid evolution of RNA genomes. Science *215*, 1577–1585.

Holmes, E.C. and Twiddy, S.S. (2003). The origin, emergence and evolutionary genetics of dengue virus. Infection, Genet. Evol. *3*, 19–28.

Holmes, E.C., Worobey, M., and Rambaut, A. (1999). Phylogenetic evidence for recombination in dengue virus. Mol. Biol. Evol. *16*, 405–409.

Holmes, E.C. (2003). Error thresholds and the constraints to RNA virus evolution. Trends Microbiol. *11*, 543–546.

Hon, C.C., Lam, T.Y., Shi, Z.L., Drummond, A.J., Yip, C.W., Zeng, F., Lam, P.Y., and Leung, F.C.-C. (2008). Evidence of the recombinant origin of a bat Severe Acute Respiratory Syndrome (SARS)-Like Coronavirus and its implications on the direct ancestor of SARS coronavirus. J. Virol. *82*, 1819–1826.

Jensen, J.D., Kim, Y., Bauer DuMont, V., Aquadro, C.F., and Bustamante, C.D. (2005). Distinguishing between selective sweeps and demography using DNA polymorphism data. Genetics *170*, 1401–1410.

Johnson, T. and Barton, N.H. (2002). The effect of deleterious alleles on adaptation in asexual populations. Genetics *162*, 395–411.

Kageyama, S., Agdamag, D.M., Alesna, E.T., Leano, P.S., Heredia, A.M., bellanosa-Tac-An, I.P., Jereza,

L.D., Tanimoto, T., Yamamura, J.I., and Ichimura, H. (2006). A natural inter-genotypic (2b/1b) recombinant of hepatitis C virus in the Philippines. J. Med. Virol. *78*, 1423–1428.

Kalinina, O., Norder, H., and Magnius, L.O. (2004). Full-length open reading frame of a recombinant hepatitis C virus strain from St Petersburg: proposed mechanism for its formation. J. Gen. Virol. *85*, 1853–1857.

Kalinina, O., Norder, H., Mukomolov, S., and Magnius, L.O. (2002). A natural intergenotypic recombinant of hepatitis C virus identified in St. Petersburg. J. Virol. *76*, 4034–4043.

Karlsson, A.C., Lindback, S., Gaines, H., and Sonnenborg, A. (1998). Characterization of the viral population during primary HIV-1 infection. AIDS *12*, 839–847.

Kim, Y. and Nielsen, R. (2004). Linkage disequilibrium as a signature of selective sweeps. Genetics *167*, 1513–1524.

Kimura, M. (1968). Evolutionary rate at the molecular level. Nature *217*, 624–626.

Kimura, M. (1983). The Neutral Theory of Molecular Evolution. (Cambridge: Cambridge University Press).

Koonin, E.V. (2006). Temporal order of evolution of DNA replication systems inferred by comparison of cellular and viral DNA polymerases. Biology Direct *1*, 39.

Kosakovsky Pond, S.L., Frost, S.D.W., and Muse, S.V. (2005). HyPhy: hypothesis testing using phylogenies. Bioinformatics *21*, 676–679.

Krakauer, D.C. and Plotkin, J.B. (2002). Redundancy, antiredundancy, and the robustness of genomes. Proc. Natl. Acad. Sci. U.S.A. *99*, 1405–1409.

Kurbanov, F., Tanaka, Y., Avazova, D., Khan, A., Sugauchi, F., Kan, N., Kurbanova-Khudayberganova, D., Khikmatullaeva, A., Musabaev, E., and Mizokami, M. (2008). Detection of hepatitis C virus natural recombinant RF1_2k/1b strain among intravenous drug users in Uzbekistan. Hepatol. Res. *38*, 457–464.

Lai, M.M. (1992). RNA recombination in animal and plant viruses. Microbiological Reviews *51*, 61–79.

Legrand-Abravanel, F., Claudinon, J., Nicot, F., Dubois, M., Chapuy-Regaud, S., Sandres-Saune, K., Pasquier, C., and Izopet, J. (2007). A new natural intergenotypic (2/5) recombinant of Hepatitis C virus. J. Virol. *81*, 4357–4362.

Leigh Brown, A.J. (1997). Analysis of HIV-1 *env* gene sequences reveal evidence for a low effective number in the viral population. Proc. Natl. Acad. Sci. U.S.A. *94*, 1862–1865.

Loeb, L.A., Essigmann, J.M., Kazazi, F., Zhang, J., Rose, K.D., and Mullins, J.I. (1999). Lethal mutagenesis of HIV with mutagenic nucleoside analogs. Proc. Natl. Acad. Sci. U.S.A. *96*, 1492–1497.

López-Labrador, F.X., Berenguer, M., Sempere, A., Prieto, M., Sirera, R., González-Molina, A., Ortiz, V., Marty, M.L., Berenguer, J., and Gobernado, M. (2004). Genetic variability of hepatitis C virus NS3 protein in human leukocyte antigen-A2 liver transplant recipients with recurrent hepatitis C. Liver Transpl *10*, 217–227.

López-Labrador, F.X., Bracho, A., Berenguer, M., Coscollá, M., Rayón, J.M., Prieto, M., Carrasco, D., Gómez, M.D., Moya, A., and González-Candelas, F. (2006). Genetic similarity of hepatitis C virus and fibrosis progression in chronic and recurrent infection after liver transplantation. Journal of Viral Hepatitis *13*, 104–115.

Lukashev, A.N. (2005). Role of recombination in evolution of enteroviruses. Reviews in Medical Virology *15*, 157–167.

McGrath, K.M., Hoffman, N.G., Resch, W., Nelson, J.A.E., and Swanstrom, R. (2001). Using HIV-1 sequence variability to explore virus biology. Virus Research *76*, 137–160.

McMullan, L.K., Grakoui, A., Evans, M.J., Mihalik, K., Puig, M., Branch, A.D., Feinstone, S.M., and Rice, C.M. (2007). Evidence for a functional RNA element in the hepatitis C virus core gene. Proc. Natl. Acad. Sci. U.S.A. *104*, 2879–2884.

Mideo, N., Alizon, S., and Day, T. (2008). Linking within- and between-host dynamics in the evolutionary epidemiology of infectious diseases. Trends in Ecology & Evolution 23, 511–517.

Milich, L., Margolin, B., and Swanstrom, R. (1993). V3 loop of the human immunodeficiency virus type 1 Env protein: interpreting sequence variability. J Virol. *67*, 5623–5634.

Minosse, C., Calcaterra, S., Abbate, I., Selleri, M., Zaniratti, M.S., and Capobianchi, M.R. (2006). Possible compartmentalization of hepatitis C viral replication in the genital tract of HIV-1-coinfected women. The Journal of Infectious Diseases *194*, 1529–1536.

Minskaia, E., Hertzig, T., Gorbalenya, A.E., Campanacci, V., Cambillau, C., Canard, B., and Ziebuhr, J. (2006). Discovery of an RNA virus 3'->5' exoribonuclease that is critically involved in coronavirus RNA synthesis. Proc. Natl. Acad. Sci. U.S.A. *103*, 5108–5113.

Moreno, M.P., Casane, D., López, L., and Cristina, J. (2006). Evidence of recombination in quasispecies populations of an Hepatitis C Virus patient undergoing anti-viral therapy. Virology Journal 3, 87.

Nei, M. and Gojobori, T. (1986). Simple methods for estimating the numbers of synonymous and nonsynonymous nucleotide substitutions. Mol. Biol. Evol. 3, 418–423.

Neumann, G., Noda, T., and Kawaoka, Y. (2009). Emergence and pandemic potential of swine-origin H1N1 influenza virus. Nature *459*, 931–939.

Nielsen, R. (2005). Molecular signatures of natural selection. Annu. Rev. Genet. *39*, 197–218.

Nielsen, R., Williamson, S., Kim, Y., Hubisz, M.J., Clark, A.G., and Bustamante, C. (2005). Genomic scans for selective sweeps using SNP data. Genome Research *15*, 1566–1575.

Nielsen, R. and Yang, Z. (1998). Likelihood models for detecting positively selected amino acid sites and applications to the HIV-1 envelope gene. Genetics *148*, 929–936.

Ohta, T. (1992). The nearly neutral theory of molecular evolution. Ann. Rev. Ecol. Syst. *21*, 263–286.

Ohta, T. and Kimura, M. (1971). On the constancy of the evolutionary rate of cistrons. J. Mol. Evol. *1*, 18–25.

Orr, H.A. (2000). The Rate of Adaptation in Asexuals. Genetics *155*, 961–968.

Pawlotsky, J.M. (2006). Hepatitis C virus population dynamics during infection. Curr. Top. Microbiol. Immunol. *299*, 261–284.

Petito, C.K. (2004). Human immunodeficiency virus type 1 compartmentalization in the central nervous system. J. Neurovirol. 10 *Suppl. 1*, 21–24.

Pfeiffer, J.K. and Kirkegaard, K. (2003). A single mutation in poliovirus RNA-dependent RNA polymerase confers resistance to mutagenic nucleotide analogs via increased fidelity. Proc. Natl. Acad. Sci. U.S.A. *100*, 7289–7294.

Pfeiffer, J.K. and Kirkegaard, K. (2005). Increased fidelity reduces poliovirus fitness and virulence under selective pressure in mice. PLoS Pathog. *1*, e11.

Rambaut, A., Posada, D., Crandall, K.A., and Holmes, E.C. (2004). The causes and consequences of HIV evolution. Nat. Rev. Genet. *5*, 52–61.

Remold, S.K., Rambaut, A., and Turner, P.E. (2008). Evolutionary genomics of host adaptation in vesicular stomatitis virus. Mol. Biol. Evol. *25*, 1138–1147.

Rico, P., Ivars, P., Elena, S.F., and Hernández, C. (2006). Insights into the selective pressures restricting Pelargonium Flower Break Virus genome variability: evidence for host adaptation. J. Virol. *80*, 8124–8132.

Roque-Afonso, A.M., Ducoulombier, D., Di Liberto, G., Kara, R., Gigou, M., Dussaix, E., Samuel, D., and Feray, C. (2005). Compartmentalization of hepatitis C virus genotypes between plasma and peripheral blood mononuclear cells. J. Virol. *79*, 6349–6357.

Rousset, F. (2000). Genetic differentiation between individuals. J Evolution Biol *13*, 58–62.

Sabeti, P.C., Reich, D.E., Higgins, J.M., Levine, H.Z., Richter, D.J., Schaffner, S.F., Gabriel, S.B., Platko, J.V., Patterson, N.J., McDonald, G.J., Ackerman, H.C., Campbell, S.J., Altshuler, D., Cooper, R., Kwiatowski, D., Ward, R., and Lander, E.S. (2002). Detecting recent positive selection in the human genome from haplotype structure. Nature *419*, 832–837.

Sanjuán, R., Codoñer, F.M., Moya, A., and Elena, S.F. (2004). Natural selection and the organ-specific differentiation of HIV-1 V3 hypervariable region. Evolution *58*, 1185–1194.

Sanjuán, R., Cuevas, J.M., Furió, V., Holmes, E.C., and Moya, A. (2007). Selection for robustness in mutagenized RNA viruses. PLoS Genet 3, e93.

Sanjuán, R., Moya, A., and Elena, S.F. (2004). The contribution of epistasis to the architecture of fitness in an RNA virus. Proc. Natl. Acad. Sci. U.S.A. *101*, 15376–15379.

Schuster, P. and Swetina, J. (1988). Stationary mutant distributions and evolutionary optimization. Bull Math Biol *50*, 635–660.

Sentandreu, V., Jiménez-Hernández, N., Torres-Puente, M., Bracho, M.A., Valero, A., Gosalbes, M.J., Ortega, E., Moya, A., and González-Candelas, F. (2008). Evidence of recombination in intrapatient populations of Hepatitis C Virus. PLoS ONE *3*, e3239.

Shapiro, B.J., David, L.A., Friedman, J., and Alm, E.J. (2009). Looking for Darwin's footprints in the microbial world. Trends in Microbiology *17*, 196–204.

Shriner, D., Rodrigo, A.G., Nickle, D.C., and Mullins, J.I. (2004a). Pervasive genomic recombination of HIV-1 *in vivo*. Genetics *167*, 1573–1583.

Shriner, D., Shankarappa, R., Jensen, M.A., Nickle, D.C., Mittler, J.E., Margolick, J.B., and Mullins, J.I. (2004b). Influence of random genetic drift on human immunodeficiency virus type 1 *env* evolution during chronic infection. Genetics *166*, 1155–1164.

Simmonds, P., Smith, D.B., McOmish, F., Yap, P.L., Kolberg, J., Urdea, M.S., and Holmes, E.C. (1994). Identification of genotypes of hepatitis C virus by sequence comparisons in the core, E1 and NS-5 regions. Gen, J. Virol. 75 (Pt *5)*, 1053–1061.

Simmonds, P. (2006). Recombination and selection in the evolution of Picornaviruses and other mammalian positive-stranded RNA viruses. J. Virol. *80*, 11124–11140.

Smith, D.B., Pathirana, S., Davidson, F., Lawlor, E., Power, J., Yap, P.L., and Simmonds, P. (1997). The origin of hepatitis C virus genotypes. J. Gen. Virol. *78*, 321–328.

Sniegowski, P.D., Gerrish, P.J., Johnson, T., and Shaver, A. (2000). The evolution of mutation rates: separating causes from consequences. Bioessays 22, 1057–1066.

Sobesky, R., Feray, C., Rimlinger, F., Derian, N., Dos, S.A., Roque-Afonso, A.M., Samuel, D., Brechot, C., and Thiers, V. (2007). Distinct hepatitis C virus core and F protein quasispecies in tumoral and nontumoral hepatocytes isolated via microdissection. Hepatology *46*, 1704–1712.

Springman, R., Badgett, M.R., Molineux, I.J., and Bull, J.J. (2005). Gene order constrains adaptation in bacteriophage T7. Virology *341*, 141–152.

Strain, M.C., Letendre, S., Pillai, S.K., Russell, T., Ignacio, C.C., Gunthard, H.F., Good, B., Smith, D.M., Wolinsky, S.M., Furtado, M., Marquie-Beck, J., Durelle, J., Grant, I., Richman, D.D., Marcotte, T., McCutchan, J.A., Ellis, R.J., and Wong, J.K. (2005). genetic composition of human immunodeficiency virus type 1 in cerebrospinal fluid and blood without treatment and during failing antiretroviral therapy. J. Virol. *79*, 1772–1788.

Summers, J. and Litwin, S. (2006). Examining the theory of error catastrophe. J. Virol. *80*, 20–26.

Taddei, F., Radman, M., Maynard Smith, J., Toupance, B., Gouyon, P.H., and Godelle, B. (1997). Role of mutator alleles in adaptative evolution. Nature *387*, 700–702.

Tajima, F. (1989). DNA polymorphism in a subdivided population: the expected number of segregating sites in the two-subpopulation model. Genetics *123*, 229–240.

Takeuchi, N. and Hogeweg, P. (2007). Error-threshold exists in fitness landscapes with lethal mutants. BMC Evol. Biol. *7*, 15.

Tenaillon, M.I., U'Ren, J., Tenaillon, O., and Gaut, B.S. (2004). Selection versus demography: a multilocus investigation of the domestication process in maize. Mol. Biol. Evol. *21*, 1214–1225.

Trifonov, V., Khiabanian, H., and Rabadan, R. (2009). Geographic dependence, surveillance, and origins of the 2009 Influenza A (H1N1) virus. N. Engl. J. Med. NEJMp0904572.

Turner, P.E. and Elena, S.F. (2000). Cost of Host Radiation in an RNA Virus. Genetics *156*, 1465–1470.

Twiddy, S.S., Holmes, E.C., and Rambaut, A. (2003). Inferring the rate and time-scale of dengue virus evolution. Mol. Biol. Evol. *20*, 122–129.

Twiddy, S.S., Woelk, C.H., and Holmes, E.C. (2002). Phylogenetic evidence for adaptive evolution of dengue viruses in nature. J. Gen. Virol. *83*, 1679–1689.

van der Walt, E., Rybicki, E.P., Varsani, A., Polston, J.E., Billharz, R., Donaldson, L., Monjane, A.L., and Martin, D.P. (2009). Rapid host adaptation by extensive recombination. J. Gen. Virol. *90*, 734–746.

van Opijnen, T., de Ronde, A., Boerlijst, M.C., Berkhout, B., and 'Donnell, K. (2007). Adaptation of HIV-1 depends on the host-cell environment. PLoS ONE 2, e271.

Viazov, S., Widell, A., and Nordenfelt, E. (2000). Mixed infection with two types of hepatitis C virus is probably a rare event. Infection *28*, 21–25.

Voight, B.F., Kudaravalli, S., Wen, X., and Pritchard, J.K. (2006). A map of recent positive selection in the human genome. PLoS Biol. *4*, e72.

Watterson, G.A. (1978). The homozygosity test of neutrality. Genetics *88*, 405–417.

Weir, B.S. (1996). Genetic Data Analysis II. (Sunderland, MA: Sinauer).

Wichman, H.A., Badgett, M.R., Scott, L.A., Boulianne, C.M., and Bull, J.J. (1999). Different trajectories of parallel evolution during viral adaptation. Science *285*, 422–424.

Wichman, H.A., Scott, L.A., Yarber, C.D., and Bull, J.J. (2000). Experimental evolution recapitulates natural evolution. Philos. Trans. R. Soc. London Ser. B. *355*, 1677–1684.

Wiehe, T. (1997). Model dependency of error thresholds: the role of fitness functions and contrasts between the finite and infinite sites models. Genetical Research *69*, 127–136.

Wilke, C.O., Wang, J.L., Ofria, C., Lenski, R.E., and Adami, C. (2001). Evolution of digital organisms at high mutation rates leads to survival of the flattest. Nature *412*, 331–333.

Wilke, C.O. (2005). Quasispecies theory in the context of population genetics. BMC Evol. Biol. *5*, 44.

Wilkinson, J., Radkowski, M., and Laskus, T. (2009). Hepatitis C virus neuroinvasion: identification of infected cells. J. Virol. *83*, 1312–1319.

Williamson, S. (2003). Adaptation in the *env* gene of HIV-1 and evolutionary theories of disease progression. Mol. Biol. Evol. *20*, 1318–1325.

Williamson, S., Perry, S.M., Bustamante, C.D., Orive, M.E., Stearns, M.N., and Kelly, J.K. (2005). A Statistical Characterization of Consistent Patterns of Human Immunodeficiency Virus Evolution Within Infected Patients. Mol. Biol. Evol. *22*, 456–468.

Wong, W.S. and Nielsen, R. (2004). Detecting selection in noncoding regions of nucleotide sequences. Genetics *167*, 949–958.

Wright, S. (1931). Evolution in Mendelian populations. Genetics *16*, 97–159.

Yang, Z., Nielsen, R., Goldman, N., and Pedersen, A.M.K. (2000). Codon-substitution models for heterogeneous selection pressure at amino acid sites. Genetics *155*, 431–449.

Yang, Z. (2007). PAML 4: Phylogenetic Analysis by Maximum Likelihood. Mol. Biol. Evol. *24*, 1586–1591.

Yun, Z., Lara, C., Johansson, B., Lorenzana,d.R., I., and Sonnerborg, A. (1996). Discrepancy of hepatitis C virus genotypes as determined by phylogenetic analysis of partial NS5 and core sequences. J. Med. Virol. *49*, 155–160.

Zárate, S. and Novella, I.S. (2004). Vesicular stomatitis virus evolution during alternation between persistent infection in insect cells and acute infection in mammalian cells is dominated by the persistence phase. J. Virol. *78*, 12236–12242.

Zehender, G., De Maddalena, C., Bernini, F., Ebranati, E., Monti, G., Pioltelli, P., and Galli, M. (2005). Compartmentalization of Hepatitis C Virus quasispecies in blood mononuclear cells of patients with mixed cryoglobulinemic syndrome. J. Virol. *79*, 9145–9156.

12 Population Genetics in the Age of Metagenomics: Impact on Investigations of Viral, Bacterial, Archaeal and Eukaryotic Microbial Communities

Jianping Xu

Abstract

Microbial population genetics examines the spatial and temporal patterns of genetic variation across diverse geographic scales and ecological niches. With the arrival of molecular biological techniques, the past 40 years have seen tremendous progresses in microbial population genetics. However, in recent years, the analyses of genetic materials directly from natural environments have revolutionized our approaches and understandings of the diversity, function, and interrelationships among microorganisms in diverse natural ecological niches. The emergence and development of this expanding new field, that of metagenomics, has been primarily driven by technical and analytical methods developed from high throughput platforms for cloning, microfluidics, DNA sequencing, robotics, high-density microarrays, 2D-gel electrophoresis, and mass spectrometry as well as associated bioinformatics software. In this chapter, I present a general overview of this exciting development, including the general approaches and common tools used for identifying the diversity and function of microorganisms in natural biological communities. Of special notes are the potential impacts of recent developments in single cell isolation, whole-genome amplification, pyrosequencing, and database warehousing on our understanding of microbial population structures in nature. These exciting developments are bringing significant opportunities as well as new challenges to the field of microbial population genetics.

Introduction

Microbial population genetics examines the spatial and temporal patterns of genetic variation among samples of microorganisms from natural environments. Unlike plants and animals where samples are relatively straightforward to identify for population genetic studies, microbial samples are rarely visible to our naked eyes and require significant resources to isolate and identify for population studies. This difficulty in microbial population studies is further complicated by our inability to culture most of the microorganisms from natural environments. The recent advances in genomics and metagenomics offer significant opportunities as well as challenges for microbial population genetic studies. This chapter introduces the broad field of metagenomics and discusses its impact on addressing various issues in microbial population genetics.

The term 'metagenomics' was first coined by Jo Handelsman *et al.* (1998) to describe the emerging field that analyses genetic materials recovered directly from environmental samples. At the fundamental level, metagenomics is synonymous with several other terms that have emerged over the last decade including Environmental Genomics, Ecological Genomics, Community Genomics, and more recently Mega-genomics. In addition, the broad-sense metagenomics also encompasses metatranscriptomics, metaproteomics and meta-metabolomics to include any investigation involving the application of modern molecular techniques to the study of biological communities directly in their natural environments, bypassing the need for the isolation, laboratory cultivation and observation of individual organisms. At present, due to their small sizes, the enormous gap in our knowledge, and the extreme functional diversity, natural microbial communities have been the primary focus of metagenomics studies. Indeed, metagenomic analyses of natural microbial communities over the last few years have revolutionized our understanding of the diversity and potential function of microorganisms in diverse ecological niches (Allen and Banfield, 2005; Xu, 2006).

While metagenomics deals with the diversity and function of natural microbial communities, microbial population genetics examines the spatial and temporal patterns of genetic variations within and among populations. Thus, microbial population genetic studies could potentially benefit significantly from metagenomics tools, approaches and databases. In the first part of this chapter, I provide an overview of the broadly emerging field of metagenomics, with an emphasis on the analyses of genetic materials from natural microbial communities. In the second section, I summarize some of the major issues and challenges of microbial population genetics. This is then followed by the description of a couple of representative metagenomic studies that addressed microbial population genetic issues using metagenomics data. At present, the impact of metagenomic studies on microbial population genetics is still relatively limited. However, there are significant potentials. Some of these potentials are discussed in the final Perspectives section.

The emerging field of metagenomics

Unlike traditional microbiology and typical microbial population genetics that rely upon pure clonal cultures, the metagenomic analyses of microbial communities enable studies of organisms that are not easily cultured in the laboratory. In addition, it allows the direct studies of organisms in their natural environments. The ability to obtain such information is essential for a comprehensive and realistic understanding of natural biological communities. For example, over 99% of microorganisms in natural environments cannot be cultured in the lab (Pace *et al.*, 1985; Xu, 2006). Even for the ones that we can grow in the laboratory, the artificial environments that we provide are unlikely to be reflective of their natural ecological niches. As a result, for the same organism, their gene expression, protein and metabolite profiles will probably be different between those observed in the laboratory and those in nature. Therefore, approaches to investigate microbial populations in their native habitats are essential for critical and realistic understanding of microbes in nature.

Early metagenomics research focused on the analyses of single genes (e.g. the small subunit ribosomal RNA gene) or random gene fragments (Pace *et al.*, 1985; Pace, 1991). In recent years, metagenomics research has developed into analysing partial or complete genomes as well as other types of molecules in environmental samples, including RNA (transcripts), proteins (translated transcripts) and metabolites (metabolic intermediates and final products). In addition, different approaches (e.g. single cell vs. community genome) have been developed to analyse different components of the microbial community and the metagenome, depending on the objectives of the research.

Fig. 12.1 presents an overview of the broad-sense metagenomics framework for analysing the structure, diversity, and function of natural microbial communities. Conceptually, the broad-sense metagenomics can be divided into four complementary fields, depending on the types of molecules as their primary targets of investigations. It should be noted from the outset that an equally important field, that of bioinformatics, is essential for interpreting the data and for integration of information from the diverse types of analyses. Briefly, the conventional, narrow-sense metagenomics targets the analyses of total DNA samples from natural biological communities; metatranscriptomics analyses the complete RNA samples from natural environments; metaproteomics investigates all the proteins from natural environments; and meta-metabolomics examines the complete collection of metabolites from such samples. A comprehensive understanding of a natural microbial community requires the integration of information from all these and other fields, bringing together by advanced bioinformatics tools.

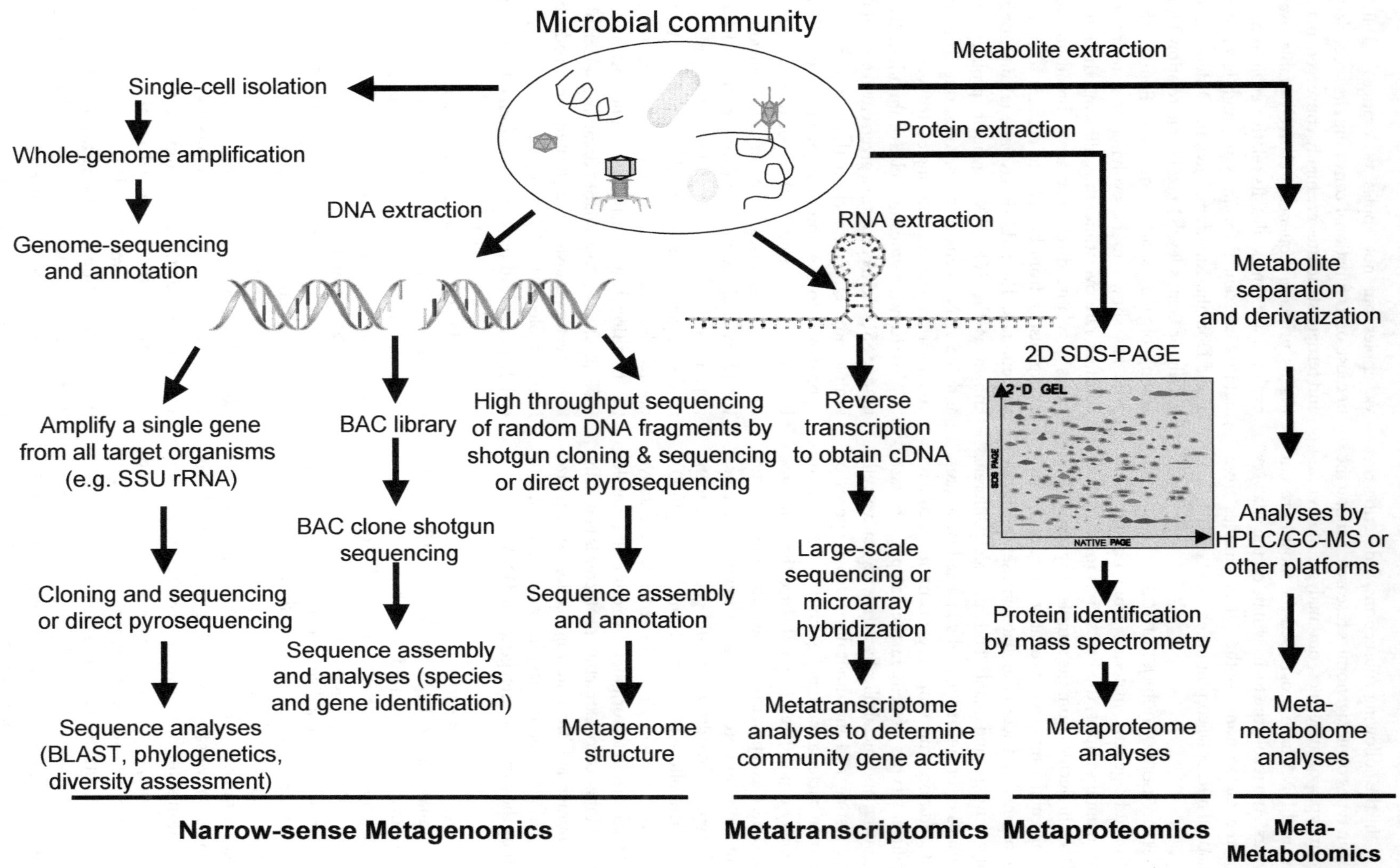

Figure 12.1 An overview of the broad-sense metagenomics: integration of information from multiple levels of high-throughput analyses is essential for a comprehensive and realistic understanding of biological communities in nature.

In the following subsections, I describe the main types of metagenomics research. Among the four types of molecules, our analyses of environmental DNA have progressed the most in recent years. As a result, the focus will be on the analyses of the metagenome – the DNA samples from natural microbial communities.

Narrow-sense metagenomics

To study genetic materials from natural environmental samples, the first step is to extract the DNA molecules from the sample. Any given environmental sample will probably contain a variety of organisms: bacteria, archaea, viruses, protozoa, algae, fungi, and plants and animals. Different groups of organisms have distinct cell structures. In addition, the same organisms may contain different types of cells in the sample, e.g. fast-growing and metabolically active vegetative cells vs. resting and metabolically inactive spores of certain bacteria and fungi. Such structural heterogeneity can differentially influence the efficiency of DNA extraction from these cells. Indeed, different extraction protocols have shown to influence the types of results you would get in subsequent metagenomic analyses (e.g. Rondon *et al.*, 2000). To ensure what's being extracted is representative of the DNA in the original environmental sample, a combination of extraction protocols is generally recommended. The following sections summarize the different approaches that have been used to study the extracted metagenomic DNA.

The single gene approach

The analyses of DNA samples directly from the environment started long before the term metagenomics came into use. Early studies examined the distribution and diversity of specific gene fragments in a given environmental sample. For several reasons, the small subunit of the ribosomal RNA (SSU rRNA) gene has been the marker of choice for this type of study. First, this gene is universally distributed across all cellular organisms. Such a universal distribution permits the potential identification, comparison and assessment of all cellular organisms in any given environment. Second, this gene is functionally homologous across all groups of cellular life forms. Such functional homogeneity ensures the comparison performed is meaningful (i.e. apples vs. apples but not apples vs. oranges). Third, because of its functional constraint, its secondary and tertiary structures are highly conserved, making its primary sequences relatively stable over evolutionary time scales. Its relative evolutionary stability is essential for analysing high-level (e.g. at the Domain, Kingdom, Phylum, Order, Class, Family or Genus levels) taxonomic relationships among organisms and sequences. Fourth, this gene has shown to be evolving at a relatively constant rate over time, thus allowing the inferences of relative divergence across broad groups of organisms, from bacteria to archaea to plants and animals. The SSU rRNA gene is commonly known as the 16S rRNA gene in prokaryotes and 18S rRNA gene in eukaryotes, due to their precipitation properties during centrifugation.

To study the types and relative abundances of unique SSU rRNA genes in a given environmental sample, after DNA extraction, the most common approach is to use universal (or group-specific) PCR primers to amplify the specific gene from the target organisms. The amplified products would generally contain a mixture of SSU rRNA sequences from diverse groups of organisms in the original sample. To obtain the DNA sequences for individual organisms in the mixture, the traditional approach is to clone individual PCR products into a vector containing selectable marker genes, transform the ligated gene products into suitable host cells (typically *Escherichia coli* cells), and spread-plate the transformed cells on selective media. Each individual colony grown from a single transformed cell should contain only one of the DNA fragments from the mixed PCR products. Individual colonies are then picked, the specific cloned fragment amplified by common primers located on the cloning vector, and then sequenced (Pace, 1991). The more recent, bead-based pyrosequencing technique allows direct sequencing of PCR products, abrogating the cloning and additional amplification steps (Edward *et al.*, 2006). Once the sequences are obtained, they can be compared to standard databases [e.g. the Ribosomal Database Project (RDP, www.rdp.cme.msu.edu), or GenBank] for their identification to various taxonomic levels, from phylum, to family, genus, and potentially species levels. The sequences can thus be used to infer the potential type and diversity of microorganisms in the specific ecological niches (Xu, 2006).

Although this methodology was limited to exploring highly conserved genes, its application supported early morphology-based observations that microbial diversity was far more complex than was known by culturing methods. Many subsequent studies followed, targeting different groups of organisms in different environmental niches. Such work revealed that the vast majority of microbial diversity had been missed by cultivation-based methods. The current consensus is that less than 1% of the Bacteria and Archaea species in a given environmental sample can be grown on existing media and incubation conditions. The non-cultivable groups include some major phyla within all microbial groups (i.e. bacteria, archaea, fungi, viruses, algae and protozoa) (Xu, 2006).

The rapid accumulation of extensive sequence information for phylogenetic signature molecules has spurred the development of high throughput microbial detection and monitoring systems. For example, based on DNA sequence information from groups of cultivable or uncultivable organisms, microarrays containing taxa-specific signature sequences have been developed to monitor the distributions of sulphate-reducing prokaryotes (Loy *et al.*, 2002), Enterococcus species (Lehner *et al.*, 2005), and other groups of organisms (Wu *et al.*, 2004; Zhou *et al.*, 2003).

Targeted partial metagenome sequencing – linking phylogeny to function

Given the relatively high cost of sequencing whole metagenomes, early efforts to study potential functional genes in natural environments were based on the analyses of partial genomic libraries containing large fragments of environmental DNA. The analyses tended to focus the clones in these libraries that contain both phylogenetic signature molecules and protein-coding open reading frames. Specifically, the genes close to the SSU rRNA gene cluster or other phylogenetic marker genes were cloned, identified and sequenced together. To achieve this, relatively high molecular weight DNA need to be extracted and used to construct clone libraries using Cosmid vectors or bacterial artificial chromosomes (BAC) so that such large fragments can be cloned and analysed individually (Beja *et al.*, 2000a). Through Southern blotting, phylogenetic probes such as the SSU rRNA gene can be used to identify specific clones in the metagenome library that contain signature gene fragments, allowing further phylogenetic inferences. These clones can then be sequenced at very reasonable cost, through either shotgun sequencing or pyrosequencing. The SSU rRNA gene as well as any other genes on the cloned fragments can be identified and annotated. Together, the information allows the inferences of the organisms and the metabolic processes in the community as well as their potential ecological functions. For example, in a classical metagenomic study of genome fragments from a BAC library of marine picoplankton, Beja *et al.* (2000b) identified a new class of genes of the rhodopsin family, named proteorhodopsin, from an uncultivated alpha-proteobacterium SAR86. At that time, this rhodopsin family was known to exist only in extremely halophilic (salt-loving) archaea and had never before been observed in cultured bacteria. Unlike the archaeal rhodopsin that does not express properly in model laboratory strains, the proteorhodopsin from SAR86 expressed readily in the laboratory model bacterium *E. coli* and it functioned as a light-driven proton pump. Later studies identified that this new type of light-driven energy generation process is in fact widespread in the ocean and that there are optimized absorption spectra of bacterial rhodopsins at different depths of ocean water (Beja *et al.*, 2001).

Random but comprehensive metagenome sequencing

The above approaches allow specific gene-based investigations of environmental samples. Recent technological advances are allowing the whole or partial metagenomes to be analysed relatively efficiently, though the costs and analyses are still prohibitive for most research labs. In this type of analyses, the DNA samples are either first cloned into a vector and then subjected to sequencing, or being sequenced directly. Because the DNA fragments are randomly sheared, overlapping fragments should exist in the sample. If sufficient numbers of DNA fragments were sequenced using either the shotgun approach through cloning and sequencing or direct pyrosequencing, it would be possible to reconstruct large DNA fragments from the randomly sequenced small pieces. In certain cases, it has allowed the reconstruction of complete or nearly complete genomes of several previously unknown bacterial species (e.g. Venter *et al.*, 2004; Tyson *et al.*, 2004). Because the collection of DNA from an environment is largely unknown and not filtered, the most abundant organisms in an environmental sample should be the most highly represented in the resulting sequence data, thus allowing assembly of large contigs of sequence information. To achieve the high coverage needed to fully resolve the genomes of underrepresented community members, large samples, often prohibitively so, are needed. On the other hand, the random nature of shotgun sequencing ensures that many of these organisms will be represented by at least some small sequence segments. Due to the limitations of microbial isolation and observation methods, the vast majority of these organisms would go unnoticed using traditional culturing techniques or microscopy tools.

For example, Breitbart and colleagues (2002) used environmental shotgun sequencing to show that 200 litres of seawater contained over 5000 different viruses. Subsequent studies showed that there were >1000 viral species in the human stool and possibly a million different viruses per kilogram of marine sediment, including many bacteriophages (Culley *et al.*, 2006). Very few of the assembled viral genomes resemble known sequenced viral genomes and virtually all the viruses in these studies represented new species. The metagenomic study of the Sargasso Sea by Venter *et al.* (2004) found DNA from nearly 2000 different species including assembling complete genome of a bacteria and revealing 148 types of bacteria never seen before. In the same year, Tyson *et al.* (2004) reported the complete, or nearly complete, genomes for a handful of bacteria and archaea in an acid mine drainage.

In the past few years, a new sequencing platform, a chip-based pyrosequencing developed by 454 Life Sciences has also been used to directly analyse environmental DNA. This DNA sequencing technique generates shorter fragments than conventional Sanger sequencing platforms. However, this limitation is compensated for by the very large number of sequence reads that can be generated, usually in the range of millions of reads, in a single sequencing run. In addition, this technique does not require PCR or cloning the DNA before sequencing, circumventing two of the main biases in metagenomics. In 2006, Robert Edwards and colleagues published the first sequences of environmental samples generated with pyrosequencing. They analysed the metagenomes from two adjacent sites that differed in chemistry and hydrogeology in the Soudan Mine, Minnesota, US. Based on the obtained metagenomic sequences, they inferred significant differences in metabolic potential of the microorganisms in each environment, including distinct biochemistry in carbon utilization, iron acquisition mechanisms, nitrogen assimilation, and respiratory pathways. In addition, these two communities were found to have very different metagenomes from other sequenced microbial communities (Edwards *et al.*, 2006).

While much progress in metagenomic studies is due to the development of high throughput sequencing technologies, the development of bioinformatics software have also played major roles and are poised to play significantly greater roles. The initial metagenome assembly and analyses used software developed for genome sequencing projects of cultured or known individual organisms. In 2007, Huson *et al.* developed and published the first stand-alone metagenome analysis tool, MEGAN. This tool was originally developed to analyse the metagenome of a mammoth sample but was later adapted for shotgun reads of metagenome sequences. In 2008, Meyer *et al.* released the Metagenomics RAST server (MG-RAST), a community resource for metagenome data analysis. As of April 2009, over 55 giga-basepairs of environmental DNA from over 1700 metagenome data sets have been analysed via MG-RAST. Other software platforms and databases for analysing metagenomic data include the IMG/M and CAMERA.

Single cell isolation, whole genome amplification and sequencing

While most current studies of uncultivable environmental microorganisms rely on the extraction and sequencing of the whole community genomes, a recent paper reported successful isolation, amplification, and sequencing of the partial genome of a single uncultured archaeon cell from the environment. In this study, Kvist *et al.* (2007) first mapped the diversity of Archaea in a soil sample by generating a clone library using group-specific primers in combination with a terminal restriction fragment length polymorphism profile. Intact cells were then extracted from the same environmental sample and probed with Cy3-labelled DNA fragment designed from the clone library using fluorescent in situ hybridization. The single cells with a bright fluorescent signal were isolated using a micromanipulator. The genome of the single isolated cells was then amplified through multiple displacement amplification (MDA) using the Phi29 DNA polymerase. The 16S rRNA gene of the single cell showed a >99% sequence identity to a soil crenarchaeotal clone SCA1170. The shotgun sequencing of the generated MDA product found its closest match to a crenarchaeotal BAC clone previously retrieved from a soil sample. In addition, this system was successfully validated using *Methanothermobacter thermoautotrophicus* as single-cell test organism (Kvist *et al.*, 2007). Such an ability to isolate and sequence a microbial cell from the environment will bypass many of the obstacles facing metagenomics research and will significantly accelerate our unbiased understanding of natural microbial communities.

Metatranscriptomics

While the analyses of DNA sequences in environmental samples can provide evidence about the diversity and function of specific groups of organisms in the analysed ecological niches, the relative importance of those genes in environmental functions cannot be determined from DNA sequences alone. To understand what genes are expressed in a specific environment, the analyses of transcripts from the communities are often needed. The total collection of all expressed genetic information of an ecosystem is called a metatranscriptome. A recent study by Gilbert *et al.* (2008) used the GS-FLX Pyrosequencing technology to study a complex marine metatranscriptome obtained from random whole-community mRNA. Through analysing over 323 million base pairs, they found that the highly expressed sequences were frequently novel. Using the microarray platform, the comparative metagenome and metatranscriptome analyses of lactic acid bacteria during kimchi fermentation identified relatively limited change in the metagenome of the community. However, the changes in the metatranscriptome profiles were more profound. Microorganisms that were less abundant in the kimchi fermentation process often show relatively high activities (Nam *et al.*, 2009).

Metaproteomics

Unlike nucleic acids (both DNA and RNA) that give indirect signals about the potential activities of genes in specific environments, proteins are promising alternative molecules as they reflect the actual functionality with respect to structural components, metabolic reactions and regulatory cascades. As a result, they can potentially provide more direct information about microbial activity than the genes and their corresponding messenger RNAs. Another advantage is that the use of proteins can help reveal the identities of the active microorganisms via database analysis using the level of homology to known protein sequences in databases of cultured microorganisms or plants and animals.

The recent growths and interests in this field have been facilitated by the development of a variety of powerful methods to visualize and identify the proteomes of pure microbial cultures and increasingly the proteomes of microbial communities. The most widely used approach in proteomics is the separation of proteins by 2 dimensional gel electrophoresis (2-DE), which combines isoelectric focusing (IEF) in the first dimension and sodium dodecyl sulphate polyacrylamide gel electrophoresis (SDS-PAGE) in the second dimension. This separation is followed by in-gel digestion of selected spots and protein identification by mass spectrometry (MS). In the analyses of proteins from fully sequenced organisms, the identification is achieved by databank comparison of peptide mass fingerprint [matrix-assisted laser desorption ionization time of flight MS (MALDI-ToF-MS)] or fragmentation data of the peptides (electrospray ionization source tandem mass spectrometry (ESI-MS/MS) or MALDI-ToF/ToF-MS). For the identification of peptides from not yet sequenced species or environmental samples, there are two general methods. In cases where proteins share high-level sequence similarity with those already published, one approach is to identify the peptides by analysing their MS/MS fragments using the approach described above. On the other hand, if the amino acid sequence similarity is low, the peptides can be fully sequenced for their putative identification and characterization.

While the extraction and analyses of proteomes of cultured microbial samples are relatively well developed, the analyses of metaproteomes from natural environments require more complicated procedures. This is because environmental samples contain mixtures of microorganisms probably in very different proportions and that proteins in some cells will be more readily extracted than others. In addition, other organic and inorganic substances in the samples can affect the quantity and quality of protein extracts. To overcome these problems and to obtain protein samples that are representative of the original community, a variety of sample type – specific extraction protocols may be needed and some have been developed. For example, Benndorf *et al.* (2007) recently developed an extraction method allowing efficient separation of proteins from the inorganic and organic constituents of the soil matrix. They used a combination of 0.1 M NaOH treatment and phenol extraction. Specifically, incubation with NaOH released humic acids and proteins from soil minerals, and simultaneously, disrupted microorganisms. The subsequent phenol extraction separated the proteins from the humic organic matter. Using spiked samples, their subsequent analyses identified the high quality protein extracts. Ram *et al.* (2005) used genomic and mass spectrometry-based proteomic methods to evaluate gene expression, identify key activities, and examine partitioning of metabolic functions in a natural acid mine drainage microbial biofilm community. They detected 2033 proteins from the five most abundant species in the biofilm, including 48% of the predicted proteins from the dominant biofilm organism, *Leptospirillum* group II (also identified directly through metagenome analyses). They found that proteins involved in protein refolding and response to oxidative stress were highly expressed, suggesting that damage to biomolecules is a key challenge for survival of microorganisms in such environments.

Meta-metabolomics

A metabolome includes all the naturally occurring metabolites (typically small molecules) in biological samples such as cells, biofluids, and tissues. These small molecules are the products of metabolism and include, for example, carbohydrates, fats (or lipids), amino acids, nucleotides, and their derivatives as well as secondary metabolites such as antibiotics. The complete complement of all metabolites in an environmental sample is called a meta-metabolome. The metabolome and meta-metabolome are the results of interactions between the genes and the environment. The study of meta-metabolome is called meta-metabolomics.

Typical biological communities have significant division of labour in metabolism. Waste products of some organisms are substrates for others. Working together they turn raw resources into fully metabolized products. In these systems, metabolite analyses can greatly accelerate comparative metagenome studies and expression experiments based on microarrays or proteomics. Such analyses would allow researchers to piece together a metabolic network that goes beyond species boundaries, enabling estimates of how metabolites are transferred and transformed through a community.

Similar to metagenomics, metatranscriptomics, and metaproteomics, the development of meta-metabolomics has benefited from the rapid development and popularization of research tools used to measure the metabolites. These tools include nuclear magnetic resonance (NMR) spectroscopy, LC/GC-coupled mass spectrometry and capillary electrophoresis coupled to mass spectrometry (CE-MS). In addition, the field of metabolomics and meta-metabolomics have their roots on hundreds of years of physiology, biochemistry, and clinical chemistry research.

However, unlike nucleic acids and proteins that are relatively homogeneous in their chemical constituents, metabolites within cells, tissues and microbial communities are very diverse in both their chemical and physical properties. As a result, different tools targeting different types of molecules may be required to reveal the complete metabolome of a cell population or the meta-metabolome of a microbial community. Because of the heterogeneity in metabolites and the tools used to analyse them, extensive efforts are needed for standardizations. In addition, unlike the widely popular and essential databases for DNA, RNA, and proteins of various types of cells, tissues, and communities, there is no GenBank-like repository for all metabolomics data. Having such a publicly available database would significantly strengthen the analyses of metabolome and meta-metabolome information. It's expected that the next few years will see great improvements in standardization of various analytical platforms, software and databases. So far, large-scale metabolome surveys have been limited to known cultures, cells and tissues (Mashego *et al.*, 2007), especially those of model organisms such as *Escherichia coli, Saccharomyces cerevisiae,* and humans.

Microbial population genetics

Microbial population genetics examines the spatial and temporal distributions of microbial genetic diversities. The distribution patterns of genetic variation are often discussed in the context of geography, ecological niches and environmental factors such as temperature, pH, salinity, pressure, the availabilities of water and nutrients, and the sources of energy and carbon. These ecological factors influence microbial activities and play very important roles in determining the spatial and temporal dynamics of microorganisms in natural environments. Below I briefly introduce the main issues in population genetics and the challenges facing microbial population genetics.

Main issues in population genetics

The patterns of variation and the potential mechanisms underlying the distributions of the observed genetic variation are among the most fundamental issues in evolutionary biology and population genetics. Information on these issues can help us not only understand the natural history of the study organism but also design effective management strategies and practical applications. For example, for plant and animal pathogens, understanding the patterns of population genetic variation can help us tracking the sources of infections and in designing effective treatment and prevention strategies. For economically and ecologically important species such as edible and medicinal plants and gourmet mushrooms, information about their population genetics could help in developing measures for conserving their natural genetic diversity. Indeed, among other applications, population genetic studies have now become an integral part of the control and prevention of infectious diseases and the conservation and germplasm management of endangered plants and animals (Avise and Hamrick, 1996; Seifert and DiRit, 2006; Xu, 2005).

Two of the specific questions addressed in typical population genetic studies are (i) the mode of reproduction within individual populations (i.e. sexual reproduction and outcrossing vs. asexual reproduction and inbreeding) and (ii) the extent of gene flow between populations (Hartl and Clark, 2007; Weir, 1996; Xu, 2005). In many plants and animals, outcrossing and sexual reproduction can often be directly observed. However, the modes of reproduction in populations of microorganisms such as bacteria, algae, protozoa, and fungi cannot be observed directly (Xu, 2004, 2005). Similarly, many plants and animals have rare sexual cycles or difficult-to-observe reproductive behaviours in nature. Under these circumstances, their modes of reproduction are usually inferred from their gene frequencies and genotype frequencies in population samples, often requiring large numbers of individuals (Weir, 1996).

The underlying principle for the currently available methods in inferring modes of reproduction is that in sexually reproducing populations, the alleles from the same locus or from different loci should randomly associate with each other (Xu, 2005). In diploid organisms, the commonly known Hardy–Weinberg equilibrium can be used to test for random associations of alleles from the same locus whereas genotypic disequilibrium is used to test the random associations of alleles from different loci (Xu, 2005). In haploid organisms, the association among alleles from different loci is used for the inferences (Xu, 2005). In general, if the observed and expected genotypes (calculated based on the assumptions of random association) were not significantly different from each other, the population is assumed to have a recombining mode of reproduction, probably resulting from outcrossing and sexual reproduction. In contrast, if alleles from different loci are significantly associated with each other, then asexual reproduction or inbreeding is assumed to be playing a significant role in reproduction (Xu, 2005). Of course, deviations from Hardy–Weinberg equilibrium and genotypic equilibrium do not mean that outcrossing and sexual reproduction do not occur at all in study populations and species. Many factors other than the mode of reproduction can cause their deviations (Hartl and Clark, 2007; Weir, 1996). These factors include biased sampling, small sample size, selection, migration, and mutations.

Similarly, the potential genetic exchange among populations is traditionally inferred from allelic data obtained from a large number of individuals in respective populations (Weir, 1996; Hartl and Clark, 2007). The tests for gene flow and population differentiation have relied mainly on allele frequency similarities and differences between populations. Based on allele frequencies, these tests partition genetic variation into those within and between populations (Xu, 2005). The greater the amount of genetic variation that is attributed to that between populations, the lower the gene flow would be between the populations (Weir, 1996; Hartl and Clark, 2007; Xu, 2005).

The challenges facing population genetic studies

There are two main challenges facing population genetic studies. The first relates to the availability of genetic markers. In model organisms or organisms with known economic, industrial, agricultural or medical implications, there is generally abundant DNA sequence information available in databases that allow relatively easy retrieval of sequences for designing and screening genetic markers for population genetic studies. However, for most other organisms, there is generally very limited genetic information in public databases. As a result, developing genetic markers can be a time-consuming process.

The second major challenge is sample size. As indicated above, all current inference methods for both the mode of reproduction and gene flow between populations rely on a large number of individuals from each population for comparison. For microorganisms, that means a large number of pure cultures or clearly defined individuals. With a small number of individuals, the confidence intervals of the obtained estimates will be large and both type I and type II errors will be significant. As a result, many species with only a few strains are not investigated for their potential modes of reproduction and gene flow between populations. These species include (i) those that are already extinct but with only a few specimens in museums; (ii) ancient populations represented by few collections; (iii) rare and endangered species with very few living representatives; and (iv) organisms that are difficult to collect because of inaccessible locations or significant risks involved in obtaining required samples (Xu, 2008). Microorganisms that are not cultivable in the laboratory also belong to this category.

Recently, an alternative conceptual approach was developed that allowed addressing population genetics questions (i.e. the modes of reproduction and the presence/absence of gene flow between populations) using one or two individuals derived from sample categories (i) to (iv) above (Xu, 2008). This approach proposed specific testable expectations of various hypotheses about population structures in the analyses of a small number of individuals. Here, I further extend that approach to address microbial population genetic issues using metagenomics data. Specifically, in metagenomic analyses of natural microbial communities, with a high rate of sequencing coverage, complete genomes may be reconstructed and multiple allelic sequences may be obtained and be compared to the expectations of various patterns of population structures (e.g. Venter *et al.*, 2004). Below I discuss the specific expectations in the analyses of small number of individuals and metagenomic data with respect to addressing population genetic issues related the modes of reproduction (sexual vs. asexual) and gene flow (frequent gene flow vs. significant differentiation). The validity of these expectations was recently confirmed in the analyses of population structure of the non-cultivable gourmet mushroom *Tricholoma matsutake* (Xu *et al.*, 2007, 2008). Specifically, the analyses of DNA sequences from multiple loci for two individual mushrooms from two geographically diverse locations identified signatures of sexual recombination and gene flow (Xu *et al.*, 2007). In a subsequent study of additional mushrooms from more geographic regions, both conclusions were substantiated (Xu *et al.*, 2008).

Expectations of no sexual recombination

In populations with no sexual reproduction, the genetic differences between individuals will only result from the accumulation of spontaneous mutations in individual lineages. Consequently, the different genes within a genome should show identical or very similar patterns of genetic variation between strains (Welch and Meselson, 2000; Xu, 2005). In a diploid species, within an individual organism, the allelic relationship from one locus should be very similar or identical in terms of sequence divergence to those at other loci in other parts of the genome. Similarly, in haploid species, the relationships between alleles from two different individuals should be very similar or identical in terms of their sequence divergence across many loci within the genome.

In metegenomic analyses of microbial communities, the multiple copies of the same gene in the dominant microbes might be present and their patterns of variation could be analysed. If the patterns of sequence variation across many gene fragments were similar or identical, it would be consistent with asexual reproduction.

Expectations of sexual recombination

In contrast to the expectations above for asexual reproduction, in populations with sexual reproduction, the genes in the genomes will be shuffled every time sexual reproduction occurs (Xu, 2005). As a result, the different genes within a genome may show very different patterns of genetic variation. This means that within a diploid organism, the allelic relationship from one locus may be very different in terms of sequence divergence to those at other loci in other parts of the genome. Similarly, in haploid organisms, the relationships between alleles from two different individuals might be very different in terms of their sequence divergence across loci within the genome.

In metegenomic analyses of microbial communities, the multiple copies of the same gene in the dominant microbes should be present and their patterns of variation could be analysed. If the patterns of sequence variation across many gene fragments were very different from each other, it would be consistent with recombination and sexual reproduction.

Expectations of no gene flow between populations

When there is no gene flow between two populations, individuals from the two populations should diverge from each other. Assuming that we have genotype data from only one individual from each population, in a diploid species, without gene flow, we should observe that at each locus, the two alleles from one individual should be more similar to each other than to alleles from the individual from a different population. Furthermore, allelic divergence and relationships should be similar across loci in various parts of the genome. In a haploid species, the later expectations should also be met (i.e. the allelic divergence and relationships should be similar between the two strains across loci in various parts of the genome). These expectations should hold for loci that are evolving neutrally and not under strong frequency-dependent selection (e.g. mating type alleles in fungi and MHC in mammals) or purifying selection. For loci that are under frequency-dependent selection, alleles have a high fitness when rare. As a result, ancestral polymorphisms might be preferentially maintained across populations and species (Hartl and Clark, 2007). For loci under strong purifying selection, there might be little accumulation of genetic mutations and a low-level divergence between populations to allow a clear distinction of alternative hypotheses.

In metegenomic analyses across several microbial communities, multiple copies of the same gene of the dominant microbes (assume that they were the same species) should be present in each population. If the allelic sequences or allele frequencies from different communities were very different from each other across many gene fragments, this result would be consistent with significant differentiation among the communities with respect to the population structure of the dominant microbial species.

Expectations of gene flow between populations

In contrast to the expectations of no gene flow between populations, when there is gene flow between two sexually reproducing populations, individuals from the two populations may share alleles and may not have completely diverged from each other. Assuming that we have multilocus genotype data from only one individual from each population, in a diploid species, due to gene flow and sexual reproduction, we might observe that at each locus, two alleles from the same individual would be less similar to each other than to alleles from an individual in a different population. Furthermore, the allelic divergence and relationships might be different across loci in various parts of the genome due to the effects of gene mixing from gene flow. In a haploid species, the later expectations might also be met (i.e. the allelic divergence and relationships might be different across loci in various parts of the genome between organisms from different populations).

In metegenomic analyses across several microbial communities, multiple copies of the same gene of the dominant microbes (assume that they are the same species) should be present in each community. If the allelic sequences or allele frequencies from different communities were identical or very similar to each other across many gene fragments, this result would be consistent with no or limited differentiation among the communities with respect to the population structure of the dominant microbial species.

Impact of genomics and metagenomics on microbial population genetics

Current microbial population genetic studies have almost exclusively focused on the analyses of cultivable microorganisms (see other chapters in this book). In typical studies, a large number of individuals are included. With the advancements in genomics and metagenomics technologies, microbial population genetics have undergone a renaissance. One area that microbial population genetics has benefited significantly is the widespread application of multi-locus sequence typing (MLST) of microbial populations. This is briefly summarized below. In addition, whole genome sequencing of cultured microorganisms have also helped reveal patterns of population genetic variations for many groups of organisms (Xu, 2006) including horizontal gene transfer and host switches and range expansions. These issues are presented in other chapters of this book and are not discussed here.

As the name suggests, MLST refers to the use of DNA sequences from multiple regions in the genome for discriminating strains in populations. The popularity of MLST is due directly to the development of highly affordable, reliable, and efficient DNA sequencing technologies that also spurred genomics and metagenomics. The low cost have allowed researchers from microbiology laboratories in virtually all countries to conduct such a research activity. Though the term was coined only in 1998 for typing human bacterial pathogens (Maiden *et al.*, 1998), its use in microbial ecological and evolutionary analyses dates back more than two decades ago. It has various other synonyms such as multiple gene genealogical analysis (MGGA) or comparative genealogical analysis (CGA) (e.g. Xu *et al.*, 2000; Xu, 2005). An added advantage of genealogical analyses over pure strain typing for unique genotypes is that genealogical analyses take into account of the specific substitutions and the relationships among alleles.

There are several advantages in analysing DNA sequences at multiple loci for strain typing, population genetic studies and phylogenetic analyses. First, sequences from multiple loci contain more information than that of a single locus, thus allowing us to obtain more robust conclusions. Second, the loci are typically located in multiple regions of the genome and thus the results are more representative of the whole genome than a single locus. This is important because, in many prokaryotes, horizontal gene transfer is very common. If only a single gene is selected and this gene happened to have been horizontally transferred, information derived from this gene will not be representative of other parts of the genome (Xu, 2005). Third, compared to other types of DNA-based strain typing methods (e.g. random amplified polymorphic DNA or RAPD, amplified fragment length polymorphisms or AFLP, restriction fragment length polymorphisms or RFLP, PCR-RFLP and PCR fingerprinting) that have been applied to analysing microbial populations, DNA sequence-based typing has several other advantages. These advantages include (i) the unambiguous nature of nucleotides in a DNA sequence; (ii) the ability to allow tracking evolutionary history more precisely; (iii) the ease of storage, sharing and retrieval from public databases such as GenBank; and (iv) the availability of many tools for analysing DNA sequences.

MLST has been used to study the population and ecological genetics of many microorganisms. It provides fine-scale measures of gene diversity and genotype diversity within and among microbial populations. These patterns of diversity have been used to infer a variety of ecological and evolutionary processes such as gene flow, cryptic speciation, hybridization, and the relative importance of clonality and recombination among analysed populations (see other chapters in this book). One issue that has been examined extensively is clonality and recombination in microbial populations and this is briefly discussed below.

All microbes can reproduce asexually and generate clones and clonal lineages. As expected, in natural populations of all microbial species examined (including viruses, bacteria, protozoa, algae and fungi), signatures of clones and clonal lineages are commonly found. These population genetic signatures include (i) limited or lack of genetic variation among strains, (ii) over-representation of certain genotypes, and (iii) significant associations among alleles located on the same or different genomic regions (Xu, 2004). However, over the past two decades, numerous studies have confirmed that genetic recombination is ubiquitous in natural populations of viruses, bacteria, protozoa, algae, and fungi (see other chapters in this book). Indeed, despite extensive investigations, no ancient asexual microbial populations or species have been found (Xu, 2004).

At present, the impact of metagenomic studies on microbial population genetics is still relatively limited. However, there are a couple of noteworthy examples. This first example illustrating the potential importance of metagenomics on microbial population genetic structures is that by Venter *et al.* (2004). In this study, they conducted an extensive shotgun sequencing of microorganisms of size ranges from 0.1 to 3.0 μm among four sites in the Sargasso Sea in the Atlantic Ocean near Bermuda (Venter *et al.*, 2004). Their study generated almost 2 million sequence reads, yielding over 1.6 billion base pairs of raw DNA sequences. Based on sequence relatedness and unique rRNA gene counts, their analysis suggested that these DNA fragments were derived from at least 1800 genomic species including 148 previously unknown bacterial phylogenetic types. While similarities were found among the four sampled sites, their data also suggested spatial variation in the richness and relative abundance of species (Venter *et al.*, 2004). Of the specific group of microorganisms identified, one stood out. This organism, most likely a member of the genus *Burkholderia*, had an estimated 21-fold coverage and comprised 38.5% of the sequence data from one of the four samples. *Burkholderia* is typically found in terrestrial environmental samples and the identification of a species in this genus in the sea at such a high frequency led the authors to suggest that terrestrial environments or coastal animals might play an important role in marine microbial community structure. However, based on several lines of evidence, Delong (2005) suggested that the abundance of highly similar *Burkholderia*-like sequences in only one of the four samples might be due to contamination of the original water sample by terrestrial organisms in the Venter *et al.* (2004) study. Such a possibility also suggests that extreme caution should be taken when conducting microbial metagenomic analysis.

As we discussed above, recombination is typically documented by DNA sequence comparisons involving a few genes from organisms obtained in pure culture. In the second example, Lo *et al.* (2007) resolved genome-wide, strain-specific expressed protein variants and provided genetic evidence for recombination in the dominant member of a natural multi-species biofilm. Specifically, they sampled pink biofilms growing in the Richmond Mine, CA that were dominated by bacteria belonging to *Leptospirillum* group II, and contained lower abundances of *Leptospirillum* group III and several archaea. Species in the genus *Leptospirillum* are chemoautotrophs, obtaining their metabolic energy by aerobic oxidation of ferrous iron. However, the metabolic by-product ferric iron can readily react with the surfaces of metal-sulphide minerals in ore deposits and mining waste, promoting sulphur oxidation and generating very acidic, metal-contaminated solutions. Such solutions can cause severe downstream environmental problems. As a result, they are of significant environmental concerns.

In the Lo *et al.* (2007) study, a mass spectrometry-based proteomic dataset was generated and compared to cultivation-independent reconstruction of near-complete genomic datasets for two related *Leptospirillum* group II bacteria from the same ecosystem. Proteins that differ by as few as a single amino acid were detected and this enabled strain-specific proteomic identifications. They detected evidence of recombination involving chromosomal regions tens to hundreds of kilobases in length that was further confirmed through multi-locus sequence typing of both pure isolates and uncultivated natural consortia.

Perspectives

In the past decade, the rapid and inexpensive genomics technologies in DNA sequencing and data analyses have revolutionized our understanding of the structures and diversity of natural microbial populations. However, most of our current knowledge has come from common cultivable microorganisms where large strain collections were available. In contrast, very limited information is available for the majority of the microbial species in our culture collections. In addition, most of the microbial diversity in nature remains to be identified.

The recent developments in metagenomics approaches and tools have the potential to push the field of microbial population genetics into new directions. Specifically, the ability to isolate and sequence the whole genome of single microbial cells will allow unprecedented access into the genetic information of microorganisms in nature. In addition, the low cost and high throughput sequencing platforms are generating large datasets for various natural microbial communities. Though still little explored, such datasets should allow population genetic issues to be addressed for at least the dominant microbial species.

Aside from benefiting microbial population genetics, metagenomics studies and the developments of high throughput platforms and analytical tools are driving a paradigm shift in modern biology: a change of emphasis from 'single gene, single protein' studies and reductionist approaches to global investigations of increasingly more complex systems of molecules and organisms. These 'systems approaches' investigate whole biological processes and enable models to be built to predict the behaviour of a system. Such systems could be at the cellular level, population level and/or community levels. Indeed, a variety of new fields have emerged, e.g. systems biology, and ecosystems biology. How biological systems at these different levels respond to environmental changes and disturbances will become increasingly important for understanding the future of the global biosphere.

Acknowledgements

Research in my lab on the structure and dynamics of natural and laboratory microbial populations has been supported by the Natural Science and Engineering Research Council (NSERC) of Canada, Genome Canada, Premier's Research Excellence Award of Ontario, McMaster University, and the Natural Science Foundation (NSF) of China.

References

Allen, E.E., and Banfield, J.F. (2005). Community genomics in microbial ecology and evolution. Nature Reviews Microbiology 3, 489–498.

Avise, J. C., and Hamrick, J. L. 1996. Conservation Genetics: Case Histories from Nature (New York: Chapman & Hall).

Béjà, O., Aravind, L., Koonin, E.V., Suzuki, M.T., Hadd, A., Nguyen, L.P., Jovanovich, S.B., Gates, C.M., Feldman, R.A., Spudich, J.L., Spudich, E.N., and DeLong, E.F. (2000b) Bacterial rhodopsin: Evidence for a new type of phototrophy in the sea. Science *289*, 1902–1906.

Béjà, O., Spudich, E.N., Spudich, J.L., Leclerc, M., and DeLong, E.F. (2001). Proteorhodopsin phototrophy in the ocean. Nature *411*, 786–789.

Béjà, O., Suzuki, M.T., Koonin, E.V., Aravind, L., Hadd, A., Nguyen, L.P., Villacorta, R., Amjadi, M., Garrigues, C., Jovanovich, S.B., Feldman, R.A., and Delong E.F. (2000a). Construction and analysis of bacterial artificial chromosome libraries from a marine microbial assemblage. Environ. Microbiol. 2, 516–529.

Benndorf, D., Balcke, G.U., Harms, H., and von Bergen, M. (2007). Functional metaproteome analysis of protein extracts from contaminated soil and groundwater. ISME J. *1*, 224–234.

Breitbart, M., Salamon, P., Andresen, B., Mahaffy, J.M., Segall, A.M., Mead, D., Azam, F., and Rohwer, F. (2002). Genomic analysis of uncultured marine viral communities. Proc. Natl. Acad. Sci. U.S.A. *99*, 14250–14255.

Culley, A.I., Lang, A.S., and Suttle, C.A. (2006). Metagenomic analysis of coastal RNA virus communities. Science *312*, 1795–1798.

DeLong, E.F. (2005). Microbial community genomics in the ocean. Nature Rev. Microbiol. 3, 459–469.

Edwards, R.A., Rodriguez-Brito, B., Wegley, L., Haynes, M., Breitbart, M., Peterson, D.M., Saar, M.O., Alexander, S., Alexander, E.C., and Rohwer, F. (2006). Using pyrosequencing to shed light on deep mine microbial ecology. BMC Genomics 7, e57.

Gilbert, J.A., Field, D., Huang, Y., Edwards, R., Li, W., Gilna, P., and Joint, I. (2008). Detection of large numbers of novel sequences in the metatranscriptomes of complex marine microbial communities. PLoS ONE *3*, e3042.

Handelsman, J., Rondon, M.R., Brady, S.F., Clardy, J., and Goodman, R.M. (1998). Molecular biological access to the chemistry of unknown soil microbes: a new frontier for natural products. Chem. Biol. *5*, 245–249.

Hartl, D. L., and Clark, A. G. 2007. Principles of Population Genetics (Sunderland, MA: Sinauer).

Huson, D.H., Auch, A., Qi, J., and Schuster, S.C. (2007). MEGAN Analysis of Metagenomic Data. Genome Res. *17*, 377–386.

Kvist, T., Ahring, B.K., Lasken, R.S., and Westernmann, P. (2007). Specific single-cell isolation and genomic amplification of uncultured microorganisms. Appl. Microbiol. Biotechnol. *74*, 926–935.

Lehner, A., Loy, A., Behr, T., Gaenge, H., Ludwig, W., Wagner, M., and Schleifer, K.H. (2005). Oligonucleotide microarray for identification of Enterococcus species. FEMS Microbiol. Lett. *246*, 133–142.

Lo, I., Denef, V.J., Verberkmoes, N.C., Shah, M.B., Goltsman, D., DiBartolo, G., Tyson, G.W., Allen, E.E., Ram, R.J., Detter, J.C., Richardson, P., Thelen, M.P., Hettich, R.L., and Banfield, J.F. (2007). Strain-resolved community proteomics reveals recombining genomes of acidophilic bacteria. Nature *446*, 537–541.

Loy, A., Lehner, A., Lee, N., Adamczyk, J., Meier, H., Ernst, J., Schleifer, K.H., and Wagner, M. (2002). Oligonucleotide microarray for 16S rRNA gene-based detection of all recognized lineages of sulfate-reducing prokaryotes in the environment. Appl. Environ. Microbiol. *68*, 5064–5081.

Maiden, M.C., Bygraves, J.A., Feil, E., Morelli, G., Russell, J.E., Urwin, R., Zhang, Q., Zhou, J., Zurth, K., Caugant, D.A., Feavers, I.M., Achtman, M., and Spratt, B.G. (1998). Multilocus sequence typing: a portable approach to the identification of clones within populations of pathogenic microorganisms. Proc. Natl. Acad. Sci. U.S.A. *95*, 3140–3145.

Mashego, M.R., Rumbold, K., De Mey, M., Vandamme, E., Soetaert, W., and Heijnen, J.J. (2007). Microbial metabolomics: past, present and future methodologies. Biotechnol. Lett. *29*, 1–16.

Meyer, F., Paarmann, D., D'Souza, M., Olson, R., Glass, E.M., Kubal, M., Paczian, T., Rodriguez, A., Stevens, R., Wilke, A., Wilkening, J., and Edwards, R.A. (2008). The metagenomics RAST server – a public resource for the automatic phylogenetic and functional analysis of metagenomes. BMC Bioinf. *9*, e0.

Nam, Y.-D., Chang, H.-W., Kim, K.-H., Roh, S.-W., and Bae, J.-W. (2009). Metatranscriptome analysis of lactic acid bacteria during kimchi fermentation with genome-probing microarrays. Int. J. Food Microbiol. *130*, 140–146.

Pace, N.R. (1991). Analysis of a marine picoplankton community by 16S rRNA gene cloning and sequencing. J. Bacteriol. *173*, 4371–4378.

Pace, N.R., Stahl, D.A., Lane, D.J., and Olsen, G.J. (1985). Analyzing natural microbial populations by rRNA sequences. ASM News *51*, 4–12.

Ram, R.J., Verberkmoes, N.C., Thelen, M.P., Tyson, G.W., Baker, B.J., Blake, R.C., 2nd, Shah, M., Hettich, R.L., and Banfield, J.F. (2005). Community proteomics of a natural microbial biofilm. Science *308*, 1915–1920.

Rondon, M.R., August, P.R., Bettermann, A.D., Brady, S.F., Grossman, T.H., Liles, M.R., Loiacono, K.A., Lynch, B.A., MacNeil, I.A., Minor, C., Tiong, C.L., Gilman, M., Osburne, M.S., Clardy, J., Handelsman, J., and Goodman, R.M. (2000). Cloning the soil metagenome: a strategy for accessing the genetic and functional diversity of uncultured microorganisms. Applied and Environ. Microbiol. *66*, 2541–2547.

Seifert, H. S. and DiRita, V. J. 2006. Evolution of Microbial Pathogens. (Washington, DC: ASM Press).

Tyson, G.W., Chapman, J., Hugenholtz, P., Allen, E.E., Ram, R.J., Richardson, P.M., Solovyev, V.V., Rubin, E.M., Rokhsar, D.S., and Banfield, J.F. (2004). Community structure and metabolism through reconstruction of microbial genomes from the environment. Nature *428*, 37–43.

Venter, J.C., Remington, K., Heidelberg, J.F., Halpern, A.L., Rusch, D., Eisen, J.A., Wu, D.Y., Paulsen, I., Nelson, K.E., Nelson, W., Fouts, D.E., Levy, S., Knap, A.H., Lomas, M.W., Nealson, K., White, O., Peterson, J., Hoffman, J., Parsons, R., Baden-Tillson, H., Pfannkoch, C., Rogers, Y.H., and Smith, H.O. (2004). Environmental genome shotgun sequencing of the Sargasso Sea. Science *304*, 66–74.

Weir, B. S. 1996. Genetic Data Analysis II. (Sunderland, MA: Sinauer).

Welch, D. M. and Meselson, M. 2000. Evidence for the evolution of bdelloid rotifers without sexual reproduction or genetic exchange. Science *288*, 1211–1215.

Wu, L., Thompson, D.K., Liu, X., Fields, M.W., Bagwell, C.E., Tiedje, J.M., and Zhou, J. (2004). Development and evaluation of microarray-based whole–genome hybridization for detection of microorganisms within the context of environmental applications. Environ. Sci. Technol. *38*, 6775–6782.

Xu, J., Vilgalys, R., and Mitchell, T.G. (2000). Multiple gene genealogies reveal recent dispersion and hybridization in the human pathogenic fungus *Cryptococcus neoformans*. Mol. Ecol. *9*, 1471–1481.

Xu, J. (2004). The prevalence and evolution of sex in microorganisms. Genome *47*, 775–780.

Xu, J. (2008). Using 1–2 Individual Organisms to Address Population Genetic Questions. In Genetic Inheritance Patterns (Editor Ren Kimura). Nova Science Publishers, Inc. pp. 9–16.

Xu, J. (2005). Fundamentals of fungal molecular population genetic analyses. In Evolutionary Genetics of Fungi (Editor J. Xu). Horizon Scientific Press, England. Pp. 87–116.

Xu, J. (2006). Microbial ecology in the age of genomics and metagenomics: concepts, tools, and recent advances. Mol. Ecol. *15*, 1713–1731.

Xu, J., Guo, H., and Yang, Z.-L. (2007). Single nucleotide polymorphisms in the ectomycorrhizal mushroom *Tricholoma matsutake*. Microbiology *153*, 2002–2012.

Xu, J., Sha, T., Li, Y.-C., Zhao, Z.-W., and Yang, Z.-L. (2008). Recombination and genetic differentiation among natural populations of the ectomycorrhizal mushroom *Tricholoma matsutake* from southwestern China. Mol. Ecol. *17*, 1238–1247.

Zhou, J. (2003). Microarrays for bacterial detection and microbial community analysis. Curr. Opin. Microbiol. *6*, 288–294.

Index

Q

R

Current Books of Interest

Salmonella: From Genome to Function	2011
PCR Troubleshooting and Optimization: The Essential Guide	2011
Insect Virology	2010
Environmental Microbiology: Current Technology and Water Applications	2010
Sensory Mechanisms in Bacteria: Molecular Aspects of Signal Recognition	2010
Bifidobacteria: Genomics and Molecular Aspects	2010
Molecular Phylogeny of Microorganisms	2010
Nanotechnology in Water Treatment Applications	2010
Iron Uptake and Homeostasis in Microorganisms	2010
Caliciviruses: Molecular and Cellular Virology	2010
Epstein-Barr Virus: Latency and Transformation	2010
Anaerobic Parasitic Protozoa: Genomics and Molecular Biology	2010
Lentiviruses and Macrophages: Molecular and Cellular Interactions	2010
Microbial Population Genetics	2010
Borrelia: Molecular and Cellular Biology	2010
Influenza: Molecular Virology	2010
RNA Interference and Viruses: Current Innovations and Future Trends	2010
Retroviruses: Molecular Biology, Genomics and Pathogenesis	2010
Metagenomics: Theory, Methods and Applications	2010
Aspergillus: Molecular Biology and Genomics	2010
Environmental Molecular Microbiology	2010
Neisseria: Molecular Mechanisms of Pathogenesis	2010
Frontiers in Dengue Virus Research	2010
ABC Transporters in Microorganisms	2009
Pili and Flagella: Current Research and Future Trends	2009
Lab-on-a-Chip Technology: Biomolecular Separation and Analysis	2009
Lab-on-a-Chip Technology: Fabrication and Microfluidics	2009
Bacterial Polysaccharides: Current Innovations and Future Trends	2009
Microbial Toxins: Current Research and Future Trends	2009

Current Books of Interest

www.ingramcontent.com/pod-product-compliance
Lightning Source LLC
LaVergne TN
LVHW062123200926
848484LV00007B/1751